Standing *on the* Shoulders *of* Giants

Standing *on the* Shoulders *of* Giants

Genesis and Human Origins

Luke J. Janssen

WIPF & STOCK · Eugene, Oregon

STANDING ON THE SHOULDERS OF GIANTS
Genesis and Human Origins

Copyright © 2016 Luke J. Janssen. All rights reserved. Except for brief quotations in critical publications or reviews, no part of this book may be reproduced in any manner without prior written permission from the publisher. Write: Permissions, Wipf and Stock Publishers, 199 W. 8th Ave., Suite 3, Eugene, OR 97401.

Wipf & Stock
An Imprint of Wipf and Stock Publishers
199 W. 8th Ave., Suite 3
Eugene, OR 97401

www.wipfandstock.com

PAPERBACK ISBN: 978-1-4982-9140-8
HARDCOVER ISBN: 978-1-4982-9142-2
EBOOK ISBN: 978-1-4982-9141-5

Illustrations: created by Luke Jeffrey Janssen or used with permission.

Manufactured in the U.S.A. AUGUST 4, 2016

Contents

Preface | vii
Acknowledgments | xi
Abbreviations | xiii

1. It's All About How You Look At It | 1
2. A Precedent-Setting Case: The Origin of the Cosmos | 20
3. Other Duels between the Church and Science | 40
4. A Basic Understanding of the Science | 71
5. Origin of Humanity: The Paleontological Evidence | 104
6. Origin of Humanity: The Genetic Evidence | 127
7. Christian Objections to the Evolutionary Model | 150
8. Adjustments to Theology | 188
9. Various Responses from the Church | 256
10. Atheist Worldviews Also Color Their Belief Systems | 275
11. Standing on the Shoulders of Giants | 284

About the Author
Bibliography | 297
Subject Index | 307
Scripture Index | 315

Figures

Figure 1: Earth rise
Figure 2: Hebrew model of cosmos
Figure 3: Aristotelian and current models of solar system
Figure 4: Nicolas Hartsoeker's perception of head of human sperm
Figure 5: Scheuchzer's "Flood victim"
Figure 6: Classic logo for evolution
Figure 7: Evolution of letters
Figure 8: UPC bar code
Figure 9: Inheritance of mitochondrial DNA
Figure 10: Mouse and human karyotypes
Figure 11: Human and great ape karyotype
Figure 12: Human eye versus octopus eye
Figure 13: Flatland
Back cover: Author's 1st book: *Reaching into Plato's Cave*

Preface

Chapter 1: It's All About How You Look At It

EVERYBODY, WITHOUT EXCEPTION, ALLOWS their worldview to color their interpretation of the world around them. That's almost as self-evident as saying everyone who is alive manifests living processes. But despite the circularity and simplicity of such a statement, many people aren't aware that they interpret things around them. Theists and atheists alike. They attribute too much certainty and fact to much of what they believe. I have three main goals for this book:

1. to point out and emphasize how we all allow worldviews to determine how we handle data, observations and experiences;
2. to equip theists to better understand the science of genetics, and then to present the data which now have tremendous impact on our theology;
3. to suggest ways to reframe our theology to be consistent with the data.

Theists from St. Augustine to Sir Francis Bacon have cautioned against allowing theology to be ridiculed because of an unnecessarily strict adherence to Scripture and a denial of scientific findings.

Chapter 2: A Precedent-Setting Case: The Origin of the Cosmos

Before addressing the impact that the new science of genetics has had on theology, it would be instructive to see how the church has dealt in the past with scientific discoveries upsetting theological applecarts. A perfect example is its response to the heliocentric theory. This chapter provides a detailed description of the original biblical/Hebrew view of the universe—a three-layered earth covered by a solid dome and a three-tiered heaven(s)—complete with abundant scriptural support. To put this strange construct in perspective, we first look at the cosmologies of civilizations that preceded that of the ancient Hebrews. In the remainder of the chapter, we look at the scientific discoveries which caused a whole reshaping of that worldview, and the various responses from the church against that paradigm shift. The main points of this chapter are twofold. First, that the church has before allowed its theology to determine how it interprets science. Second, that it has been able to discard major theological viewpoints, based upon a faithful reading of Scripture, when faced with an abundance of conflicting scientific evidence.

Chapter 3: Other Duels between the Church and Science

To show that the conflict addressed in chapter 2 is not a unique, one-off experience in church history, we consider several other less well-known examples of the church and theists allowing theology to drive understanding of the way the world works, despite scientific evidence to the contrary. These include several examples from biology, from archaeology and history of Palestine, from geology and even from meteorology.

Chapter 4: A Basic Understanding of the Science

The central goal of this book is to equip a believer having little or no scientific training with the means to handle the overwhelming findings from genetics which have the potential to destroy faith. But first it will be necessary to provide that reader with a basic understanding and vocabulary to handle the science itself. In this chapter, I will summarize the tools and strategies used by paleontologists (those who study fossils) and geneticists to explore the world around them, and will also provide a rudimentary understanding of the scientific method and of the theory of evolution.

Chapter 5: Origin of Humanity: The Paleontological Evidence

Most theists have a rudimentary understanding of the paleontological evidence for "cavemen" and Neanderthals; for many, this understanding is so vague that they're able to completely overlook how they might fit into the creation narrative. This chapter provides an overview of the abundant findings of various hominid species, including Neanderthals, *Homo erectus*, *Australopithecus*, *Floresiensis*, Denisovans and *Homo naledi*. The oft-cited hoax known as Piltdown Man is also summarized.

Chapter 6: Origin of Humanity: The Genetic Evidence

When nineteenth-century archaeologists found clay tablets describing a Babylonian version of a global flood, the parallels with the Biblical version in detail after detail forced them to wonder whether one copied from the other. Today, geneticists have learned how to read the genetic code, and when they make comparisons between the human and great ape sequences, once again the parallels between the two are so staggering that it is becoming impossible not to wonder whether one has been copied (and modified) from the other. And they've found much more than that. Other topics to be addressed include: Mitochondrial Eve and Y-chromosomal Adam; inversion of chromosome #2; proviral sequences; and pseudogenes. So much of what they're finding threatens to completely shatter the Genesis version of the origins of humans.

Chapter 7: Christian Objections to the Evolutionary Model

Chapters 5 and 6 present an incredible volume of scientific evidence which speaks against the traditional Christian understanding of the origin of mankind. That evidence instead undergirds a model in which humanity evolved over millions of years from an ancestor we share in common with the primates. Understandably, many Christians react vigorously against this model. In this chapter, fifteen objections often made by Christians against this model are presented and rebutted. The first seven of these objections happen to be the exact same ones raised against the heliocentric theory, which the church and Christians have now come to accept and even embrace.

Chapter 8: Adjustments to Theology

The worlds studied by paleontologists and geneticists were made by God. As Sir Francis Bacon first put it, the book of God's words should say the same thing as the book of God's works. In the sixteenth century, discoveries made by astronomers forced the theologians to rethink their thoroughly scripturally based cosmology (chapter 2). In the same way today, the discoveries made by paleontologists and geneticists now make it incredibly difficult to hang on to the traditional biblical version of the origin of mankind, no matter how much Scripture can be called up to support it. But adjusting this worldview necessarily leads to a variety of tensions in our theological frameworks. This chapter delves into some of the important theological issues directly impacted by the new discoveries. These issues include: the historicity of Adam and whether that impacts our view of Christ; the fall and original sin; the atonement; world religions; our view of the Bible (its origins; inspiration; inerrancy; infallibility; authority; purpose). The chapter also provides some strategies to resolve the tension without throwing the baby out with the bathwater.

Chapter 9: Various Responses from the Church

The responses of theists to these new data from the world of genetics have not been uniform. Some, unfortunately, feel it necessary to give up faith entirely. Others dig in deeper and push back harder, even to the point of proposing bizarre ad hoc explanations for the threatening evidence. As is inevitably the case, some take a middle ground and attempt to accommodate the science into their worldview, which others interpret as compromise. This chapter addresses the variety of responses, their strengths and weaknesses, and offers suggestions on how to rationalize faith and science.

Chapter 10: Atheist Worldviews Also Color Their Belief Systems

Theists are not the only ones to allow their belief systems to determine how they view the world and interpret new data from scientific endeavors to fit their worldview. Atheists do so as well. Many atheists aren't aware that they even hold belief systems, or won't admit to doing so. This chapter explores this phenomenon of belief in nonbelievers.

Chapter 11: Standing on the Shoulders of Giants

Everything humans know is founded, at least in part, on the observations and conclusions of other humans that preceded them. Sir Isaac Newton, to whom three major scientific advances can be traced, is often credited for the expression "If I have seen further, it is by standing on the shoulders of giants." It is equally true for theology. Humans across all ages, all corners of the globe, and all strata of societies have sought to understand the "Great Being." and have developed all kinds of religions and theologies, often borrowing ideas from other groups and developing them further in response to their own journey. We can certainly see this in the Judeo-Christian religion. We benefit greatly from the unique perspectives of JHWH had by Moses, or David, or Isaiah. Jesus brought yet an entirely different perspective, although one clearly rooted in the former perspectives. The apostles brought a whole new and greater level of understanding of the teachings of Jesus, especially Paul and his ideas about the "first Adam" and the "Last Adam." But today, we see much further than Paul: we have since learned that there likely may not have ever been a "first Adam." If our goal is to seek truth in all forms, then it seems to be time for us to rethink some of our theology.

Acknowledgments

THERE ARE SO MANY I want to thank for their input as this book evolved, and in some cases for taking the time to give me feedback on excerpts from the book. There are indeed so many.

In particular, a large and growing number of people with whom I meet semi-regularly over coffee and in social settings to hash out various theological issues. This includes several local pastors, ministry leaders and otherwise critically-thinking active believers: Rev. Lane Fusilier, Rev. Jimmy Rushton, Rev. Peter Cowley, Rev. Ken Styles, Paul Almas, Marvin Kuehn, Dr. Malcolm Sears, Dr. John Harvey, Drs. Bonnie and Sean Marshall, Harold Laser, Rick Bradford, Lawrence Howe, and Don Corry.

Some of those mentioned above have helped me form a theology/philosophy group that meets semi-regularly to "open some pints and opine some points." and with whom I've explored many of the questions raised in this book: in addition to the ones mentioned above, key among these are Drs. Janet Warren, Sarah Woods, and John Seaman.

Thanks to the various faculty and students at the McMaster Divinity College for illuminating discussions and courses which have been truly mind-opening. Meeting other believers from such diverse backgrounds, and hearing their various stories, has helped me to see that the theology I grew up with and which I thought represented the full scope of all things Christian was in fact just a narrow and superficial slice of the entire spectrum of Christian belief.

Thanks also to colleagues at McMaster University through whom I could gain entirely different perspectives on life issues, whether that be an agnostic/atheistic worldview on matters which often have a very spiritual/religious impact, or others for being open to sharing their experiences coming from a non-Christian religious paradigm, including the Islamic, Sikh, and Buddhist faiths. These include Drs. Wolfgang Kunz, Param Nair, Shyam Maharaj, Paul Forsythe, Warren Foster, Subhendu Mukherjee, Mozibur Rahman, Nabeel Ghayur, and Taran Singh.

I feel I owe a huge debt of gratitude to Justin Brierley and the podcast / radio show that he hosts called "Unbelievable?" which gave me innumerable insights into many of the questions addressed in this book; I highly recommend all readers to check out the diverse list of podcasts available on their website.

Closest and dearest to me are my whole family, who are a feisty bunch with no end of diverse opinions and a long history exploring who God is: my son, Ryan, my parents, Harry and Grace, my siblings, Allan, Margot, Renee and Ivan, and paternal aunt, Rennie Janssen. Of course, my wife, Miriam, who has heard me carry on at great length at many of the discussions alluded to above, or just between her and me.

Finally, I thank God for guidance in preparing this manuscript, which I'm sure will ruffle some feathers and stir up debate in his children. Many times while writing I've prayed, "Lord, honestly: what did you mean by that?" or, "What am I supposed to do with that?" and I've done my best to be open to his leading. I have sincerely sought to learn who God is and what he wants from me: my first book describes that journey, and this book represents where I believe I've landed. That journey has included long hikes through a variety of spiritual valleys-of-the-shadow-of-death, and I feel he has led me to this point where I find myself now. He inspired the hunger to pursue these questions, and I believe he's inspired the answers I've found.

Abbreviations

AAAS	American Association for the Advancement of Science
AD	Anno Domini
AiG	Answers in Genesis
ANE	Ancient Near East(ern)
ASA	American Scientific Affiliation
ATP	adenosine triphosphate (energy molecule)
BC	before Christ
CE	Common Era
DNA	deoxyribose nucleic acid
DoSER	Dialogue on Science, Ethics, and Religion
ID	Intelligent Design
KJV	King James Version
NAE	National Association of Evangelicals
NIV	New International Version
NT	New Testament
OEC	Old Earth Creationism
OT	Old Testament
RATE group	Radioisotopes and the Age of The Earth group
RNA	ribonucleic acid
RTB	Reasons to Believe
SETI	Search for Extraterrestrial Intelligence
SNP	single nucleotide polymorphisms
UV	ultraviolet
YEC	Young Earth Creationism
YHWH	Yahweh

1

It's All About How You Look At It

As a university professor, part of my job involves marking long assignments from students. Sometimes my reward is seeing evidence of brilliant, original thought. Other times, not so much.

Let's just say I'm marking essays on American history. Admittedly, this is a strange thing for me to do, given that I'm a professor from a department of medicine in Canada teaching a practical physiology lab course. But bear with me: the reader will relate to this form of the analogy better than if I turned it into one of marking pharmacology lab reports.

One student writes on the Civil War: the circumstances that led up to it, the main details of the war itself, and the societal changes that followed after it. Another student goes through a similar exercise around the Great Depression. A third student steps back and takes a bigger picture approach by covering two hundred years of American history, which means it includes many of the same points and details found in the first two essays. But I'm not concerned: there's bound to be some overlap in the essays if they're all dealing with basically the same set of historical facts.

It's the next two essays that make me feel really uneasy. The fourth essay covers yet a different aspect of American history, but bounces back and forth through the historical timeline, sometimes in directions you wouldn't expect, and even draws a bizarre parallel with a particular episode of the TV series *The Simpsons*, and has two unfinished paragraphs where the student started developing something but didn't finish.

Sure, I'm going to have to be harsh with this fourth paper, but that's not what unsettles me. Instead, it's the fact that essay number five does exactly the same thing. Same historical event. Same hop-scotching around the timeline, and in the same sequence. Same reference to *The Simpsons* episode. Same half-finished paragraphs. It even has the very same spelling and grammatical mistakes in the very same places.

What am I to conclude? I could be open-minded and nonjudgmental, and say to myself: "Perhaps by chance they just happened to pick the same topic, and when you're dealing with a limited number of historical facts, they're bound to include some of the same points. Besides, it might hurt their feelings too much if I actually accused one or both of copying from the other. And it might take too much effort to prove that one of them did: I'd have to interview the students, ask them for their early drafts and preparatory notes, and look into their computer hard drives and recycle bins."

I could choose to ignore my suspicions and try to convince myself of what I might hope is true: it's entirely possible that these two just happened to be on the same wavelength and wrote two very similar papers completely independently.

On the other hand, I could be realistic and principled: I should stand up for what is right and not be meek or cowed into complicity. The facts speak for themselves: clearly one of these two authors copied from the other. Agreed: this is a serious matter with major consequences for the students, ranging from having to rewrite the entire essay, to getting a failing mark to even getting expelled. But that shouldn't cause me to dance around the undeniable inference, or deter me from concluding the obvious.

This analogy sets the stage for one of the main goals of my writing this book: there's been some plagiarism going on which now calls into question a number of fundamental theological tenets that the Christian church has held for millennia. But you'll have to wait till chapter 6 to find out what plagiarism I'm talking about. First, I need to cover a few important concepts.

Escaping the Matrix

In the 1999 blockbuster movie *The Matrix*, a character named Cypher (played by Joe Pantoliano) speaks to the star of the movie (Neo, played by Keanu Reeves) about life in the Matrix. (For those who aren't familiar with the story, the "Matrix" is a simulated reality created by sentient machines to imprison humans and extract their heat and electrical energy.) Cypher holds up a piece of steak and says, "I know this steak doesn't exist. I know that when I put it in my mouth, the Matrix is telling my brain that it is juicy and delicious. After nine years, you know what I realize?" [Cypher bites into the steak and savors it.] "Ignorance is bliss."

We all do it. We allow our worldview to dictate or at least color our interpretation of the facts before our eyes. Sometimes we're conscious of it; other times, not so much.

We deny factual evidence because it doesn't fit our understanding of things, or at least how we want to believe things should be. When watching a news report on television about a crime committed by an old high school friend or a colleague at work, the response is, "No, it can't be true. They're not like that." Sometimes the family of a loved one who died will deny that the latter is actually dead: they'll cling to the belief that one day their loved one will return home and all will be well.

We interpret data presented to us according to our preconceived bias. Lawyers on both sides of a case they're defending or prosecuting may call for the replacement of a potential juror simply because their experience tells them that two different people can see the same facts quite differently. A certain event in the Middle East can be interpreted completely differently by the ethnic groups involved.

We grow up for decades with a certain sense of propriety, only to be told by our teenagers that "it's not like that anymore": hairstyle, clothes, lyrics in music, sexual norms. What is plain and simply true in one era isn't necessarily the case in another era.

Sometimes the consequences for getting it wrong are insignificant. I still don't understand why it's such a faux pas to wear socks with sandals or white after Labour Day, and I've even been known to flaunt these rules. But other times the stakes are enormous, particularly within those subjects that one should never broach during a dinner party: religion and politics.

This book addresses the former of those two taboo subjects. In particular, it focuses attention on our tendency to allow our theology to drive our interpretation of the world around us, even to the point of believing things which defy the facts. Mark Twain is credited for defining faith as "believing what you know ain't true."

"–isms"

No matter who you are, where you live, or how you've been raised, you have a carefully defined worldview. A set of values, and a way to understand the world around you. Sometimes we don't really know what our "–ism" is, and so we have to go out and "find ourselves": this is especially the case for teenagers who have grown up for almost two decades under the –ism(s) of their parents and they've reached the stage where they're ready to be their own person. Often, we hold several of these worldviews simultaneously. –isms have all kinds of dimensions:

- Religious: Buddhism. Catholicism. Zoroastrianism.
- Social: Feminism. Humanism. Libertarianism.
- Economic: Capitalism. Communism. Socialism. Materialism.
- Political: Liberalism. Republicanism. Conservativism.
- Perspectival: Optimism. Pessimism. Nihilism.

The problem with –isms is that we can allow them to become too rigid. The –ism doesn't allow the facts to speak for themselves: it colors the interpretation of the facts. There will be stark differences in how the actions, motives and life story of a successful white male CEO are assessed from the perspective of a feminist, a capitalist, a devout Buddhist monk, and a poor person from a non-Caucasian non-Western background.

A different problem, but one equally as bad, is that we allow our community and peers to define our –ism and how we should interpret the world. To impose a zeitgeist upon us. We may feel strongly about a certain topic—say, gun control—and that automatically defines how we are expected to feel about gay rights, national fiscal responsibility, and international policies. Choosing to support the Republican candidate crystalizes my stance about global warming. And as we navigate our way through life, and start to see things from a new angle, –isms seem to force us to have to choose a side.

The Eyes See Only What the Mind Is Prepared to Comprehend[1]

I grew up with a very distinct Christian worldview. One which allowed the Bible to define and evaluate everything around me and about me. One phrase I heard repeated many times was: "If the Bible says it, I believe it, and that settles it." Another one which I've come across far more often, especially in the more recent past when reading or listening to a discussion about some aspect of apologetics, is: "a plain reading of Scripture." That phrase will come up often in this book.

That brand of Fundamentalism defined where I stood on social issues, fiscal issues, lifestyle choices, world history, scientific discoveries and many other matters. One particular tenet it forced me to take on was the idea that the universe was created approximately six thousand years ago, over a period of six days, as described in the book of Genesis. This view is often referred to as Young Earth Creationism (which I'll abbreviate as YEC through the remainder of this book, in contrast to Old Earth Creationism or OEC).

But that imposed a tension in my life: I often had to hold two or more contradictory ideas together as fact, a phenomenon which also goes by the label "cognitive dissonance." (Musically speaking, two notes may clash and so we call them dissonant: the same thing can happen at a cognitive level.)

As a child, I would read about dinosaurs that lived millions of years ago, but I "knew" that wasn't so, because the Bible told me they were created six thousand years ago. When I heard about evolution producing all kinds of new species in response to changes in habitat, I again "knew" that wasn't the case because the Bible said that God created all animals, each according to its own kind.

But eventually the evidence became too overwhelming, and I began to entertain other possibilities. Maybe God could have used the processes of genetic mutation and natural selection to create changes in the growing and expanding world of living organisms (which I'll otherwise refer to as the Tree of Life). I didn't realize it at the time, but this began a whole reconsideration of all the basic tenets of my belief system. Initially, those changes centered round the book of Genesis, particularly the stories of creation, the flood, the Tower of Babel, and the Nephilim. The catalyst for those changes was simply allowing the facts to speak for themselves, rather than bend the facts or the explanations in order to blend with my belief system. Learning to recognize cognitive dissonance and call that what it was.

1. Davies, *Tempest-Tost*.

One very unsettling outcome of that paradigm shift was that I then had to recalibrate my understanding about the inspiration of Scripture. Others around me were walking down the same path, and their response at various steps along the way was to let go of their entire belief system. All too often, peers around them (us) would condemn that choice, sometimes harshly. But those doing the condemning didn't recognize the irony in that it was an honest, sincere and deliberate seeking for truth that led us to reject what the other felt was the truth. It wasn't motivated from a desire to indulge in behaviors or lifestyle choices which had previously been denied us. It wasn't because we wanted to stop giving money to the church. It was simply being open to other interpretations of the same facts around us.

I nearly made the same mistake of giving up entirely on my faith. However, I recognized there's a difference between letting go of the Bible and letting go of a certain *interpretation* of the Bible. I found out that many other theists interpreted the Bible differently. They didn't rewrite the Bible, or like Thomas Jefferson did, rip out pages and cut out passages in order to create a new, more acceptable Bible. They just looked at the same wording from a different perspective. "Is the glass half-full or half-empty?"

This is the first goal in my writing this book. To call attention to and highlight a flaw in the thinking of certain parts of the Christian community which allows or even insists upon bending and rejecting truths in order to hold on to truth. Some examples of the ways in which the church has done this are summarized in chapters 2 and 3 below. This is a particular problem for those coming from a YEC point of view, as I did. But this problem is not exclusive to Fundamentalists. I'll elaborate later how this is also a problem for those who accept OEC, either confidently or cautiously, but who, in the words of Robertson Davies, haven't taken the time to prepare their minds to reflect on how their theology is nonetheless very much based on a YEC perspective. This will be the focus of chapter 8.

I should add, though, that this problem is not unique to the Christian church in particular, nor to theists in general. In the course of researching what and why I believe, which has entailed listening to the points made by many atheists, I've come to see that they too allow their worldview to color how they interpret data. I'll explain this a little further at the end of chapter 3 and in chapter 10.

Life Jackets or Surfboards?

You're in a boat and a tidal wave is coming that will likely roll the boat over: do you want a life jacket or a surfboard? The one will keep you afloat but leave you stranded. The other could be a lot of fun and give you mobility and control of your situation.

The church is facing a tidal wave. Many don't know it. Those that do and who recognize its proportions and potential to do harm (as they see it) aren't equipped to deal with it.

What is this tidal wave? It's a barrage of new data coming from the relatively new science of genetics. We humans have developed an understanding of genes and how they control our lives. They define our very being, our history, our strengths and weaknesses, our susceptibility to diseases (at least in part: a new science referred to as epigenetics is informing us of how other factors also play into these matters). And we've learned how to "read" our genes like a book. When we do read the genetic book, we find things about ourselves and other animals which conflict with another book: the Bible.

The natural response of many theists is to then declare the genetic book, or the science of genetics, as wrong, simply because the Bible has to be right. That was the problem I referred to in spelling out the first goal for writing this book.

But I've found it possible to read the Bible a little differently—same words, but a different interpretation—and the cognitive dissonance melts away. In order to be able to do that, though, one has to have a reasonable understanding of genetics and how the Bible came into our possession. Many theists don't have either of those tools.

So another goal of this book is to provide a basic level of understanding of genetics, as well as of another relatively older science which corroborates much of the new findings of genetics: paleontology and anthropology, the study of fossil bones and stone artefacts. Both of these are given in chapter 4. Next, I summarize the findings of anthropology (ch. 5) and of genetics (ch. 6) which directly impact Christian theology.

Chapter 7 teases apart how these new findings might be perceived to threaten a Christian belief, and provides some alternative ways to interpret certain biblical passages and ideas which seem to conflict.

The typical Western way to deal with a threatening force is to meet it with an equal or greater force. In contrast, a practitioner of the martial arts of East Asia will often take the force and simply redirect it. She may even take a punch that is thrown at her and pull it toward her, subtly shifting it over her shoulder in order to use the assailant's own strength and momentum to flip him head-over-heels. Likewise, trees will bend to the wind, and sea-weed will slosh with the tides.

An architect, on the other hand, will just make a building stiffer and stronger in order to stand up against a hurricane wind or earthquake. That is all fine until the wind or quake exceed the breaking point of the hardened material and the latter cracks and snaps.

The same thing happens when theists brace themselves against facts of science which seem to conflict with their theology. They may be able to stand firm for a while, but eventually their convictions will splinter. And the result is often a catastrophic failure. This is what I've seen many times in believers around me, but especially when students head off to the campuses of high school, college and university. Introductory Biology is mandatory for a great many students, and in those courses they are hit square in the face with evolution theory. Their faith system shatters and they end up rejecting it entirely rather than allow it to mature. Again, chapter 8 aims to provide theists with alternative interpretations of the book of God's Word which are more consistent with the book of God's Works.

Declaration of War

The church has often clashed with society over a variety of matters: morality; art; economics; international affairs; law; and so on. The specific conflict being addressed in this book is primarily the one over science. Their disagreements have flared up frequently, sometimes violently; history is filled with examples. Hollywood has capitalized greatly on it: their adaptations of Carl Sagan's *Contact* and Dan Brown's *Angels and Demons* are classic examples of this age-old battle. The next chapter of this book will focus on one particularly well-known (but grossly misunderstood) clash, but also show how the two parties were able to come into complete agreement: my hope is that this might set a precedent for the present growing conflict between the church and science over the new data coming from the science of genetics.

Too often, the underlying reason for the dispute is that the two use the same word in very different ways: examples of this which will be highlighted later in this book are the words: theory, myth, and random.

Words which are grossly overused and too often misused are: literal, inerrant, and infallible.

Some words have too much baggage and emotion attached to them: Creationism, faith, skeptical, and Intelligent Design, for example.

Or the groups will use different words to describe the same thing. An example of this would be the creationists using the phrase "each according to its kind" while scientists refer to species, genotype and phenotype (these terms, and this confusion, will be elaborated upon later in this book).

Sometimes they will create their own definitions for concepts which lie outside of their areas of expertise. For example, creationists referring to microevolution versus macroevolution, and creating the term "baramin" (derived from the Hebrew words *bara* [to make or create] and *min* [kind], and used in the sense of the word species), or atheists referring to Creationism or Intelligent Design. Similarly, some may create definitions within their camps, such as those who divide between Theistic Evolution versus Evolutionary Creationism.

One of the most important reasons for the friction between the church and scientists is that they defer to two different authorities on a given subject. It should come as no surprise to the reader that the primary example of this is the church using the Bible as a final authority on scientific matters. This too will be considered in great depth later in this book.

Finally, a major source of conflict and confusion arises from the different approaches used by the two parties, as explained in the next section.

Many have tried to resolve the conflict between these two camps, or at least have tried to understand and describe it. Some of these have authority in both camps: that is, they are at the same time bone fide scientists and accredited theologians. The attempts of some of these have been reviewed recently,[2] and from this the authors identified five different conceptual frameworks which people use to understand issues having a theological and/or a scientific basis. As befits many a sermon, alliteration is used to name these five frameworks to help remember the scheme. And to further aid in recall, they also refer to well-known individuals who exemplify the particular strategy (in order to help put a face to those names):

I. Conflict: Theology over Science. If the two camps disagree on the matter in question, the theological view is taken to be superior. Exemplars: Ken Ham, and the apologetics organization he directs (Answers in Genesis); Kurt Wise.

II. Conflict: Science over Theology. When the two disagree, the scientific view is taken to be superior. Exemplars: Richard Dawkins; Daniel Dennett; Sam Harris; Christopher Hitchens.

III. Compartmentalism. The two camps occupy completely separate realms, so there can be no agreement or disagreement. Exemplar: Steven Jay Gould, who coined the terminology non-overlapping magisteria (NOMA).

IV. Complementarism. Either camp can describe or explain the matter in part, but together they provide a more complete understanding. Exemplars: Denis Lamoureux; Francis Collins.

V. Concordism. The authors found it difficult to describe this concisely and distinguish it from complementarism. Concordism first assumes that the scientific and theological explanations will find agreement and harmony. There is also the sense that the two overlap in that biblical texts will reveal or contain certain elements of modern science: in this context, it resembles accommodationism (God having to dumb down the science in order for the ancient audience to get it). This view that scientific truths are hidden within the pages of the Bible and are only now in the modern era being recognized and explained is applied not only to the stories of creation and the flood, but also to various laws found in the Pentateuch: certain laws which might sound bizarre to modern ears are explained as being based on sound medical and scientific principles that only YHWH would have known when those laws were written and we are only just now coming to understand their rationale. Exemplars: Hugh Ross, and the apologetics organization he directs (Reasons to Believe).

---

2. Tenneson et al., "New Survey Instrument," 200–22.

The authors of this review also developed a questionnaire which aims to identify the viewpoint(s) that any particular participant might employ, and this survey instrument was then tested on five relatively different groups of people (1,491 in total):

1. A diverse group of science professors in the United States (n = 312);
2. A group of educators, pastors, and students in the Assemblies of God (n = 117);
3. A group of college undergraduates at a large Christian university in the South (n = 551);
4. Protestant pastors, educators, and students who attended a faith and science conference (n = 109);
5. Faculty and students from [Assemblies of God] higher education institutions in the United States (n = 402).

Their overall findings are interesting, although perhaps not too surprising: "The favored approach of all groups we studied was Complementarism. Three groups with strong religious commitment also used Concordism to a great extent. In some populations, a large number of people did not use any science-theology paradigms to evaluate theology and science propositions. YECs predominantly used Conflict: Theology over Science and Complementarism. OECs and evolutionary creationists relied mostly on Complementarism."[3]

Apologetics

The church has long employed apologetics in attempting to reconcile new scientific discoveries with existing theology. The term may be new for some readers, and misleading for many more because of its similarity to the word apologize. So a definition at this point would be helpful. Apologetics are defined as "reasoned arguments or writings in justification of something, typically a theory or religious doctrine." The word comes from the classical Greek legal system in which the prosecution made accusations (*kategoria*), and the defendant replied with an *apologia*. Apologetics (derived from that latter Greek word), then, are a defence or counterargument, and certainly not a statement of regret (in the sense of the English word apology).

The past few decades have seen the emergence of several apologetics organizations staffed by bone fide scientists. Four of these organizations merit brief introduction, given their worldwide influence in the science-faith dialogue today.

The oldest of these four is the Institute for Creation Research (ICR), founded in 1970 by Dr. Henry Morris (PhD in hydraulic engineering and author of many papers and books including the very popular book *The Genesis Flood*).[4] Their team now in-

---

3. Ibid., 200.
4. Whitcomb and Morris, *Genesis Flood*.

cludes several with a variety of advanced scientific degrees. Their mission statement is "to conduct scientific research within the realms of origins and earth history, and then to educate the public both formally and informally through graduate and professional training programs, through conferences and seminars around the country, and through books, magazines, and media presentations."[5] They hold the books of the Protestant Bible[6] to be divinely inspired, "infallible and completely authoritative on all matters with which they deal, free from error of any sort, scientific and historical as well as moral and theological."[7] As such, they hold a YEC point of view and promote the traditional interpretation of a historical Adam and Eve and the fall in the garden.

Answers in Genesis (AiG) was founded in 1987 by Ken Ham, who holds no advanced degrees but was previously a high school biology teacher. It is a rebranding of an older creationist organization—the Creation Science Foundation—which in turn arose by merging yet two other Australian creationist organizations in 1980. In 2005, a dispute led to the splitting off of the American and British branches of AiG, who retained the brand name, while branches based in other countries took the name Creation Ministries International. The staff at AiG include an MD and others with advanced scientific degrees in cell biology, molecular biology, astronomy and geology. Their statement of faith also refers specifically to the Protestant Bible: "The 66 books of the Bible are the written Word of God. The Bible is divinely inspired and inerrant throughout. Its assertions are factually true in all the original autographs. It is the supreme authority in everything it teaches. Its authority is not limited to spiritual, religious, or redemptive themes but includes its assertions in such fields as history and science."[8] Again, they are firmly within the YEC camp, and hold to the special creation of Adam and Eve and their subsequent fall in the garden roughly six thousand years ago. In 2007, AiG opened its Creation Museum in Kentucky, full of exhibits designed to promote a YEC perspective and to vigorously challenge Darwinism: some of these show humans and dinosaurs living together. Their latest addition is a Noah's ark–themed amusement park.

Reasons to Believe (RTB) was founded by Dr. Hugh Ross in 1986. Their relatively small team includes staff with advanced scientific degrees in astrophysics, biochemistry, and theology. They too hold the Bible (and again refer specifically to the Protestant Bible) to be "verbally inspired and completely without error (historically, scientifically, morally, and spiritually) in its original writings"[9] (the latter four words inserted here by RTB, and above by AiG, presumably to help address contradictions

---

5. Institute for Creation Research, "Who We Are."

6. Although "Protestant" is not explicitly stated, they do refer specifically to 66 books of the Bible (39 OT and 27 NT). Other versions have 73 books (the Catholic Bible), 76 books (the Greek Orthodox Bible), and 81 books (the Ethiopian Orthodox canon).

7. Institute for Creation Research, "Principles of Scientific Creationism."

8. Answers in Genesis, "Statement of Faith."

9. Reasons to Believe, "Our Mission: Engage & Equip."

and flaws, which some claim have since entered into the copies that we now have today). Their scientific training has enabled and even encouraged them to integrate into their theology those scientific data which conflict with the traditional Christian teaching on origins. As such, they are OEC in the sense that they accept the universe to be approximately thirteen billion years old and the earth approximately four billion years old, and they are more willing to use evolutionary terms and mechanisms when trying to explain various biological phenomena. However, their view on the origin of humanity is a hybrid between YEC and the standard Darwinian model. In particular, they hold that humans are a *de novo* creation of God, "distinct in kind from all other life on earth," and that "Adam and Eve, the first human beings" fell in the garden less than a couple hundred thousand years ago.[10] It seems they make this concession not because of any scientific data which lead them in that direction. Admittedly, a limited amount of data can be bent to conform to this idea, but the bulk of the data clearly go against it. Instead, this concession is made primarily to retain certain theological interpretations of Scripture. This will be explored in much more detail in chapter 8.

Perhaps in response to the recently increased acceptance within the Christian community for OEC views and a less Fundamentalist view of Scripture, the BioLogos Foundation was established in 2007 by Dr. Francis Collins, the former director of the Human Genome Project and current director of the National Institutes of Health (as of 2016). The organization affirms that "the Bible is the inspired and authoritative word of God" and "that God created the universe, the earth, and all life over billions of years . . . that the diversity and interrelation of all life on earth are best explained by the God-ordained process of evolution with common descent."[11] Their relatively large team includes some with advanced degrees in astrophysics, biology (genetics), ecology and evolutionary biology.

All four organizations publish a wide variety of articles on various subjects within this area (and their staff have collectively written a large number of books), and make them freely available on their websites and/or through periodic journals. Claims are made that those articles go through peer review, but it should be kept in mind that the group of peer reviewers are carefully selected for being consistent with a certain worldview.

The Purposes and Target Audiences of This Book

I have three primary goals for writing this book:

First, to point out the mistake of believing in things despite evidence to the contrary, or insisting on a certain belief which has no evidence behind it simply because it supports a worldview one chooses to hold. The tendency we all have toward allowing

---

10. Ibid.
11. Biologos, "What We Believe."

our presuppositions to interpret our science. I'm particularly sensitive to this coming from theists, but I will also show that atheists are prone to this as well.

Second, to show that we the church have in the past been quite able to jettison aspects of our theology when the science challenging it became overwhelming. This is important, because I feel the time is now that we need to reconsider certain strongly held theological views which are being challenged by new and provocative findings from the field of genetics.

Third, I want to prepare readers who are not particularly well-founded scientifically for this barrage of new data from genetics. I want to first provide a basic understanding of genetics and some of the strategies used by geneticists to study DNA. Hopefully this will help those readers as they wade through chapter 6, which provides overall summaries of some of those key findings from genetics which seriously call into question certain theological views.

I'm not necessarily trying to convince the reader to change their views on creation or the origin of mankind. Although they may not like the direction that some of these new discoveries are taking us, those readers who hold very different views about Scripture and the creation accounts nonetheless need to be made aware of the new discoveries if they want to stay relevant to their children who are facing these things daily on campuses all across North America. Or they will need this information as they provide spiritual leadership to their youth groups, or their churches, or to confused individuals who come to them for help in navigating through these new waters.

With these three goals in mind, I'm aiming this book primarily at an audience whose theology is based largely on YEC principles. Surprisingly, this includes many people who would describe themselves as having an OEC theology.

On the one hand, there will be those who hold a YEC view based on "a plain reading of Scripture." Generally speaking, these will refer to the Bible with adjectives such as inerrant, infallible, and final authority, and will hold firmly to a six thousand–year-old cosmological perspective. Some might refer to such people as Fundamentalists. Let me be clear and honest: I don't have an ax to grind against this group, but I do feel an over-exaltation of Scripture, which is a form of idolatry, has led to many problems for Christianity. It is this group which has given us snake handlers in certain eastern parts of the United States, Westboro Baptist Church in Kansas, and faith-healers all across the map (admittedly, all three groups and many others represent an extreme fringe of the spectrum of Fundamentalist beliefs). People who make rash decisions because they pin certain dates to the second coming of Christ (and then have to backpeddle when their predictions prove to be wrong). The ones who say hurtful or even hateful things about people from other religions and/or about people having views on human sexuality distinct from their own.

I would hope that in reading this book, they might be able to see that the Bible was not delivered to us in completed form, lowered down from heaven on a platter, nor whispered into the ears of the authors the way the Muslims believe about their

Quran and some Mormons believe about their Scriptures. It was written by humans who let divinely inspired ideas percolate through their own human limitations: their cultural biases and ancient Near East (ANE) and premodern Hebrew worldviews. Furthermore, I hope they come to see that those writings have been manipulated, edited, censored, translated and retranslated by other humans who inevitably added their fingerprints to the final product.

On the other hand, there will be those who hold an OEC view, and can accept much of evolution theory, even human evolution, but have not thought through what this does to their theology, which was handed down to them through a church history steeped in YEC tradition. Consciously or otherwise, their theology is based on a literal reading of Genesis, including the origin of all humanity from a common pair (which Genesis names as Adam and Eve). So while they might confidently and proudly declare that they do not sense any tension between their Christian faith and science, they are unaware that key aspects of their faith are rooted in YEC principles which clash against their OEC worldview. If they wish, this subgroup can actually skip over chapters 4–6 of this book, since those three chapters are intended to inform readers of the evidence supporting the theory of evolution, and go straight to chapter 8, which deals with the theological implications of an OEC view on their theology.

The demographics of these two subgroups can be broken down into three general categories.

First, I'm particularly concerned about students stepping onto campuses of high schools, colleges and universities, especially the ones who come from a religious background which avoided any aspects of the science-faith discussion, and more so those coming from backgrounds which took an adversarial view on science. Too many of them are simply not prepared for the inevitable encounters with Big Bang and evolution theories, other world religions, and diverse moral issues which are found behind every corner of those campuses (except private schools, where those influences are carefully excluded, which still leaves the individuals unprepared for those encounters later in life). Far too many end up rejecting their faith, or letting it go dormant, simply because they're unable to hold those ideas up in juxtaposition. The cognitive dissonance overwhelms and they relegate the Bible to being a fairy tale and their religious upbringing as entirely antiquated. The latest Pew Forum on Religion and Public Life has shown that Americans between the ages of eighteen to thirty-four are by far the least religious among the various age groups, and 34 percent of them claim no religious affiliation whatsoever[12] (up from 25 percent in 2007), despite the fact that America is generally seen to be a very religious nation and that 75 percent of these survey respondents were raised in houses that were self-identified as Christian.

Second, I'm also hoping that anyone of any age who was once a believer but found it necessary to "throw the baby out with the bathwater" will realize it wasn't necessary to do so. Many people who thoroughly accept evolution theory, who may

12. Pew Research Center, "America's Changing Religious Landscape."

be employed in jobs which are based on such acceptance (researchers; teachers; professors; physicians), are committed believers. This group may also include motivated adults actively engaged in the faith-science dialogue, particularly as it pertains to the scientific and theological questions of human origins.

Third, there are those authority figures to whom people will turn for answers to questions arising from science-versus-faith discussions. My advice to youth leaders, campus ministry leaders, and parents is this: teens and twenty-somethings will likely go to college or university, and those that get into any academic stream in the life sciences will almost certainly take Biology 101, which will positively confront them with the theory of evolution. Are they ready for that? Pastors: you can be sure there are people in your congregation who read books and newspapers and watch TV, and who are struggling to reconcile those inputs with their religious views (as well as your own, which may not be the same). These authority figures may balk because they feel they're not scientifically trained, and don't know what to say or think about some of these issues. This book attempts to prepare them for those discussions.

Overall, the target audiences will be ones that are well-read and well-educated, but which may not have a strong scientific foundation. More importantly, the target audience will be one which seeks to understand the new findings from the anthropological and genetic sciences which point to a gradual evolution of mankind over millions of years, and how those may affect our theological beliefs. This work comprises an in-depth review of that new evidence, and includes two chapters which will provide the reader with a basic grounding in the scientific principles and understanding needed to evaluate the genetic and anthropological data.

Why Is This Important?

In conversations had while researching this book, some theists would express the sentiment that this question regarding the historicity of Adam and Eve is not all too important to them. They would often affirm, "I believe that Adam and Eve are literal, but even if God chose to use the process of evolution, I still believe that mankind is fallen and needs redemption." However, I would suggest that many of those theists have not carefully thought about what they believe and why. Jesus told two parables in the context of "counting the cost" of deciding to follow him—one of a man who started to build a tower, the other of a king about to go to war[13]—and highlighting the mistake made if one doesn't carefully consider the basis of making any such large commitments. Likewise, Paul instructs Timothy: "Do your best to present yourself to God as one approved . . ." (the King James translates this passage as "Study to shew thyself approved unto God"), ". . . a worker who does not need to be ashamed and who correctly handles the word of truth."[14]

13. Luke 14:25–33.
14. 2 Tim 2:15.

At the most fundamental level (if one will pardon the use of that adjective), this question is important for anyone who holds the Bible to be inerrant and infallible, and they interpret it literally. They find the evidence that the universe has been around for billions of years, and that life evolved on earth approximately four billion years ago, gradually unfolding to reveal the ecosystems we see today, to be completely unacceptable. For them, the Bible clearly describes a very different explanation of origins, and if that description is shown to be false, the foundation for their faith is shattered. If they can't take every word in the Bible at face value, they feel forced to have to reject the whole thing. And as the evidence for the evolutionary paradigm grows all around them, their faith buckles. Like the metaphorical house built upon the sand, "The rain came down, the streams rose, and the winds blew and beat against that house, and it fell with a great crash."[15]

Even for those who do not describe themselves as Fundamentalists, who are able to accept a less stringent (some say "literal") interpretation of the Bible, a great deal of Christian theology is founded upon the fall in the garden approximately six thousand years ago.

As will be elaborated at great length later in this book, Paul builds a tremendous cornerstone of theology on the concept of the first Adam and the Last Adam:[16] and yet science tells us there was no first Adam.

Christians often refer to being "fallen creatures, living in a fallen world"; and yet "the fall" is looking very much like a metaphorical event.

What does one do with the concepts of original sin, and inheriting guilt and a sinful nature? How far back through the human line do we extend these? At any arbitrary point in real earth history where we might declare that God revealed himself to humans who then rebelled, those humans would have had predecessors, so-called "prehumans." Where then do we draw the line? Should that line include other hominids, such as Neanderthals? Or should it only include *Homo sapiens*, and if so, on what basis? Why should humans be treated so differently from the animals if we evolved alongside them? What does it mean to be created "in the image of God"?

What about the key theological word *redemption*, which the *Oxford Dictionary* defines as "the action of regaining or gaining possession of something in exchange for payment, or clearing a debt"? This word is entirely appropriate in the context of the Genesis account in which the primordial pair take part in an act of defiant betrayal or rebellion against God, and in fact many dictionaries include that context in their definitions for this word. Do these terms *possession*, *payment* and *debt* take on new meanings if one accepts the idea that humans simply descended through an evolutionary process over millions of years along with other animals?

If humans have wandered the face of the globe for eons, and have always had this inner sense of a Great Being, and in so doing developed a wide variety of religions

15. Matt 7:27.
16. 1 Cor 15:45.

very different from the one introduced by the Semitic race in Mesopotamia three or four thousand years ago, should we continue to be so adversarial or judgmental of those other attempts to find the Divine?

For these reasons, and many more, I think some homework is warranted in this area.

Fair Warning

I heard another author being interviewed about his own book directed at an overly literalistic interpretation of Scripture, and in the process being cautioned that his book might hurt people's faith. His reply was that neglecting to write the book would also hurt people's faith. I can relate to that dilemma. I do feel that the status quo with respect to the relationship between science and faith in some Christian circles is an unhealthy one, and people are getting hurt either way.

This book is likely going to stretch your faith. It will ask hard questions for those who insist that the Bible is inerrant and infallible, who otherwise find it hard to accept that it's just as much a human document as a divine one, and/or who insist that it doesn't just *contain* God's word, but that it *is* God's word (in the sense of a dictation). But before quitting too quickly or easily, ask yourself another important question: if your faith is so fragile that it cannot withstand being exposed to mere scientific facts, is it really a strong faith? Is it built on rock or on sand? Stretching it, probing its depths, and letting go of those parts that actually hinder it might be a good thing.

Making the decision to avoid facing this hot new issue will make one irrelevant to today's society and put one in the same camp as the Flat Earthers and the Geocentrists standing with their heads in the sand. Even worse, those who do that become stumbling blocks for others coming to faith.[17] By not adequately preparing their children, or even intentionally instilling in them a poorly substantiated theology and anti-science worldview, they risk watching their children reject any faith when the latter venture into an academic setting (high school, college, university) or otherwise encounter the facts in their everyday lives. By presenting themselves as willfully ignorant or defiant deniers of this emerging science, they compromise their ability to be ambassadors of Christ and contribute to the rejection of the gospel message by the hearers of their testimony.

On the other hand, if you are open to the possibility of finding a faith that is true to the words on a page as well as to the observations from a test tube, and want to know how it's possible to fully maintain a Christian faith (in fact, one that I think can be stronger because it fully embraces a larger set of facts), then let's move forward together.

---

17. 1 Cor 8:9.

Mea Culpa

In the arena of Christian apologetics, there are a great many fully accredited theologians weighing in on the biggest origins questions which science is trying to answer (the origins of matter, life, and humans), and interpret scientific findings despite lacking scientific qualifications. Even in just the past decade or two, many of these theologians have written specifically on the subject of Adam and Eve, including: John Walton;[18] C. John Collins;[19] John MacArthur;[20] Peter Enns;[21] Alvin Plantinga;[22] J. Richard Middleton,[23] and William D. Barrick;[24] (also see a collection of essays from theologians on this subject).[25] I have no problem with them voicing their theological arguments, but become uncomfortable when they attempt to incorporate science into their arguments and in the process distort and mishandle the data to fit their ideas, or if they make theological declarations which completely ignore contradictory data.

I ask them, and the reader, to accord to me the same kind and degree of leniency vis-à-vis being a scientist stepping into theological waters. Although I myself am not formally accredited in theology, defined in the *Oxford Dictionary* as "the study of the nature of God and religious belief." I have spent decades in that pursuit (see About the Author, p. 295). In this book, I share how I've successfully married my world of science with my world of theology. This has worked for me, where it wasn't working before. My voice adds to those of other accredited scientists who have also written on the mutual interactions between their own worlds of science and of faith: Francis Collins;[26] Hugh Ross;[27] Fazale Rana;[28] Denis Alexander;[29] Deborah and Loren Haarsma;[30] Michael J. Behe;[31] David Montgomery;[32] and John Lennox,[33] to name a few. And yet others who, in addition to having scientific credentials, also have advanced degrees in theology or

---

18. Walton, *Lost World of Genesis One*; Walton, *Lost World of Adam and Eve*; Walton, "Historical Adam."

19. Collins, *Adam and Eve Really Exist*; Collins, *Science & Faith*; Collins, "Adam and Eve as Historical People," 147–65.

20. MacArthur, *Battle for Beginning*.

21. Enns, *Evolution of Adam*.

22. Plantinga, *Conflict*.

23. Middleton, *Liberating Image*.

24. Barrick, "Historical Adam."

25. Madueme and Reeves, *Adam, Fall and Original Sin*.

26. Collins, *Language of God*.

27. Ross, *More Than a Theory*.

28. Rana and Ross, *Who Was Adam?*

29. Alexander, *Creation or Evolution*.

30. Haarsma and Haarsma, *Origins*.

31. Behe, *Darwin's Black Box*; Behe, *Edge of Evolution*.

32. Montgomery, *Rocks Don't Lie*.

33. Lennox, *Seven Days*.

philosophy, including: John Polkinghorne;[34] Denis Lamoureux;[35] William Dembski;[36] and Stephen C. Meyer.[37]

Each of those authors have presented slightly different perspectives on the questions of origins, and mine will be somewhat different yet again. Even among these scientific colleagues, I sometimes see examples of molding the data and ignoring outliers in order to maintain some congruence with long-held theologies and church traditions. I've struggled against this myself in the recent past, but am trying to purge myself of this kind of presupposition. Which means asking some pretty tough questions about the theological implications, as I do in chapter 8.

Instead, I advocate simply spreading the scientific data and the scriptural writings on the table, without any preconceived biases, and looking at what they tell us. And if they challenge any particular dogma, even concepts as central as original sin and the atonement, we don't reflexively throw a blanket across that part of the table and pretend that evidence isn't there, or bend the data and manipulate them for the sole purpose of hanging on to the dogma or the church tradition.

On the other end of the scholarly spectrum, the internet and print media are also chock full of the views of those who have no academic credentials whatsoever and who sometimes choose to not even look at the data. I have heard them say that "the Bible is so simple, even a child can understand it." On one level, that may be true. But the Bible deals with such complex issues—ones that span the millennia of human history in every corner of the globe, and which strike at the very core of our being—that simple phrases such as "God did it," "the Bible says it, I believe it, and that settles it," and "a plain reading of Scripture" are hopelessly unhelpful, to say the least. The Bible, and indeed many other Scriptures, contain the insights of countless authors representing a diverse array of peoples and times, on the three big questions in life which everyone must answer: "Who am I?" "Where did I come from?" and "How must I live?" And we need to mine all the Scriptures as well as the growing mounds of scientific data for all they're worth for answers.

In this book, I will ask a number of questions which the reader may find quite challenging. I will also provide alternative interpretations for the reader to consider, ones which may be seen to be incompatible with traditional ideas. I would ask the reader to take careful note when these are posed as legitimate questions and suggestions, rather than as statements of fact, and consider them on that basis. My purpose is to stimulate thought and discussion, as well as a search for a theology which is fully compatible with the scientific data.

---

34. Polkinghorne, *Quarks, Chaos and Christianity*.

35. Lamoureux, *Evolutionary Creation*; Lamoureux, "No Historical Adam"; Lamoureux, "Ancient Astronomy"; Lamoureux, "Ancient Biology"; Lamoureux, "Beyond the Cosmic Fall."

36. Dembski, *Intelligent Design*.

37. Meyer, *Signature in Cell*.

Conclusion

We all have worldviews which shape our experiences in life. Sometimes, those perspectives act like sunglasses which exclude certain aspects of reality: we wear them because they feel comfortable, and make it easier to look out at the harsh glare of reality. That's alright as long as we remain able to remove the shades when the occasion calls for doing so. Unfortunately, theists as a group haven't demonstrated the best track record in this respect when it comes to the intersection between faith and science, especially on the questions of the origin of humanity. In the next chapter, I'll use a different experience in church history as an example of how we previously resisted vigorously the advances of science but eventually learned how to embrace those and keep our faith intact.

2

A Precedent-Setting Case
The Origin of the Cosmos

FOR SOME READERS, THIS chapter might seem at first to be irrelevant for a book aimed at meeting the challenge brought on by the field of genetics against theology. Bear with me. I want to use this chapter to set a precedent: the conflict unfolding today is certainly not the first time that the church has had to wrestle with new scientific findings, and come out on the other side having massively overhauled its fundamental theological tenets.

Most people in the Christian world don't know this, but for thousands of years, Judeo-Christians believed in a three-layered universe surrounding a three-layered earth built on pillars which also supported a large overturned bowl over our heads.[1] As bizarre as this might sound to our ears now, it stems directly from direct observations of the world around us and from passages in Scripture. But before describing how we got that model of the universe from Scripture, I need to set the stage.

This book will begin from an OEC point of view. There is just far too much evidence to indicate that the cosmos is billions of years old, not merely ten thousand. This is not the place to rehash that evidence, nor will I explore the origin of life in general or the evolution of the various species which have inhabited earth: a great number of other books are available for that. Instead, the focus of this book will be the origin of humanity. As an OEC believer, I am convinced that human history did not begin six thousand years ago: *Homo sapiens* go back many hundreds of thousands of years. No one would dispute that the Old Testament is a product of ancient Hebrew culture, although there is much discussion about exactly who wrote it and when, and about what exactly is meant by divine inspiration. It is also a fact that the Hebrew culture developed within a Mesopotamian culture comprised by the Akkadian, Sumerian, and Egyptian empires which ruled the region for many millennia. Some believe Moses is

---

1. Lamoureux, *Evolutionary Creation*; Lamoureux, "Ancient Astronomy"; Lamoureux, "Beyond the Cosmic Fall."

the primary author of the Pentateuch: Scripture says he was raised in Pharaoh's household, which would have given him an extensive Egyptian education and a gold-plated card to the massive libraries of the Egyptian empire. Others believe the Pentateuch was written by Jewish scribes during the Babylonian captivity: the latter too would have had regular access to the libraries of the Babylonian empire. In both cases, the writer(s) would have been quite familiar with a scientific framework that pervaded the thinking of the world at that time. Those authors were steeped in the cosmology of their day. And unless one holds to the view that the author(s) of the Pentateuch merely transcribed the dictated words of God—a phenomenon I've heard referred to as divine ventriloquism—then there's no way to deny that those other ANE cultures had an influence on what the authors of Genesis wrote and believed.

Many features clearly show a Mesopotamian or ANE influence on the writing of the first half of the book of Genesis.[2] The first few chapters of Genesis use the word *elohim*—the plural form of the deity—to refer to God: "let *us* create man in *our* image" (my emphasis). The creation and flood accounts in Genesis parallel suspiciously closely certain stories in the *Epic of Gilgamesh*. The lists of patriarchs and their unbelievably long lives in chapters 5 and 11 of Genesis echo the Babylonian kings lists which named royal figures who ruled for thousands of years. The garden of Eden is described as being situated between two Mesopotamian rivers, the Tigris and the Euphrates (whether that garden actually existed or not isn't the point here; instead, the context given is clearly a Mesopotamian one). The Tower of Babel story in Genesis describes "the whole world" as having settled down in the plain of Shinar, which is in Mesopotamia. Many of the cities which were said in Genesis to be settled by the descendants of Adam are Mesopotamian ones. Abram and Sarai are called out of Mesopotamia[3] where their ancestors had worshipped other gods.[4]

So before delving into the Hebrew thinking on how the universe was organized, we need to understand the milieu in which that Hebrew thinking emerged: the well-developed cosmologies of those other empires. And for that, we need to see how the latter in turn bubbled out of a human psyche that had been evolving for millennia.

Ancient Cosmology

Take a walk with me back in time. Let's get into the mind of someone looking up at the sky thousands of years ago, long before modern scientists began to explain to us the things we observe up there. But it's absolutely essential that you take this walk without any of our modern baggage.

It's harder than you might think. It's like trying to enjoy a joke or figure out a riddle in the same way that you did the first time. When you first listened to that joke

2. Hooke, *In the Beginning*, ch. 5.
3. Gen 15:7.
4. Josh 24:2.

being told, or first tried to understand that riddle, it just did not make any sense at all. It held your attention because of the mystery and the novelty. You entertained all kinds of possible outcomes, but couldn't be sure if you had figured it out. And it was that uncertainty which kept driving you to try.

After a long pregnant pause, the joke-teller finally delivers the punch line. Or once you'd finally given up on the riddle, you turn to the back page and look for the answer. At that moment, the answer is so plainly obvious. It all makes sense now! And the next time you hear the joke being told, you nod your head and hurry the teller along because you already know where she's going with it. The next time you consider the riddle, you immediately see the answer: now it's child's play.

The same thing happens when you watch a magician perform an amazing trick, and then watch it again from the side or from behind and you can see exactly what they're doing with their hands: once you've learned their secret, the trick will never amaze you again.

It's hard to go back and recapture that sense of confusion and bewilderment you had when you confronted these problems for the first time. But that's what I'm asking you to do now. To look at the sky without any of our modern scientific explanations of what you're seeing.

When the ancient humans looked up, they had no telescopes, binoculars, satellites or any other tool to tell them how far away the sun, moon and stars were. There was absolutely no reason for them to not see those as just lanterns a few miles up at best, rather than being unfathomably bigger than earth and many millions of miles away: the stars and planets might otherwise look like the campfires of other human settlements far off on other hillsides. They couldn't know that the sun's diameter was four hundred times bigger than that of the moon: to the uninformed person they look like they're identical in size, which is why we can experience both a solar eclipse and a lunar eclipse. They couldn't know that many of the stars were millions of times bigger and hotter than the sun. How could they? They had no concept whatsoever of "billions" or "light years" or things so big that they could swallow up thousands of earths.

Stand in the place of an ancient farmer, five or ten thousand years ago, tilling his small plot of land. He sees the sun make its appearance on the horizon and he basks in its warmth. As the day progresses and he labors through the field, pulling weeds and squishing bugs, he's aware that the sun climbs higher in the sky, reaches its zenith, and then approaches the opposite horizon, eventually disappearing behind it. The next morning, just

Figure 1. "Earth-rise" over the lunar horizon, a paradigm-changing perspective that no human could have conceived of when the Bible was being written. Taken in 1968 by William Anders. Used by permission from NASA.

like Bill Murray's *Groundhog Day*, it all happens again. And again. Over and over and over. At some point, the farmer might ask his partner what she thought the sun did at night: how it returned to the same spot every morning. "It doesn't just return the way it came, because we'd see that. And it can't simply plow its way through the dirt like a gopher." (Remember, they don't have our understanding of the shape of earth. They've never seen a picture of earth as a globe rising over the lunar horizon, as we have (figure 1). In fact, they likely would never have heard anyone talk about it as a sphere. From their point of view, they look in all directions, and all they see is a flat horizon. Mother Earth must be flat, and they had no comprehension of how deep her foundations went, or upon what they were laid.)

"Perhaps," she says, "there's some kind of underground cavern through which the sun returns to its point of rising every morning."

He shrugs his shoulders.

At night, our farmer would look at the sky and gaze with wonder at the stars, awestruck by their number, their subtly different colors and brightness, how they twinkled and seemed to form patterns in the night sky. He notices that they, too, slowly trace a path across the sky which begins at the same part of the horizon as the sun's rising and ends at the opposite horizon. Except the moon did something completely different. The part of the sky in which it appeared was different from one evening to the next, and its shape always varied from being a round ball to a sharp crescent; and yet there was still some sense of regularity in both these changes. Unlike the sun and the stars, it might be visible during the day or the night, as if it had a mind of its own, but it was still somewhat comfortingly predictable.

They might have given names to the moon, simply because it seemed to have a mind of its own and therefore had a sense of aliveness. And to the stars because they formed patterns in the sky that looked like people or animals. And to the sun because of its power and warmth and the good feeling and life-giving energy it provided.

But they certainly would have invoked ideas of great celestial powers when the sky did strange things.

Like eclipses. Sometimes the sun would suddenly disappear in the middle of the day for no apparent reason. They would not have known that the moon was now in front of it, because whenever this kind of thing happens only the dark side of the moon is facing earth, so they wouldn't at all realize that the moon was even there. The only conclusion that made sense was that some Great Being had overpowered the sun! They would interpret the relative totality of the eclipse to indicate which of these two great powers was winning the conflict at that moment, but the ultimate reappearance of the sun would confirm its absolute power over everything.

On very rare occasions, the moon might suddenly become orange-red in color, and go dark, and might even disappear altogether, again for no reason. This always happened at night, when *Homo sapiens* tend to feel the most insecure and jump at every noise that comes out of the dark. (Remember, they would not think about the sun being behind a spherical earth, but instead would think that the sun was somewhere in its underground tunnel making its way to its starting point for tomorrow's sunrise.) Eventually, though, the moon would return, and all things would be back to normal.

Otherwise, nearly any other night that the clouds didn't obscure the sky, the ancient observer would notice that one or more of the stars suddenly streaked toward the ground. They had no reason to think that these "stars" had originated billions of miles away in a vacuum of space and had only become visible when they entered earth's atmosphere. How could they ever understand the concept of atmosphere, or gases, let alone vacuum? Instead, it was as if the star(s) had been struck down by one of the great sky powers. During a meteor shower, this might seem like a cosmic battle being waged. The humans had no way to see the combatants: perhaps they were invisible spirits, or were hidden behind some kind of heavenly fortress walls. But clearly something fearsome was going on up there, and one did not want to be in the line of fire. So they would retreat to their tents and talk about it and perhaps offer up some kind of sacrifice.

That kind of thinking would have dominated for thousands of years, and all kinds of religions would be built up around that: Babylonian; Egyptian; Zoroastrian; Mayan; and Stonehenge.

Sumerian/Babylonian Cosmology

The Babylonians had a very sophisticated cosmology by 1800 BC. They seem to be the first to have applied mathematics to predict astronomical events, such as the changes in the length of the days during a year, and the rising and setting of several of our planets which can be observed without telescopes. Surprisingly, they seem to have done this without understanding fully the concept of planets orbiting around the sun because, in the oldest piece of human literature we have today—the *Epic of Gilgamesh*—they describe an underground tunnel through which the sun returns every night to its point of rising.

They, in turn, derived their model from a Sumerian cosmology which was developed one or two thousand years before them. The earliest recorded Sumerian history we have goes back to approximately 2900 BC, but scholars agree that the Sumerian civilization began around 4000 or 4500 BC.

The ancient Sumerians believed that earth was surrounded by a salt-water sea and the universe was a giant dome placed on top of earth (I'll explain in the next section why this would be the case). Underneath earth was an underworld and a freshwater ocean. Of course, they had a whole pantheon of gods. Even the earliest of Sumerian literature—the most ancient of recorded human history we have—referred to the deity of heaven (Anu), of the air (Enlil), of freshwater (Enki), and of earth (Ninhursag), plus many other lesser gods.

All of this literature and cosmological knowledge would have been found in the Egyptian libraries at the time that Moses is said to have been raised by the daughter of Pharaoh (the ruler of the Egyptian empire), as well as in the Babylonian libraries during the Israelite captivity and exile approximately a thousand years later. Can there be any doubt that the author(s) of the Pentateuch, whoever they were, were not influenced by this well-developed cosmology of two major civilizations that ruled the world for thousands of years?[5]

Hebrew Cosmology: "A Plain Reading of Scripture"

There is debate within the biblical scholarly community as to when the Pentateuch, which includes the book of Genesis, was written and by whom. Was it collated by different groups of priests around the time of the Babylonian exile in the middle of the sixth century BC by gathering together other documents a few hundred years older? Or was it Moses himself after the Exodus from captivity in Egypt at some point in time between 1500 and 1000 BC? Or did Moses collate writings from other authors long before him? I've heard at least one theory that Moses took writings from Adam, writings from Noah, and writings from several other patriarchs and stitched them together into one book. The most recent Old Testament (OT) manuscripts we have today have been dated to just a few centuries BC, so they can't be used to settle the question of who wrote the stories which seem to stretch back several thousand years BC. Nor can it be decided without careful dissection of the wording, which in turn involves extensive interpretation. Advocates for dates of authorship occurring around the time of the Babylonian exile point to the mixture of writing styles and repetitions of stories, sometimes with details altered, as evidence of the work of many different original authors over an extended period of time and/or from different isolated parts of the region.[6]

The Jews trace their origin to Abraham and Sarah,[7] who came from the city of "Ur of the Chaldees,"[8] which was of Sumerian origin and even older than the Assyrian and Babylonian cultures. Joshua reminds the Israelites that "long ago your forefathers, including Terah the father of Abraham and Nahor, lived beyond the river and worshipped other gods."[9] Thomas Cahill[10] gives a very readable summary of the changes in thinking as this family transitioned from Sumerian polytheists in Babylon to Semitic monotheists in Canaan. It was those wandering nomads—Abraham's family,

---

5. Hooke, *In the Beginning*, 3.
6. Hooke, *In the Beginning*.
7. Named in the earlier part of their story as Abram and Sarai.
8. Gen 15:7.
9. Josh 24:2.
10. Cahill, *Gifts of Jews*.

who were thoroughly Mesopotamian in their language and culture—that gave us the earliest stories in the Bible.

Like the Egyptians, Babylonians and Sumerians who lived thousands of years before them, the descendants of Abraham and Sarah developed their own cosmology, and wrote about it not just in the first few chapters of Genesis, but also throughout the other books of the OT and the New Testament (NT). That cosmology was presumably influenced by Divine inspiration, but it was undeniably also influenced by the cosmology that the Babylonians developed millennia before them. Irrespective of the degrees to which the reader believes that the Hebrew cosmology was divinely inspired, borrowed and adapted from their ANE neighbors, and/or developed by the Hebrew authors themselves, most of the Christian world has long ago discarded that ancient Hebrew cosmology.

Let us examine further the biblically based cosmology that contemporary Christianity has almost universally discarded. In the opening paragraphs of this chapter, I referred to a three-layered heaven and a three-layered earth.[11] This bizarre cosmology is directly and unequivocally rooted within "a plain reading of Scripture."

Right in the very first verse of the Bible, we read about God creating the heavens. Note the plural. There are many other passages in which the plural form of this word is used; in fact, too many to list here. In other places, this plurality is emphasized by phrases like "the heavens, even the highest heavens."[12] Paul talked about knowing someone who had been caught up into "the third heaven."[13] When they did this, the Hebrew authors were thinking that: the nearest heavenly layer in front of the firmament was the one in which moved the birds, sun, moon, and stars; the firmament held back a second layer comprising "the waters above"; beyond that second layer was a third celestial one in which God and the heavenly hosts resided (Paul's "third heaven"[14]).

The stars appeared to the ancient observer to be simply little pinpoints of light—like the campfires of bands of people a few miles away—hanging somewhere high in the sky and moving together in an arc across the sky. They had no way to explain how these lights might simply float in mid-air a couple miles up and move together in such precise unison, so they envisioned them hanging from a big dome or upside-down bowl placed on top of the flat earth, held up by the mountains and curving down to the seas.

When raindrops fell down from the sky, they had no understanding of the water cycle, including the concept of evaporated water condensing out of thin air. The only way for them to comprehend this mysterious water from the sky was to propose

---

11. Lamoureux, *Evolutionary Creation*; Lamoureux, "No Historical Adam"; Lamoureux, "Ancient Astronomy."

12. Deut 10:14; 1 Kgs 8:27; 2 Chr 2:6; 2 Chr 6:18; Neh 9:6.

13. 2 Cor 12:2.

14. 2 Cor 12:2.

that there was a large body of water beyond the dome (the "waters above" in the first chapter of Genesis) and holes in the dome which let the water drip through. How else would a completely scientifically illiterate people explain water falling out of the sky? Noah's flood is said to have resulted, in part, when "the floodgates of the heavens were opened."[15]

When they explored the springs and rivers around them, they found that water seemed to just continually flow out of the ground: in many cases, this flow seemed to never stop, even when there hadn't been rain for several months and all the surrounding fields had long since dried up. "Where does all this water keep coming from?" they might ask. It just kept pouring out from somewhere underneath ("the waters below" in Genesis), so there must be yet another large body of water deep underneath the earth, gushing to the surface. This underground body contained fresh water, very different from the salt-water seas that surrounded the lands.

These simple misinterpreted observations led the ancient Hebrews to see the dome in the sky as something solid (the Complete Jewish Bible uses the word dome, while other translations of the Bible use the words firmament or expanse or vault). When describing God's control over various aspects of weather, one of Job's friends says: "Can you join him in spreading out the skies, hard as a mirror of cast bronze?"[16] This solid dome was thought to hold back the waters above (from which the rain trickled out) and separate them from the underground waters below the earth (from which the rivers came forth).

With these interpretations of observed nature (which was the leading science of their day) in your mind, reread the creation story in Genesis from the viewpoint of a premodern author and pay particular attention to the words firmament or expanse or vault. In particular, notice the placement of things *within* the firmament (sun, moon and stars) or *beyond* the firmament (the "waters above"). Isn't that placement *opposite* to what you would expect, based on today's understanding of the cosmos? That is, we would place the clouds of water within that bowl (representing our atmosphere), and the cosmological bodies beyond the bowl.

This solid dome, and even the flat earth itself, were described as being set on pillars.[17] Those pillars provided a firm foundation: several other passages stated that earth is established and will never be moved.[18]

And finally the third layer of earth—the one under the land and "the waters below"—was the place of the dead. The Hebrew authors, as did everybody throughout the ANE, firmly believed that souls went to some kind of place deep in earth: "Sheol" in OT Hebrew; "hades" in the Greek NT; and "hell" in modern English. In the Complete Jewish Bible, David talks about how God "lifted me *up* from Sheol" (Ps 30:3), Job

15. Gen 7:11.
16. Job 37:18.
17. 1 Sam 2:8; Job 9:6; Ps 75:3.
18. 1 Chr 16:29–30; Pss 93:1; 96:10; 104:5; 119:90.

describes "the limits of the Almighty as being *higher* than the heavens above . . . and *deeper than* Sheol,"[19] and God talks about his anger "that burns *down to the depths* of Sheol" (my emphasis).[20] Isaiah and Ezekiel refer many times to people going *down* to hell, or hell being brought *up* against nations. Jesus referred to a city being "brought *down* to hell" (my emphasis).[21] Paul referred to the spirits in the underworld.[22] Even the Apostles' Creed, written in the first century AD, says that Jesus "*descended* into hell" (my emphasis).

Scripture also made it clear that earth did not move.[23] The sun, on the other hand, did move over earth: there are numerous repetitions in the Scriptures of phrases like "the rising of the sun" and "the going down of the same."

Some are quick to point out that this is simply phenomenological language: phrasing based on what something looks or feels like even though the author doesn't truly believe it literally (like saying "my heart sank when I heard the news" or "my car was flying down the highway").

However, the ancient Hebrews (and others) *did* believe the sun physically rose and set. The Teacher in Ecclesiastes specifically described the sun's orbit: "The sun rises and the sun sets, and hurries back to where it rises."[24] Likewise, the psalmist describes the sun stepping out of its tent in the heavens "like a bridegroom coming out of his chamber, like a champion rejoicing to run his course. It rises at one end of the heavens and makes its circuit to the other."[25] There were also interesting passages which describe interruptions to the forward movement of the sun: it "stopped in the middle of the sky and delayed going down about a full day" at Joshua's command,[26] and another one in which the sun went "ten degrees backwards" as a miraculous sign for King Ahaz.[27]

Put this all together and you have the Bible full of references to a three-layered cosmos, a solid firmament or dome separating "waters above" from "waters below," and a flat three-layered earth (including the lands we walk on, the "waters below," and Sheol) built on pillars and foundations, over which the sun and planets move (figure 2).[28] Their revolution around a central spherical earth came later, when the church melded together this Hebrew model with Greek thinking.

19. Job 11:8.
20. Deut 32:22.
21. Matt 11:23; Luke 10:15.
22. Phil 2:10.
23. 1 Chr 16:29–30; Pss 93:1, 96:10, 104:5, and 119:90.
24. Eccl 1:5.
25. Ps 19:5–6.
26. Josh 10:13.
27. Isa 38:8.
28. Hooke, *In the Beginning*, 20.

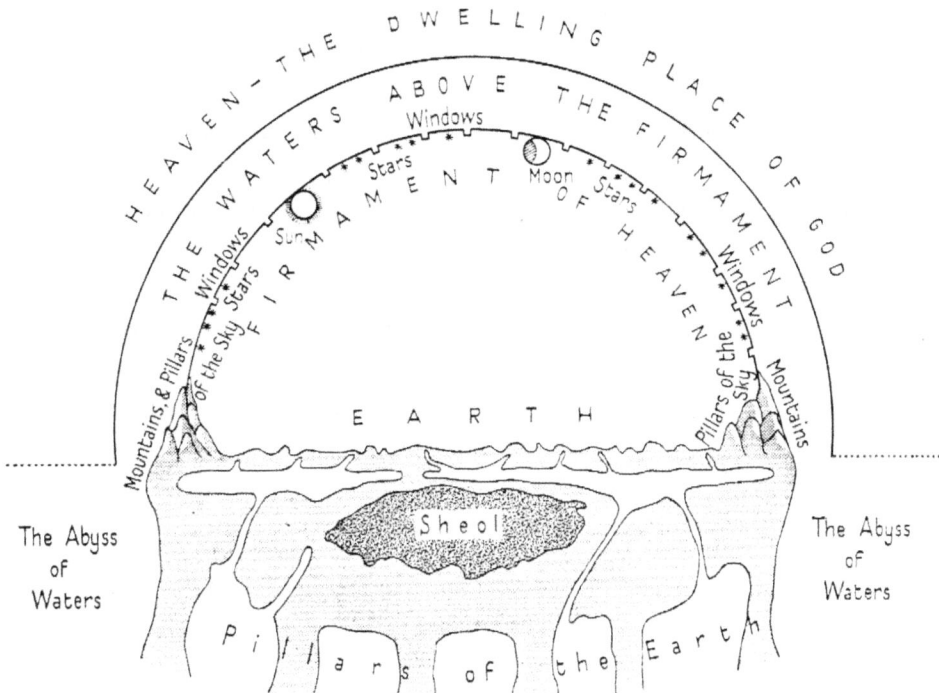

Figure 2. Hebrew model of the cosmos. From In the Beginning, page 20, by S. H. Hooke. Used with permission from Oxford University Press.

If you proclaim *sole scriptura* as a rallying anthem, then you should be pushing for this Hebrew cosmological model to still be taught in grade schools today (and for the Hebrew understanding of weather to be included in our college and university courses on meteorology: see p. 63). Needless to say, we just don't think that way anymore: this point will be important when we come to an entirely different conflict between "a plain reading of Scripture" and other scientific discoveries (chapter 6).

Ancient Greek cosmology

While it goes without saying that the cosmology of the early Christian church had one foot firmly planted upon ancient Hebrew cosmology (which itself was clearly influenced by Babylonian cosmology, and that in turn derived directly from the Sumerians), the other foot was firmly planted on ancient Greek cosmology.

The ancient Greeks are noted for their pursuit of truth, knowledge, and understanding. They used reason to distill many scientific explanations from their observations of nature, and had a highly developed mathematics which they used to explain the movement of the sun, moon, planets and stars. But they too let their value systems—both theological and philosophical—shape the conclusions they came to.

For example, around 500 BC, Pythagoras (the mathematician who will forever be remembered for the equation that all elementary grade students must learn even today) tried to model the universe in terms of whole numbers simply because something as grand as the universe should not have any numerical fractions or unsightly remainders to be carried. In this model, they also saw earth as the center of the universe, as did the Sumerians, Babylonians and ancient Hebrews before them, but they concluded it must be a perfect sphere (rather than a flat surface), with the sun and planets orbiting it in perfectly circular orbits, and the rest of the universe comprising a perfect sphere. This is because they considered the circle to be a perfect geometric shape described by a simple and perfect mathematical formula, and a sphere to be a perfect three-dimensional extension of that, and the Divine would only work in/with perfect shapes.

This model satisfied their philosophical biases (with respect to whole numbers and perfect shapes) and also explained certain observations like the rising and setting of the sun, and the arc-like movement of the stars. However, there were problems. The most bothersome of these included the observations that:

1. some planets at times seemed to slow down in the sky and even do a small loop-the-loop before returning on their original paths to the west (a phenomenon we now call retrograde motion);

2. the planets seemed to be brighter and closer at certain times but smaller and more distant at other times; and,

3. the sun also seemed to be speeding up or slowing down depending on the season.

A solution to these problems was provided by Aristarchus of Samos, who proposed that earth orbited a tremendously large sun while spinning on its axis. Even though this model explained everything quite well (and is remarkably close to the model we have today), it was utterly discarded by one no less than Aristotle himself because it seemed to raise other unexplainable questions. "If earth is spinning, why don't we feel it moving beneath our feet, and why don't we feel a strong wind in our face? Why does an object thrown straight up fall down to the exact same spot? If earth is indeed rotating in one direction while the thrown object travels up and down unattached to the ground, the object must of necessity land at some distance from the thrower. In fact, if a rock is thrown as hard as possible in one direction and then again as hard as possible in the opposite direction, it must necessarily travel further in one direction than the other because of the movement of earth underneath it while it was in flight." These unfulfilled predictions made perfect sense to Aristotle.

Aristotle's followers continued to work on the model, and eventually began to congregate around one which still had the centrally placed spherical earth surrounded by a series of circles upon circles, or epicycles. More specifically, there were very large circles centered around earth itself, and a number of very tight circles centered on the paths of those larger circles (figure 3). The sun and planets travelled on these circles, which explained their arc across the sky while also explaining retrograde motion and the other paradoxical findings. It was an incredibly complicated model, but it explained a lot of observations. In fact, it stood unchallenged for over one-and-a-half millennia until Copernicus and Galileo began to shake things up again.

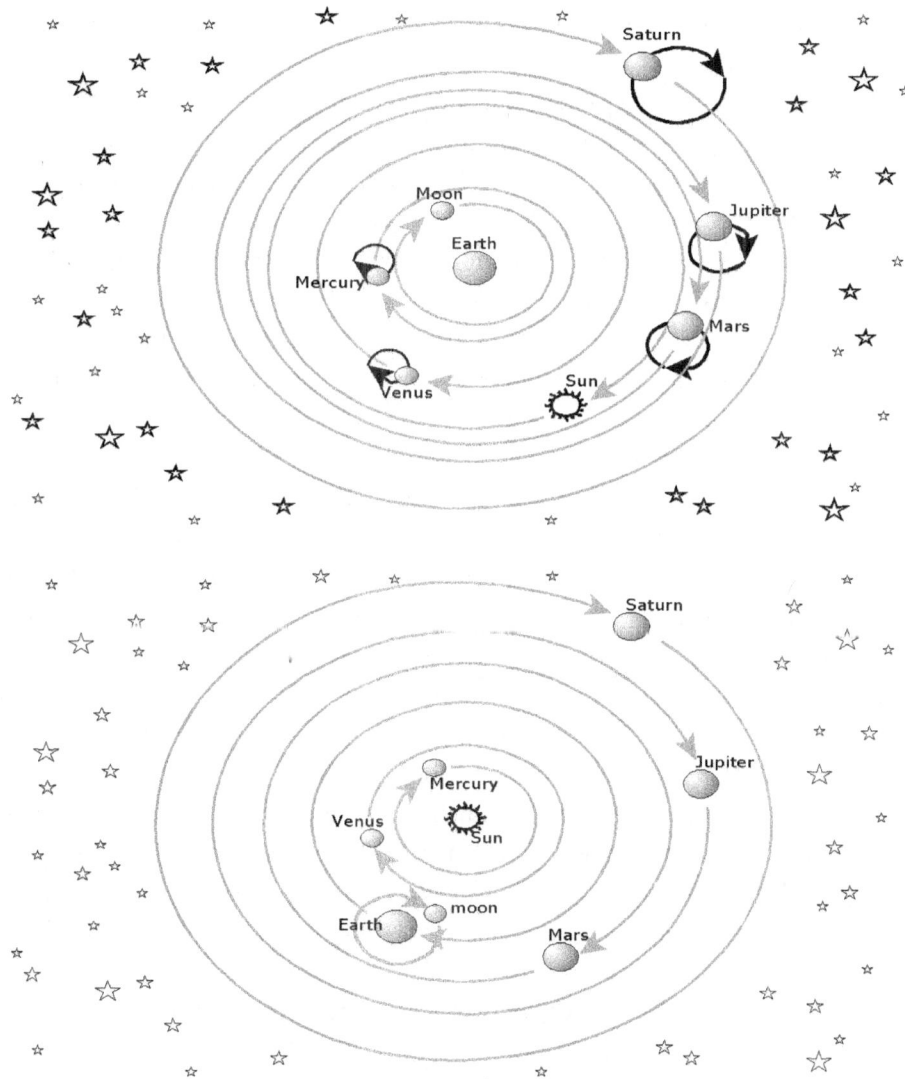

Figure 3. Top, Aristotelian model of our solar system. Bottom, Galileo's model of our solar system.

The Clash between the Church and Science over Cosmology

Many readers will have a vague understanding of the Christian church hanging on to either a flat-earth model or to one with the sun revolving around the earth (these are two completely different models of earth, since the sun would arc over a flat earth, not rotate around it). And most readers will be aware that the church resisted newer models from astronomers, who described a spherical earth revolving around a centrally placed sun, because that was seemingly incompatible with Scripture. The reader's memory of this conflict may also have the church leaders torturing and even martyring those astronomers, probably by burning at the stake.

The truth is, much of this history has been distorted by the media and misunderstood by the public.[29]

One early major challenge to the Hebrew cosmological model, which the church had embraced for several millennia, had to do with the shape of earth. Clearly it was flat. All you had to do was walk with the religious leaders to the highest balcony turret in the church cathedral or the king's palace and look all around you and you would see a flat horizon. Even if you climbed to the highest mountain peak available and rotated yourself 360 degrees, all you would see is a flat horizon (occasionally obscured by other mountains). There may have been some lack of clarity over the actual shape of the borders of this flat earth. Isaiah described it as a circle above which God placed his throne,[30] but in another passage he also wrote about the four corners of earth,[31] as did other biblical writers.[32]

Clearly the idea of a flat earth eventually disappeared, but not due to any overt intellectual pressure from scientists. How could there be? There was no way for scientists to actually prove that earth was round, and there wouldn't be until humanity developed the ability to lob satellites in orbit and return selfies. Instead, this idea itself fell flat by its own irrelevance to daily life. It had absolutely no practical benefit for most of the human population who didn't travel very far, and it contradicted the common sense of those that did. Ferdinand Magellan, the famous Portugese explorer who organized the Spanish sailing expedition to the East Indies that achieved the first circumnavigation around earth, commented: "The church says Earth is flat, but I know that it is round, for I have seen the shadow of the Moon, and I have more faith in a shadow than in the church." Mariners could watch mountains or even ship's masts dip below the horizon before actually shrinking out of sight. Sailors shook their heads and testified that no one has ever sailed over the edge of earth, no matter how far they sailed. And if you sailed far enough, you eventually returned to your starting position!

The flat-earth dogma died a slow death of neglect (although every now and then it seems there are still some people who resurrect it). Earth is clearly round. Not only could they concede to that, but they echoed the Greek claims that it must be a perfect sphere. An oblong shape is a distorted or imperfect circle, and it was felt that God would only create something perfect (reader: please note the theology driving the science here).

---

29. Brooke and Cantor, *Reconstructing Nature*; Danielson, "Great Copernican Cliché," 1029–35; Lindberg, "Galileo, the Church, and the Cosmos," 33–60.

30. Isa 40:22.

31. Isa 11:12.

32. Ezek 7:2; Rev 7:1; Rev 20:8.

A PRECEDENT-SETTING CASE

The relative positions of the spherical earth and spherical sun and spherical planets, on the other hand, was much more contentious because of the perceived gravity of the theological implications, and the newer arrangement proposed by astronomers and mathematicians was actively opposed for those reasons (that is, not for any scientific or logical reason). Many readers will understand Galileo to be at the heart of this conflict, and that he was burned at the stake for pitting science against "a plain reading of Scripture." However, this whole chapter in church history has been grossly misrepresented.

First, it was not simply a church-versus-science conflict. Almost all scientists at the time were churchmen, and many theologians were quite knowledgeable about the science of that era, including astronomy.

Second, the dispute was not solely founded on Christian theological presuppositions, but also upon Greek philosophical ones. The Greeks believed there were four basic elements—earth, water, air, and fire—and that nature sought ardently to organize these four elements in certain relative positions.[33] If one puts some earth into water, it immediately sinks down. If one releases air underneath the water, it immediately bubbles straight up. Likewise, fire apparently "wants" to ascend above the air (just picture the smoke and sparks from a campfire streaming straight up high into the sky when the air is otherwise still and calm). For this reason, they viewed earth as the least noble element, water more noble, air more so yet, and fire as the most noble element of the four. Some would even claim a fifth element—the celestial—which was the most noble of all elements and arose even above the other four. This philosophical paradigm told those Greeks that earth was actually at "the bottom" of creation, with the more noble seas resting on top of the earth (the lands we live on are actually intrusions through that nobler aquatic element), the yet nobler atmosphere above both land and water, and the sun furthest above all. This philosophy fit with Christian theology in that it saw sinful mankind relegated to earth, and our salvation/redemption involving being washed in the waters and entering into the fifth element: the third or highest heaven.

Third, it was not merely a debate between two competing models. In fact, there were several different models at that time describing earth's relationship to the sun, moon, stars and other planets. The debate was not simply over which celestial body was at the center, but also the shape of the orbits involved.

Some of the models had earth at the center; as such, we refer to them as geocentric models (the Greek word for earth is *gē*). This mistaken belief was based on many passages in the Holy Scriptures about earth being firmly established and immoveable, and other supporting passages describing various movements of the sun.

Other newer models placed the sun at the center, and are therefore referred to as being heliocentric (the Greek word for the sun is *helios*). Copernicus advocated a heliocentric model, decades before Galileo was born, because it seemed to better explain

---

33. Lindberg, "Medieval Church Encounters," 16–17.

some of the movements of the planets through the night sky. It should be pointed out, though, that he borrowed the idea from Greeks many centuries before him who had their own heliocentric models.

The models also differed with respect to the shape of the orbits. There were both geocentric and heliocentric models that drew these orbits as ellipses (oblong or egg-shaped) because they better explained certain observations, while other models in both camps required them to be circular (perfectly round), for the theological reasons already given above.

The relative merits of all these various models were debated not only with respect to their differing scriptural support and theological implications, but also with respect to their ability to explain scientific observations. Some argued against the heliocentric models on the basis that we should observe the various stars from different angles, depending on whether earth was on this side or on that side of its orbit around the sun. For example, if you stand in a forest and shift your head from side to side, you'll notice the changing relative positions of those trees closest to you and less so for those trees far off in the distance. The same phenomenon happens when you drive through mountain ranges and you take note of the relative positions of mountains near versus far as the journey progresses. In the same way, it should be possible to precisely measure the relative positions of the stars at two different times of the year, particularly when earth is on opposite sides of the sun, and find differences depending on how close the stars actually are to earth. Opponents of the heliocentric models argued that such differences, a phenomenon we call "stellar parallax," were simply not observed. Keep in mind that this debate was occurring at a time when almost nothing was known about the stars, including whether they were "just lanterns a few miles up," or were instead tremendously gigantic bodies billions of miles away. It would take a number of centuries before astronomy developed sufficiently to actually document stellar parallax (in 1838).

The geocentric models had a long history (thousands of years) and church tradition advocating for them, and they seemed to more closely adhere to scriptural passages. When Copernicus published his book (again, decades before Galileo) describing his heliocentric theory, it immediately attracted proponents and opponents. Protestant theologians—particularly Martin Luther and John Calvin—were the most vehemently against it. Not so much because this put the sun at the center, but because this new theory set earth in motion—both in the sense of orbiting around the sun and by spinning on its own axis—in clear contradiction with the Scriptures which explicitly described it as established and immoveable.[34]

The Catholic Church, on the other hand, was willing to consider and debate all the cosmological models, and was in discussions with several astronomers, including Johannes Kepler, Tycho Brahe, and Galileo Galilei. In fact, the Jesuit Order claimed Galileo as one of their own because he had Jesuit training. Cardinal Robert Bellarmine

---

34. 1 Chr 16:29–30; Pss 93:1; 96:10; 104:5; and 119:90.

was appointed to look into this controversy and after much discussion and deliberation declared that he could accept a heliocentric model if proof for it could be given, and that "one would then have to proceed with great care in explaining the Scriptures that appear contrary, and say rather than we do not understand them than that what is demonstrated is false."[35] But he also added that "this is not a thing to be done in haste, and as for myself I shall not believe that there are such proofs until they are shown to me." In other words, he couldn't just arbitrarily overturn such a long church tradition until there was hard evidence on the table.

Galileo fabricated his own telescopes and added many observations to the field of astronomy. Some of the latter included the waxing and waning of Venus, which transitions from being spherical to crescent-shaped exactly like the moon does (and for the same reasons). This observation of the phases of Venus is indeed consistent with a heliocentric model but is not definitive proof for it: it is entirely possible that the sun has a wide orbit around earth (geocentrism) and that Venus has a tight orbit around the sun (explaining its phases). He also described the moon as not being perfectly smooth, but pockmarked with craters and valleys and mountains. This also upset the sensibilities of the religious leaders because they insisted that anything created by God should be perfect: in the case of the moon, it should be smooth as a billiard ball.

Because Galileo pushed persistently and dogmatically for a heliocentric model (particularly one with circular orbits, rather than elliptical), he was invited to write a book which evaluated both sides of the debate. His *Dialogue concerning the Two World Systems* had two characters discuss the relative merits of the two models. However, he didn't do so in an even-handed manner, but instead gave much more favorable treatment to the heliocentric model. To make things worse, the character defending the geocentric model (and therefore representing the church in this debate) he named Simplicio, which in Italian means fool. As if to say: "In your face, Cardinal Bellarmine!"

This insult was too much, and Galileo was taken to court. Throughout his trial in Rome, he was the houseguest of various wealthy supporters who kept him in their palaces and extravagant apartments. He was never under any threat of execution, and in the end was sentenced to house arrest in his villa in Florence for the rest of his life (another nine years), where he continued his studies of astronomy.

Looking Back on It All: Lessons to Be Taken

This story of Galileo's trial by the church over the heliocentric theory has long been distorted and misunderstood. It was not a simple church-versus-science conflict, and it did not conclude with the church martyring the scientist(s). Many of Galileo's challengers were themselves scientists, and many of those supporting him were

---

35. Brooke and Cantor, *Reconstructing Nature*, 114.

theologians. Although not being allowed to ever again leave your villa in Florence could be described as some form of punishment (some would call this a comfortable retirement), it is certainly a far cry from the claim that he was tortured and burned at the stake. Galileo was perhaps somewhat a victim of political maneuvering between the Protestant and Catholic churches and professional rivalries among his peers, but he also exacerbated his situation with his own defiant and disrespectful attitude.

However, the point of this chapter was not to rehabilitate Galileo's wrongly maligned image, nor to emphasize that this was not a straightforward church-versus-science conflict, nor to simply summarize the ancient Hebrew cosmology. Many others have already done an excellent job of this, and I have borrowed extensively from them.

Instead, I wanted to first present a clear example of how the church, and indeed also the non-Christian Greeks, interpreted scientific findings and theories on the basis of their preconceived ideas. But more importantly, I hope to use this as an illustration of how the church was eventually able to move past its firmly entrenched theologically based science and to fully embrace an evidence-based science which contradicted the former.

At the time, the church was firmly opposed to the heliocentric theory. For them, the scientifically based model raised many problems for their theology. Their objections included:

Objection #1: "It contradicts Scripture." First and foremost, paramount above all, the scientific model conflicted directly with Scripture. This cannot be stressed enough. There are many, *many* scriptural references supporting the Hebrew cosmological model described above (only some of which are cited in that section, to avoid redundancy and monotony). Scripture paints a very clear picture of the way the solar system is constructed, they would say, and it specifically describes earth as established and immoveable. And the Scripture is God's word. There can be no debate on that. If science contradicts it, then science is wrong.

Objection #2: "It goes against church tradition. We've always held the old way to be true: our leaders, called by God himself to shepherd us, have always taught us these things." This is similar to, but not the same as, the point raised above. We have many other documents—not included within the canon of Scripture—written by church fathers (sorry ladies) throughout the ages, during OT times and more recently, which attest to the existence of the third heaven, the firmament, hell being deep underground, and the passage of the sun over an immoveable earth. How could they all be wrong?

Objection #3: "God would not work that way." Both the ancient Greeks and the early Christian church felt that the orbits of the celestial bodies must be perfectly circular because the divine would only create perfection. They felt Galileo could not have seen craters and fractures on the moon because, again, the Divine would not have made imperfect things: some of them claimed that the craters and mountains on the moon's surface must instead be smudges and scratches on Galileo's telescope lens.

Objection #4: "It means we are not unique among all created things." Then, as now, we saw ourselves as being at the very center of all of God's attention: literally, the universe revolves around us. But this new scientific model said that earth was just one of several planets. Not even the first one, or last one, or biggest one, or the one with the most moons. Just a medium-sized planet occupying a middle position and having only one moon, while Jupiter had four (they didn't know that Jupiter actually has at least sixty-three moons).

Objection #5: "It challenges the idea that we are created in the very image of God." In this new heliocentric model, earth is actually only a middling planet among many other planets which were more glorious than our own in different ways. Other planets were larger (Jupiter) or had more moons (again, Jupiter) or more beautiful (Saturn with its rings) or faster (Mercury) or closer to the sun (again, Mercury). This is important, because the relative glory of the planet must reflect the relative glory of the creatures inhabiting that planet, in the same way that they would have viewed the lands of civilized Europeans to be more glorious or noble than the lands of uncivilized barbarians, which in turn would be more so than wild areas inhabited only by animals, and much more so than uninhabited desolate wastelands. And yet humans were supposed to be the pinnacle of creation. "Why would God place us on a middling planet, rather than on one more glorious? This would suggest that we are not the pinnacle species: the more glorious planet(s) must have more glorious pinnacle species, expressing more of God's image. This can't be." It doesn't matter that we're now pretty confident that no advanced beings exist on those other planets: the possibility existed in their minds at that time, and it ruffled their theological feathers.

Objection #6: "On the sixth day, God declared everything 'very good.' How can such a distortion of that which is right be very good?" Again, the prevailing idea was that, among the four basic elements, earth was the least noble and fire the most noble. But this new heliocentric model placed fire—the sun—at the base or bottom of creation, and placed earth above it. This new model inverted the natural order of things: it was an upheaval of all that is right.

Objection #7: "Jesus and the apostles clearly believed in the accuracy and historicity of Genesis." Jesus referred indirectly to Adam and Eve,[36] and directly to the murder of Abel[37] and to Noah's flood.[38] Peter also referred to that flood,[39] while Paul builds a whole theological case around Adam.[40] So obviously they took Genesis to be historical and literal. Furthermore, Jesus referred to the rising and/or setting of the

---

36. Matt 19:4–6; Mark 10:6–8.
37. Matt 23:35; Luke 11:51.
38. Matt 24:37–38; Luke 17:26–27.
39. 1 Pet 3:20; 2 Pet 2:5.
40. Acts 17:26; Rom 5:12–19; 1 Cor 15:21–22; 1 Tim 2:13–15.

sun,[41] as did Mark,[42] Luke,[43] Paul,[44] and James.[45] They also referred to hades being under the earth,[46] and to the third heaven.[47] So they all clearly took the ancient Hebrew cosmological model to be accurate and true. "Who are we to say they were wrong?"

Objections such as these sound ridiculous to us now, but the advocates raising them even just a few centuries ago were dead serious about these. To them, this was a battleground. Theological positions such as these were vigorously defended. And yet the church has learned to give up that outdated and flawed cosmological thinking. In fact, nearly everyone now laughs at much of it and has no problem whatsoever with attributing those ideas to the antiquated science of that era and relegating them now to the dust-bin of history.

This point is the second goal of my writing this chapter: I wanted to demonstrate that we, the church, were not always closed-minded about scientific data which seemed to challenge "a plain reading of Scripture," but rather sought hard to rationalize faith and science in its search for truth. Admittedly, the church advocated proceeding slowly and carefully. But when the scientific evidence eventually overwhelmed the theological paradigm, the church conceded, though with the proviso that "one would then have to proceed with great care in explaining the Scriptures that appear contrary."[48] Likewise, one of the greatest early church fathers, St. Augustine of Hippo, in 415 AD wrote: *"Be on guard against giving interpretations of scripture that are far-fetched or opposed to science, and so exposing the Word of God to the ridicule of unbelievers."* These points will come up again in chapter 6 when we look at another more recent scientific challenge of long-held theological viewpoints.

Standing on the Shoulders of Giants

Our knowledge about the universe, and about life itself, has been continually changing and growing. We humans have been constantly taking previous forms of knowledge and building upon them.

Sir Isaac Newton is the one who brought to us calculus (although some dispute that that recognition should be shared with his contemporary Gottfried Wilhelm Leibniz), the laws of motion (Newtonian mechanics), the theory of gravity, and the wave theory of light, not to mention several other lesser contributions and his many

---

41. Matt 5:45; Mark 4:6.
42. Mark 1:32; Mark 16:2.
43. Luke 4:40.
44. Eph 4:26.
45. Jas 1:11.
46. Phil 2:10.
47. 2 Cor 12:2.
48. Brooke and Cantor, *Reconstructing Nature*, 114.

writings on theological matters. Altogether, this is an incredible contribution to the body of knowledge owned by *Homo sapiens*.

But when attention was drawn to him and his work, he deflected the praise by saying, "If I have seen further, it is by standing on the shoulders of giants," thus recognizing and acknowledging the contributions of many others before him to his own great achievements.

For example, with respect to calculus, we find elementary forms of that branch of mathematics in the writings of the fourteenth-century Indian mathematician Madhava of Sangamagrama, and even simpler forms in Egyptian and Greek writings.

The same point could be made for his other powerful achievements. This intellectual giant built on top of work from previous giants, who in turn were founded on yet other earlier giants, and so on. Even in using the phrase alluding to standing on the shoulders of giants, he was borrowing the words from another luminary, Bernard of Chartres, a French Neo-Platonist philosopher who developed that allusion centuries before him.

Throughout this book I will present new ideas building on other older ideas: hence, the title. The next chapter presents other examples of how the church needed to let go of ideas which were subsequently shown by scientific giants to be inferior. Chapter 6 will take this one step further yet, showing how it is now time to reevaluate our understanding of the origin of humanity.

3

Other Duels between the Church and Science

THE DIVISION OVER THE geocentric and heliocentric theories certainly didn't mark the only time that the church has stood up against science. There have been many disputes since then over "a plain reading of Scripture" and scientific evidence; many times at which we allowed our theological glasses to color our interpretation of the data held up to us. Some of these have been inconsequential, others are of major impact. We're still navigating our way through several of them: for example, flood geology and the theory of evolution. However, none of these have had or will have as much impact on theology as the one which is now looming on the horizon and which is the main subject of this book (chapters 5, 6 and 7): the origin of mankind. But first, let's look at a few of these previous incidents of theological division to drive home the point that theists have frequently insisted on ideas that conflicted with a plain reading of data, but eventually softened their resistance and gave in.

Please bear in mind that my goal in this chapter is not to argue for or against any particular theological belief. It may appear that I am advocating a certain side and trying to convince the reader as I stack up the evidence against certain theological ideas. Instead, I'm merely giving examples in which there is a large body of factual evidence against a given interpretation of the scriptural texts, but believers have chosen to ignore that evidence, or to bend the facts to fit around their belief. This is the very definition of a phenomenon referred to as cognitive dissonance.

A second, equally important goal is to demonstrate how the church has eventually been able (sometimes) to accept the scientific evidence, incorporating it into its theological paradigm and in so doing essentially change its thinking 180 degrees, despite loud claims that the people doing the changing are leaving the faith. If we've done it before, we can do it again.

I'll start with a few relatively insignificant and light-hearted examples, before getting into other ones that seriously rocked the church.

Ribs and Penises?

One humorous story involves Andreas Vesalius, a Belgian anatomist and physician at the University of Padua during the sixteenth century who overturned many ideas about human anatomy which had until then held strong for over a thousand years (from the teachings of the Greek-speaking Roman physician Galen). One of those ideas that he overturned pertained to the number of our ribs: he showed irrefutably that men had the same number of ribs as did women.[1] This stoked a minor storm of controversy because certain religious leaders were adamant that men had one fewer rib, given the account in Genesis in which God took one of Adam's ribs to make Eve. Although many words were spoken and written, the religious leaders didn't actually physically mistreat Vesalius for poking this hole in their balloon.

There's no need to correct me by pointing out that removal of a rib from Adam would constitute nothing more than a surgical injury, and we know that such injuries are not passed on genetically to our children and grandchildren; even those church fathers should have known that the accidental loss of a body part like a finger wasn't forever transmitted on through the family line. That's not the point. Instead, it is simply to show that this finding of Vesalius was challenged simply because it conflicted with the story in Genesis (at least in their minds it did). The church resisted an unequivocal scientific finding because it felt the latter simply must be wrong based on theological principles. And why must it be wrong? For no other reason than that it conflicted with "a plain reading of Scripture."

I was surprised to learn that the dispute over Adam's rib was being resurrected once again today, and even more surprised to learn of the entirely new direction it had taken.

Ziony Zevit, a distinguished professor of biblical literature and northwest semitic languages, and a member of faculty at the American Jewish University, promoted the idea that the material God took from Adam to make Eve was not one of Adam's ribs, but his *os baculum*, or "penis bone."[2] The *baculum* is a bone found in the penis of many placental animals, and provides stiffness for the act of sexual reproduction. His reinterpretation was based in part on an exhaustive linguistic study of the Hebrew word which has otherwise been translated (incorrectly, he claimed) as rib, but also on the fact that the *baculum* is present in many primates but is conspicuously not present in humans. Accordingly, his interpretation is that when YHWH created Adam from some ancestor in common with the primates, he took the *baculum* to create Eve, forever after depriving their male descendants of this bone.

Once again, Zevit seemed to misunderstand that, irrespective of whether it was a rib or a *baculum*, this change would have been surgical in nature, not genetic, and therefore would not have been transmitted down to all of Adam's progeny.

---

1. Andreas Vesalius, *De Humani Corporis Fabrica*, bk. 1.
2. Zevit, *What Really Happened*, ch. 12.

Zevit has since backed away from this claim, but not because he realized children don't inherit surgical injuries. Instead, he did this because it was pointed out to him that the text uses this word in the plural sense—YHWH took "one of" whatever it was that he took—and animals only ever have one *baculum*.[3]

I recognize that this thesis posited by Zevit was a flash in the pan idea. Nonetheless, I included this vignette in the ongoing debate over human origins because it illustrates powerfully and embarrassingly the risks taken by theologians when they speak to scientific questions (in this case, why humans don't have a *baculum*, let alone the origin of human females in particular, and of humans in general), and justify their statements on wholly theological grounds.

I've Got a Gut Feeling

In another example of mistaken biology: the ancients, including the Israelites, believed that our intelligence, emotions and personality arise from the very center of our being: from our kidneys, liver, and intestines, rather than our brain. That is the language that the Hebrew texts used. For example, when David wrote in Psalm 40:8, "Your law is in my heart," the text is literally written, "Your Torah is in my guts."

That thinking and language continued into the NT era, and into the Greek language. In Philippians 1:8, Paul literally longs for his readers "in the bowels of Jesus Christ," and this is how the King James Version translates the passage.

Needless to say, we've long since changed our thinking about the process of thinking.

This is not to say that there was any kind of conflict between the church and science over this matter. Instead, it's just another example of how the ancients understood things differently than we do now, and the words in the Scripture reflect that premodern understanding. We don't have to endorse that ancient thinking, nor to insist that biology classes be taught that way, just because it's found in the pages of Scripture.

Babies Come from Where?

The ancient Hebrews certainly comprehended at some level that it takes both a man and a woman to make a baby. However, their understanding of the details behind reproduction was quite different than how we see it today. As we explore this question, please remember to set aside any modern scientific ideas you might have about sexual reproduction and simply approach this from the point of view of an ancient Mesopotamian trying to understand this mysterious process.

---

3. Zevit, "Was Eve Made from Adam's Rib," 33–35.

As explained by Lamoureux,[4] their understanding was based on the phenomenological science of their time. The people of that Mesopotamian region and era were very familiar with agriculture. They worked closely with the soil to produce their crops. They would marvel at the magical properties of this dirt, acted on by their hands, to generate living things. Is it surprising, then, that many ANE texts describe the creation of humanity using references to clay, including those of the Sumerians (Song of the Hoe; Hymn to E'engura; Enki and Ninmah; KAR4), the Akkadians (Atrahasis; Enuma Elish), the Egyptians (Pyramid texts; Coffin texts; CT spell; Instruction of Merikare), and of course Genesis?[5] Enki and Ninmah refer to clay. Some Egyptian Pyramid texts refer to clay on a potter's wheel, as do Job, Isaiah, and Jeremiah.[6] Seeds are planted and humans break out from the ground like crops in the Song of the Hoe, and in the Hymn to E'engura.[7]

Farmers planted seed on the fields, and seed *always* sprung up unless there was some kind of problem with the land. Unproductive fields were either too wet, too dry, too hard, or did not have enough organic matter, for example. They also knew that all males had seed: every man could demonstrate, if need be, that they could produce an ejaculate. All that was needed, then, was for that seed to find a good place to spring up. Women, on the other hand, did not produce anything that looked like seed, although the men would correlate a similarity between the woman's menstruation and a wound acquired by the uprooting of a partially developed seedling from the earth's soil. So if a couple were unable to produce a child, they concluded that the woman was the limiting factor: in the biblical texts, it is only ever women who are said to be infertile, using the word "barren" or referring to the Lord having "closed her womb," or "opened her womb," or "enabled her to conceive." This is the case for Sarah (wife of Abraham, mother of Isaac),[8] Abimelek's wife and all his female slaves,[9] Rebekah (wife of Isaac, mother of Jacob),[10] Rachel and Leah (wives of Jacob, mothers of several tribes of Israel),[11] the unnamed wife of Manoah (mother of Samson; according to Rabbinic tradition, her name was Hazelelponi),[12] Hannah (wife of Elkana, mother of the prophet Samuel),[13] Michal (wife of King David),[14] Elizabeth (wife of Zechariah;

---

4. Lamoureux, "Ancient Biology."

5. Lamoureux, *Four Views*, 57–59 and 98–100.

6. Job 10:8–9; Isa 29:16; 64:8; Jer 18:1–6.

7. The Psalmist writes about being "woven together in the depths of the earth" before being birthed (Ps 139:15).

8. Gen 11:30; Heb 11:11.

9. Gen 20:17–8.

10. Gen 25:21.

11. Gen 29:31; 30:1–3, 22.

12. Judg 13:2.

13. 1 Sam 1:2, 6.

14. 2 Sam 6:23.

mother of John the Baptist),[15] and in Isaiah's prophecy of the coming of the Messiah.[16] They also observed that men were capable of fathering children far into old age, whereas women have only a brief window of opportunity before the Lord would close her womb. I am not aware of any biblical passage which pins the problem of infertility specifically on the male in the relationship.

Those same ancient farmers would open up the seeds of certain legumes, nuts or fruits, and in between two relatively large lobes (which we now call cotyledons) would be found what looked like a tiny little plant: tiny little leaves and a single rudimentary root (the reader can easily confirm this with seeds from peanuts, beans or peas). That observation would lead them to believe that even inside the seeds from men would be found tiny little babies, which just needed good "soil" to fully develop. That idea persisted for millennia, even into the modern era. In 1694, Nicolas Hartsoeker, one of the first biologists to use a microscope, sketched what he thought he saw inside the head of human sperm: a tiny little baby nicely curled up and waiting to unfurl like a plant once it found good soil (see figure 4).

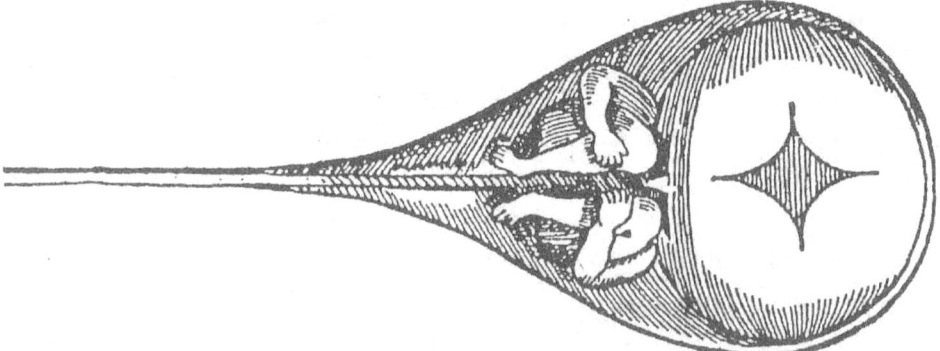

Figure 4. Nicolas Hartsoeker, view inside the head of a human sperm.

Likewise, the ancients had no idea quite how fetal development occurred. How could they? Instead, they resorted to analogies. "For you created my inmost being; you knit me together in my mother's womb."[17] In the opening chapter of Jeremiah, God says, "Before I formed you in the womb I knew you."[18] Job also acknowledges that God formed him the way a potter forms something out of clay.[19] We now know that embryogenesis—the development of a fetus—is a preprogrammed sequence of biochemical events which can be repeated and sustained in a test tube. Most contemporary readers of the Bible will admit that the allusions to knitting needles and clay are just poetic devices rather than scientific explanations. But when it comes to the formation of Adam, some will insist that he was literally shaped by God's hands from

15. Luke 1:7, 36.
16. Isa 54:1; Paul also quotes this passage in Gal 4:27.
17. Ps 139:13.
18. Jer 1:5.
19. Job 10:8–9.

the dust of the earth because that is how it is described in Genesis. They maintain this, even though Job uses the same language to describe his own formation, yet they don't feel compelled to believe that Job also was formed out of handfuls of dirt by God's own literal hands. When pushed on this matter, they might admit that Job is speaking metaphorically, and yet they won't allow the text of Genesis to be metaphorical about Adam and Eve.

Evolution

Although most of the church has put these matters of ribs, guts, knitting needles, soil, and potter's wheels behind them, the same cannot be said when it comes to the origin of life and of species. The theory of evolution has divided theists ever since Charles Darwin published his theory in 1859, and it continues to pit theist against theist, and leave atheists laughing and pointing on the sidelines. Biologists have amassed mountains of data—multiple lines of evidence—which all corroborate the idea that the various species of life arose through gradual modification of preexisting forms.

The fossil record may have been sparse in Darwin's day, but today we have far more fossil specimens than can be studied in detail. Due to a lack of financial and human resources, many fossils are simply quickly notarized and then filed away in cabinets for future study when time and funding permit. And more fossils continue to be discovered at an ever accelerating pace.

The usual, reflex response to this from certain theists has often been a reference to the many and large "missing links." Unfortunately, in most cases, those that make this accusation have not done enough homework, and are simply parroting phrases they've heard others utter or that they just want to believe are true. The fossil record is overwhelming, and gaps continue to be filled in. Just the same, some will doggedly persist in denying the evidence, insisting that every single branch and twig of the Tree of Life must be found and put in place before they will accept the theory. Actually, I suspect some would still deny it even if we reached that point. And once again, the reason for this opposition is simply that it conflicts with "a plain reading of Scripture."

Some theists are willing to concede some ground. They will accept that plants and animals might have come to be through the processes of evolution, but will draw the line when it comes to the creation of mankind. In their minds, humans must be different. But this too is a presupposition. A bias. Another example of forcing the science to fit a certain theology. I have much more to say about all of these points, but will do so in chapter 7.

Radioisotope Dating

A major focal point in the battle between faith and science has been the dispute over ages: of the universe, of the earth, of fossils, and of artefacts which are unearthed.

Many lines of evidence indicate that these are much older than the six thousand-year timeline which seems to be delineated by the biblical text, but perhaps the strongest of these has been radioisotope dating. Very few scientists familiar with this technique question the very long ages which arise.

The section below explains how this technique works. Although some readers will feel this particular topic has already been written about by so many others in the past, the fact remains that even to this day many YECs continue to vigorously dispute those dates and how they are obtained. This is largely because they simply do not understand the techniques. The informed reader is free to skip over this section. However, I hope such readers first make sure that they are sufficiently informed so as to be able to explain to those believers who continue to dismiss radioisotope dating as unreliable.

For thousands of years, scholars had believed the universe was approximately six thousand years old. This fundamental point was based solely on "a plain reading of Scripture." Anyone could go through several genealogies provided in the book of Genesis, add up all the ages at which a long list of genetically related men sired what is implied to be their first son, then match up key individuals within that family tree to the years associated with other unique individuals or events in recorded history. I've described this exercise in much more detail in my first book.[20] Suffice to say, the generally accepted age-of-everything was much less than ten thousand years.

And the justification for this conclusion is simply "because the Bible says so."

Then came science. Which told us a very different story.

Physicists discovered the process of radioactive decay of certain long-lived elements, a process whereby the latter change from one form to another by spontaneously and naturally rearranging their atomic structure.

For example, uranium is a highly radioactive element. It has a unique atomic structure, distinct from all the other elements with respect to the number and arrangement of protons (positively charged particles) and neutrons (non-charged particles) in its nucleus. Physicists have organized all the different elements according to those numbers, and arranged them from smallest to largest in a "spreadsheet" called the periodic table. This table is an amazing achievement of science. It explains all kinds of chemistry, and, more importantly, even led to the predictions of elements which were previously unknown but which were later verified years later.

Uranium occupies position number 238 on this periodic table, and so we refer to it as uranium-238. But these atoms of uranium are constantly jiggling and vibrating, and every now and then, a couple protons and a couple neutrons fly off, leaving behind a somewhat smaller element called thorium, which occupies position 234 on the periodic table (so we refer to it as thorium-234).

The amount of time needed for this change to occur in any given atom of uranium is completely unpredictable: it might be one second, it might be one trillion

20. Janssen, *Plato's Cave*.

years. But if you have a handful of uranium containing trillions of uranium atoms, you can be sure that half of those atoms will have decayed to thorium-234 within four and a half billion years. It's like rolling a pair of six-sided dice: you can never know for sure whether or not the next roll will produce a twelve, but you can be sure that if you rolled them a thousand times, twelve would come up roughly twenty-eight times (the chance of rolling the twelve is one in thirty-six).

So uranium-238 has a "half-life" of four and a half billion years. But thorium-234, through a couple radioactive steps with a half-life of twenty-four days, rearranges itself into uranium-234. That in turn decays (half-life of 240 thousand years) through a cascade of rearrangements to produce several other lighter elements with varying half-lives, including: thorium-230 (77 thousand years), radium-226 (1.6 thousand years), radon-222 (3.8 days), polonium-218 (3.1 minutes), lead-214 (27 minutes), bismuth-214 (20 minutes), polonium-214 (160 microseconds), lead-210 (22 years), bismuth-210 (5 days), polonium-210 (140 days), and finally lead-206, which as far as we can tell no longer decays or rearranges.

Those half-lives, and the half-lives of other elements, have been measured many times and shown to be constant to a very high degree of precision and confidence. So much so that we can build incredibly complex things like nuclear reactors and know exactly how much energy we can extract from a kilogram of uranium-235 (a different form of uranium from the two mentioned above), and exactly how much of those other elements we'll have after running the reactor for a couple decades.

Getting back to the point of this chapter, geologists would crack open rocks that had been unearthed and measure the various elements within them—uranium-234, radium-226, bismuth-210 and lead-206, for example—and after plugging in the various amounts and half-lives and going through the math, they kept coming up with ages for those rocks numbering in the billions. Repeatedly. After thousands of measurements done by thousands of different laboratories all over the world, measuring different combinations of elements and their radioactive decay products, we keep coming up with numbers in the billions. Four and a half billion years to be a bit more precise.

It's hard to argue against that consistency. But some do, because it threatens their worldview. They'll emphasize that this approach to dating earth is based on assumptions, one of which is that radioactive decay occurs at a constant, unchangeable rate. For the critics, an assumption is a weakness: it gives pause for concern. Some even take it as license to set aside all those measurements as unreliable. But we've got no reason to think otherwise. The rates of decay have been measured thousands of times and found to be unaltered by changes in temperature, pressure, pH or other chemical factors, electrical fields, magnetic fields, or gravitational fields. And yet some theists will dispute this overwhelming evidence for no other reason than that it leads to a conclusion that conflicts with their theology.

Another assumption which is often criticized is the starting concentration of the radioisotope being measured. It's easy to measure how much of that radioisotope we have today, but to determine how old that is, we need to know how much there was to begin with, which in turn will tell us how many half-lives have occurred. For example, if we measure an object and find in it one kilogram of an element with a half-life of ten thousand years, and if we can somehow know that there had originally been four kilograms of that element in the object, we can conclude that the latter is twenty thousand years old (the original four kilograms decays to two kilograms after ten thousand years, and to one kilogram after another ten thousand years). But how can one know (or assume) the original amount of that element so long ago?

In the case of carbon-14, it is possible to make a very reasonable assumption (and one which is frequently the target of critics of radioisotope dating). The assumption is based on what we know about how carbon-14 is generated and how it disappears.

Carbon-14 is generated high up in the atmosphere when neutrons emitted by the sun hit nitrogen atoms (roughly 78 percent of the atmosphere is nitrogen). Just like one billiard ball can be used to knock another billiard ball into a pocket of a billiard table, the neutron smacks a proton out of the nucleus of the nitrogen atom and takes the place of that proton. This change in the composition of the nucleus—loss of a proton and gain of a neutron—converts that element from nitrogen-14 to carbon-14. The process is therefore referred to as neutron capture.

That carbon-14 atom then undergoes radioactive decay to carbon-12 with a half-life of 5730 years.

So carbon-14 is constantly being generated in the atmosphere via neutron capture, and constantly breaking down via nuclear decay. Eventually, these two processes of generation and decay come into equilibrium and you are left with a constant proportion or concentration of radioactive carbon-14 atoms relative to the much more abundant and non-radioactive carbon-12 atoms.

The situation is like a large lake which receives water from several streams, and in turn empties into some other lake downstream. Even though water is constantly coming in (analogous to neutron capture) and flowing out (analogous to radioactive decay), the depth in the middle of that lake might always be approximately three hundred feet (give or take a few feet depending on the time of year).

Once the processes of constant generation and constant decay of carbon-14 come into equilibrium, the concentration of carbon-14 in the atmosphere is very low: on the order of one part carbon-14 to one trillion parts carbon-12 (one trillion is also represented as 10^{12}). Having measured the current rate of solar emission of neutrons and the half-life of carbon-14, and assuming that both have been constant over the past thirty thousand years (several half-lives of carbon-14), then we can determine what the concentration of carbon-14 had originally been thirty thousand years ago. Yes, there are assumptions built into that, but we've got no reason to think otherwise.

Living organisms incorporate carbon during their lifetimes. Plants take up carbon dioxide from the atmosphere (containing that background concentration of carbon-14) and convert that into hydrocarbons (sugars and starches), herbivores eat those plants, carnivores eat the herbivores, omnivores eat both, and scavengers eat the dead remains of everything. The concentration of carbon-14 in the bodies of all these organisms while they're still alive will be determined by what they've eaten (this point is absolutely crucial, and will come up again in a few pages when I discuss "discordant dates"). For two reasons, their bodies will contain the same proportions of carbon-12 and carbon-14 as their food sources. First, their metabolic systems can't tell the difference between carbon-14 and carbon-12. Second, their own lifetimes are profoundly less than the half-life of carbon-14, which means that essentially none of the carbon-14 in their bodies will decay to carbon-12 *while they're still alive* and interacting with the atmosphere and the environment.

When the organisms die, however, the situation changes. At that point, they stop taking in new carbon (the rivers upstream of our metaphorical lake stop refilling the lake), but the carbon-14 they've incorporated continues to decay (the rivers downstream keep emptying the lake). If the remains of those organisms are analyzed thousands of years later, the remains will contain far less carbon-14 than they had originally contained when the organisms were alive and constantly exchanging their carbon with the atmosphere and environment.

So it is possible to measure the relative proportions of carbon-12 and carbon-14 in an object today, and from that determine its age, given the assumptions behind the constancy of the rates of neutron capture and radioactive decay. Both processes have been measured and not yet found to vary.

Furthermore, we've compared the ages determined by radiometric dating with the ages obtained by other dating techniques and found them to agree.

One other dating technique that has been used in this way to calibrate carbon-14 dating has been measuring the number and varying thicknesses of tree rings. Trees generate one new ring every year. If it's been a particularly good year for the tree—one with lots of sun, rain and warmth—that tree ring will be quite thick. In a bad year that ring will be quite thin. Only rarely will it appear that two rings are generated in one year (that year would have to have produced an unusually large amount of growth in the first few months as well as in the latter few months, but have included an unusually bad drought and/or cold snap during the middle of that summer). Or equally rarely it may appear that no rings were generated in a given year (one with spectacularly cold, dry and/or sunless summer weather). In that way the rings in the tree(s) will form a unique pattern which reflects the varying goodness and badness of each successive growth year.

Some trees can live for hundreds of years, and thus be used as historical weather records. Bristle cone pines still alive today in the Western United States are five thousand years old! By examining the tree rings in several different dead stumps which lived at different times, and matching up the unique sequences or patterns of rings within them, it is possible to overlap the timelines displayed within all of them and thus build up a calendar going back twelve thousand years. Getting back to the main point of this paragraph, carbon-14 testing of the tree rings gives ages which match up with those obtained by simply counting the rings. Likewise, if an object is found buried in a soil layer together with such tree stumps, then similar ages are determined by matching up the tree rings within the calendar referred to above and by carbon-14 dating the object.

A similar approach can be used in counting the ice layers going back a hundred thousand years in cores obtained from ice sheets in Greenland, with each layer reflecting the distinct patterns of summer's melt and winter's deep freeze.[21] The exact same thing has been done with the sedimentary layers at the bottom of Lake Suigetsu going back fifty thousand years.[22] Carbon-14 dating of objects found within those sedimentary layers also gives corroborating ages.

In other cases, one can find layers of ash or lava put down by major volcanic disruptions which were described by human observers, or which can be dated in other ways: radiocarbon dating of objects found immediately above or below those layers also produce dates which are consistent with that timeline.

These three calibration techniques—tree rings, ice layers and sedimentary layers—confirm the original assumptions made and the reliability of the carbon-14 dating technique.

Several measurements will be made from different parts of an object and produce different estimations of ages, simply because of what we call experimental error. This admission of error disturbs some people, but only because they don't understand the precision of science. If you take a ruler and measure the length of a stick several times, you would expect to always come up with the same answer: say, 17 inches. But if you use a carefully calibrated scientific instrument called a micrometer, you might at one time get a measurement of 17.072641 inches, while a second measurement gives 17.072894 inches. Most reasonable readers will not be the least bit concerned about these differences in the readings.

Nonetheless, sometimes a strange result is obtained which merits careful consideration. For example, let's say carbon-14 dating of ten samples of a particular object produces the following ages:

- Sample #1: 27,000 years
- Sample #2: 30,500 years
- Sample #3: 29,000 years
- Sample #4: 3,500 years

---

21. Ramsey et al., "Terrestrial Radiocarbon Record," 370–74.
22. Reimer, "Atmospheric Science," 337–38.

- Sample #5: 26,900 years
- Sample #6: 28,100 years
- Sample #7: 30,900 years
- Sample #8: 29,900 years
- Sample #9: 31,200 years
- Sample #10: 27,900 years

The scientists doing these measurements will omit the date obtained from Sample #4—because it's clearly far outside of the range of measurements—and determine that the object was 29,040 years plus/minus 1,655 years. That omitted value they will call an outlier, an artefact or a sampling error. Critics of carbon-14 dating, however, will cry foul and will focus attention on the date obtained from Sample #4, perhaps even insisting that the short date obtained by one of the ten samples must be the correct one (because it fits better within a YEC frame of reference). But let's look more carefully at this.

Can an unusually short or long date result from sampling error? Given that the starting concentration of carbon-14 in the sample at the time that it died was incredibly low—only one part per trillion—and that its concentration only gets smaller as time goes on, it's entirely possible that a particular slice off the original object is taken which has a few more or a few less of these atoms. This may not make a difference for a large sample, but can be very problematic for a small sample. To better appreciate this, let's return to my analogy above of measuring the depth of a lake with water flowing in and out at both ends. If that lake were an exceedingly large and deep lake, then measurements will likely be highly reproducible throughout the year because changes of even a few feet represent only a small fraction of its full depth (of hundreds of feet). But if it's just a tiny pond, there will be times in the year that the depth is six inches, and other times that its depth is six feet: a change of 1200 percent!

Can a strange date be due to contamination? This technique assumes that the carbon within the specimen prior to death was in equilibrium with background atmospheric carbon, and that upon death the accumulation of carbon-14 stopped whereas radioactive decay continued. However, other living organisms could have invaded that specimen at various times in the past: bacteria, fungi, worms, tree roots, and many others. When the specimen is later unearthed and tested, the carbon-14 from those invading organisms will be included in the tally. If they invaded the specimen close to the time of death of the animal, this will make very little difference because they also would have the same background concentration of carbon-14 as the specimen which had just died. But if they invaded the specimen in the relatively recent past, long after the specimen had died, they will throw off the measurement (skewing the measurements toward the date of those invading organisms).

It is very common for this kind of contamination to occur. Many specimens which one would want to get a radiometric dating on—such as a set of bones, or the wooden part of a weapon or tool—were already close to the surface and are often discovered after being exposed by erosion, landslides, earthquakes, frost upheaval and other physical processes, or when people crawl into caves and trip over them. Being so close to the surface, or even being exposed, those long-dead specimens will inevitably be invaded by living organisms (again, bacteria, fungi, worms, tree roots, pollen). The contribution from the living organisms will shorten the age determined, not lengthen it, but the magnitude of this error depends on how much have penetrated the sample and how recently the penetration occurred. If the specimen/sample were in reality one hundred thousand years old, but gathered some modern bacteria, fungi or pollen at the time that it was unearthed and brought to the lab, radiocarbon dating could easily produce an age of only a few thousand years old. So unexpectedly short dates are not uncommon, and require multiple samplings and determinations to be done, with careful precautions taken to avoid or remove contamination.

Then we come to those discordant dates which YEC proponents love to publicize: estimates of ages which on the surface defy explanation. One category will be ages which are impossibly long, and the other will be ages which are impossibly short.

Before proceeding with the examples which appear to be impossibly long, I need to remind the reader to keep in mind that the original carbon-14 content of a specimen will be entirely dependent upon the sources of carbon-containing material from which it ate while alive.

Several studies have determined that shells from *living* freshwater mussels appear to be over two thousand years old according to carbon-14 dating. The interested reader can find some of these studies cited in a follow-up investigation done by Keith and Anderson, who sought to explain this surprising anomaly.[23] This strange result is peculiar to freshwater mussels (comparisons were made with marine mussels), and is related to the environment they live in. More specifically, the freshwater mussels live in ground water which has percolated through/over humus and limestone rock beds and become saturated with the background carbon of those materials rather than the carbon of the atmosphere.

The humus comprises very old decaying vegetative matter, and gives off its carbon content in the form of partially digested materials, metabolites, waste products and CO_2 produced by the bacteria eating away at it and breaking it down.

Adding to that, the ground water bearing the carbon from that long-dead vegetative matter flows over and through limestone rock, which itself is made up of the shells of very ancient animals: ones that may have died hundreds of thousands or even millions of years ago. That water will continually erode that limestone away, especially if the water is slightly acidic. For example, the CO_2 from the decaying humus

---

23. Keith and Anderson, "Radiocarbon Dating," 634–37.

spontaneously produces carbonic acid, the fizz in carbonated drinks, which then accelerates the dissolution of the limestone.

The bottom line is that freshwater mussels do not derive their carbon from the atmosphere, which contains the background amount of newly generated carbon-14, but instead derive their carbon from ancient sources whose endogenous content of carbon-14 has long ago decayed to carbon-12. "You are what you eat." The same strange outcome was found when a different research group carbon-dated the shells of living snails from three different freshwater sources having three very different levels of limestone content (high, medium, low), and compared them to snails which were grown in the lab and fed fresh lettuce (and therefore had zero limestone content).[24]

This is an example of what is now known as the marine reservoir effect, and it addresses the fact that one needs to take into account the nature of the sources from which the organisms are drawing their carbon, since carbon-dating is based on the assumption that the original carbon content of the organism before it died was in equilibrium with the atmosphere (where carbon-14 is generated). The bottom line is that carbon-14 dating works best on plant materials (because they draw their carbon almost exclusively from the atmosphere) and animals which live solely on eating those plants, and is least accurate with marine organisms that survive on ancient carbon sources. Similar results have been obtained with living penguins dated to being thousands of years old, and for similar reasons.

It is also essential to consider how the samples being carbon-dated were handled prior to their testing. Much ink has been spilt by anti-evolutionists over radiocarbon test results of dinosaur bones which indicated the latter to be only 20,000 years old,[25] and this story continues to bubble to the surface from time to time. In the early 1990s, a YEC group presented themselves surreptitiously to the Carnegie Museum of Natural History as chemists seeking to obtain fossilized dinosaur bones with the intent of analyzing their chemical composition. They sent samples to the University of Arizona's Laboratory of Isotope Geochemistry for radiocarbon dating, without telling them the nature of the samples or how they were obtained. The Arizona Laboratory in turn reported a carbon-14 date of only 20,000 years. That date could not have been owing to the fossilized bone itself, since fossils normally contain little detectable carbon: in the process of fossilization, the original living materials of the sample are replaced by minerals.[26] Consistent with this, the Arizona lab also told them that the sample contained no detectable collagen, which should otherwise have given the YEC group pause for concern given that bone is almost exclusively calcified collagen. Instead, the carbon detected in the sample came from the preservative coating with which the

---

24. Rubin and Taylor, "Radiocarbon Activity," 637.

25. Lepper, "Radiocarbon Dates for Dinosaur Bones?"

26. Exceptions to this rule are living materials sequestered in inaccessible parts of the bones, such as collagen and red blood cells found in dinosaur claws, and DNA found within teeth; both examples will be revisited later in the book.

fossils were treated: the Carnegie Museum warned the YEC group posing as chemists that the samples were "covered heavily in shellac," which is a resin secreted by a certain insect and is made up of complex carbon chains. Despite all this misrepresentation of themselves and misinterpretation of the data obtained, this claim of dinosaur bones being dated to only twenty thousand years ago—concurrent with humans—continues to this day to be resurrected in the YEC literature.

Another anomaly that YEC proponents will frequently point to are measurements of carbon-14 in coal carried out by a group of scientists associated with the ICR. The latter refer to themselves as the RATE group (Radioisotopes and the Age of The Earth), and its members have PhDs in atmospheric science, physics, geophysics, and geology (and, interestingly, one in Hebraic and cognate studies).[27] So they're admittedly a well-credentialed group. They obtained the coal from a coal sample bank maintained by the US Department of Energy. So far, everything in the story sounds reassuring.

Coal is comprised of thick layers of vegetative matter which presumably contained the background amount of carbon-14 existing at the time that those plants had been alive and which had subsequently been compressed and heated deep beneath the earth's surface over long periods of time. These samples were taken from ten different coal layers which were determined by the US DOE to have been deposited during the Cenozoic, Mesozoic and Paleozoic eras; in other words, many millions of years ago. After millions of years, that coal should contain no measurable amounts of carbon-14.

To their surprise, the RATE group measured the samples and derived dates as recent as one hundred thousand years old, and determined an average number for all the layers of approximately fifty thousand years old. Similarly, others have been able to detect carbon-14 in diamonds, which also are assumed to derive from ancient vegetative matter compressed and heated deep in the bowels of the earth over millions of years.

Taken at face value, these findings are astounding. How can coal and diamonds contain carbon-14 if they're supposed to be millions of years old? Given that the half-life of carbon-14 is only 5,730 years, half a million years represents just over eighty-seven half-lives, so the amount remaining will be one part in a number with thirty-six zeroes behind it. To put this in perspective: one thousand has three zeroes, one million has six zeroes, one trillion has twelve zeroes, and so on. So a number with thirty-six zeroes is an incomprehensibly large number (and the corresponding ratio an incomprehensibly small number).

To ICR, the RATE group and YEC proponents, being able to measure carbon-14 in coal or diamonds, and to pin the date to fifty or a hundred thousand years, is scientific evidence that the earth is not millions or billions of years old. In fact, they push the point even further by saying: "Using a more realistic pre-flood 14C/12C ratio reduces that age to about 5,000 years." (They don't explain or define what they mean

27. Riddle, "Chapter 7."

by a "more realistic" ratio. The cynical side of me can't help but wonder if they simply represent numbers chosen to produce a determined age that perfectly supports a YEC view.)

But let's look more closely at their claim that they were able to obtain measurements which indicate ages of fifty to one hundred thousand years.

One hundred thousand years represents seventeen or eighteen half-lives for carbon-14. After the first half-life (5,730 years), there will only be 50 percent, or one part carbon-14 per two trillion (2×10^{12}) parts of carbon-12. After the second half-life (11,460 years), only 25 percent. After the third, only 12.5 percent. And so on till you get to the eighteenth half-life, at which point the samples contain only 0.000381 percent of their original carbon-14 content. Given that the original content was already only one part in a trillion (10^{12}), this means that after eighteen half-lives, it will contain one part carbon-14 per 2.6×10^{17} parts carbon-12. In case this isn't immediately obvious to the reader, this means one atom of carbon-14 for every 2.6×10^{17} atoms of carbon-12. That's an incredibly lonely carbon-14 atom!

And that's where the claims of the RATE group become suspect: the number of carbon-14 atoms becomes so small that even minute traces of contamination become a huge problem. The contamination can only shorten the measured ages of the sample, and it only takes a few carbon-14 atoms to make a carelessly small sample look substantially younger than it actually is. When we're talking about this level of sensitivity, then the risk of picking up even just a handful of contaminating carbon-14 atoms make the measurements meaningless.

Even though the RATE group claimed that "careful precautions were taken to eliminate any possibility of contamination from other sources," it takes very little contamination to throw off a measurement of a single carbon-14 atom for every 2.6×10^{17} atoms of carbon-12. This contamination could come from breathing, coughing or sneezing done by the researchers, since this will contain faint but detectable amounts of saliva, which in turn contains bacteria, enzymes, and DNA. It can also come from floating dust particles (which are predominantly particles of skin), pollen grains and bacteria (which are easily lifted by gentle breezes raised by colleagues walking into a room). These represent sources of contamination from the handling of the samples by the RATE group members themselves, or by the DOE who gave them the samples, or the employees of the companies involved in digging the coal out of the ground. There is also the possibility that present day atmospheric carbon-14 becomes introduced via pollen grains and bacteria in the outside air during the extraction of the coal, or if ground water carrying bacteria or algae made its way into the mine and the fractured seams of coal. Likewise, contamination could have occurred during the many years which the coal sat buried deep underground, with ground water percolating through its seams.

Then there's an entirely different source of contamination which needs to be taken into account. The carbon-14 dating technique assumes that the carbon-14 in the sample comes *only* from that generated in the atmosphere. But carbon-14 can come about by mechanisms which operate underground.

Other atoms can absorb neutrons and be converted into carbon-14, including carbon-13, oxygen-16 and oxygen-17. All three of those elements, as well as nitrogen-14, are also present in the vegetative matter which was then buried deep within the earth to produce the coal. These elements are present within the very molecules that make up the vegetative matter, not as free gases which could be squeezed out by heat and pressure. Likewise, diamonds are nearly pure carbon-12, but also contain a variety of impurities (including carbon-13, nitrogen-14, oxygen-16, and oxygen-17): it is these small amounts of impurities within them that give them their unique colors and/or produce flaws which lessen their market value.

So both the coal deposits and the diamonds will contain atoms of carbon-13, nitrogen-14, oxygen-16, and oxygen-17: all potential substrates which could be turned into carbon-14. It's true that, being buried underground, they're protected from the sunlight which provides the neutrons that convert nitrogen-14 into carbon-14 in the atmosphere. But radioactive substances deep underground also emit neutrons and other radioactive particles. So both the coal deposits and the diamonds could continuously pick up additional carbon-14 simply from this radioactive bombardment underground, right up until the day they are unearthed. This may occur at a very slow rate, but over millions of years it can produce a significant number of carbon-14 atoms. And I've already gone through the mathematics above to show how few carbon-14 atoms it takes to throw off the readings from objects which are millions of years old. These other sources of carbon-14 need to also be considered when one is talking about readings so close to the limits of measurement error (let alone far below that practical limit).

But the RATE group did not do this, and of course their YEC adherents don't know or appreciate this important fact. And these anomalies are then picked up and widely publicized by nonexperts who mislead the masses. Such distorted and misinterpreted stories keep coming out of the YEC camp, often led by individuals with superb speaking abilities and just enough scientific understanding to be dangerous, but who lack scientific degrees from accredited institutions: high profile individuals include Ken Ham and Kent Hovind who are high school teachers lacking any graduate degrees in science. Please don't misconstrue this as simply an anti-YEC attack: I am simply stating facts here.

On the other hand, some YEC proponents who will accept the validity of measurements by other techniques which produce ages in the millions or billions of years will rationalize this by claiming things only look old because God created everything to look old, a concept first posited in 1857 by the British biologist and preacher Henry

Philip Gosse[28]—"apparent age." But God would have had to follow through on this ruse in many ways, because many other things look very ancient: many features on the surface of the moon; the light from distant stars; the fossil record; the genetic record (I'll explain this in more detail in ch. 6). Why would God go through so much trouble to make everything look so old? The only explanation ever given is that he does it to test our faith. To force people to make a choice: him or science? Faith or fact? But this turns God into a deceiver. And a disturbing one at that. I used an analogy in my previous book which I want to use again here.

Imagine that I decide to test my wife's trust in me by making it look like I was having an extramarital affair. I plant clues all over the house which can only be interpreted in that way. I write a letter from someone other than her, leave it in one of my shirt pockets and then intentionally put that shirt in the laundry for her to discover. Leave half-composed email messages on my computer screen. PhotoShop my credit card bills to show expenses for tickets to movies and theater presentations that my wife had never been to, or to show purchases for jewelry that she never received. When I return from business trips, I leave things in my luggage that clearly don't belong to me. And many other such tests of her faith. When she finally does confront me, my response is an emphatic denial and an insistence that she has to simply trust me. And if she can't "just believe" me despite the mountain of evidence to the contrary, then I end our relationship.

How can I not come out of that exercise looking like a pathetic husband, or even worse? And yet that's what we turn God into, if we hold on to the idea of apparent age.

Noah's Flood

Geologists also spend their time looking at layers in the rocks, and how those reveal a history of earth. Many of the layers look like outflows of lava. Others look like they were produced by a steady accumulation of layers of particulate matter over millions and billions of years. Many of the latter incorporate a fossil record which documents the steady evolution of life on earth. Altogether, these pieces of evidence have helped paleontologists develop a detailed history of earth over the past four and a half billion years.

Some theists on the other hand, attribute these observations to a massive worldwide flood which occurred a few thousand years ago. I highly doubt that any reader doesn't already know the church has been engaged in an acrimonious argument with scientists over the flood account in Genesis. Many books have already been written on that subject. One which I highly recommend to the interested reader is *The Rocks Don't Lie: A Geologist Investigates Noah's Flood*, by David R. Montgomery. It is written

---

28. Gosse, *Omphalos*. (Note: Gosse's book is entitled *Omphalos*, which is the Greek word for "navel," because a central argument in his book is whether or not Adam had a belly button, given that he never had an umbilical cord attached to a mother).

by a trained geomorphologist at the University of Washington who specifically compares what he learns in the lab with what he learned in Sunday school. The summary of the situation from this highly trained expert is the following: "Whatever you may think of evolution, the creationist belief in a several-thousand-year-old Earth shaped by Noah's flood is as scientifically illiterate as the idea that the sun circles us. Both have been known to be wrong for centuries. And to embrace the creationist view of Earth history is to deny Earth's autobiography inscribed on pages of stone."[29]

Mott Greene has a written an excellent synopsis of "Buckland's Dilemma," referring to the nineteenth-century geologist and Reverend William Buckland, who found himself torn between following truth to where it leads even if that took him far outside of his Anglican belief system. Greene goes on for two pages recounting Buckland's many achievements, honors, and recognitions, leading up to Buckland's taking up the first ever faculty position in geology at Oxford University in part because of his recent geological discovery. Greene writes in the first person, asking the reader to imagine themselves in Buckland's shoes:

> This discovery is going to make you the leading scientist of your generation. It will make students and faculty jostle for seats in your lecture hall . . . make you a best-selling author . . . newspaper and magazine profiles . . . prizes, offices, money, honors, and fame. . . . However, you have a problem. Your fantastic discovery in the Yorkshire cave, if you pursue it, is going to undo your diluvial theory . . . will show that the biblical Deluge is *not* responsible for a great many phenomena attributed to it. . . . Your discovery is certain to appall and enrage biblical realists. . . . Finally, you will not easily escape the charge that you are contradicting the account of the Creation given in the Book of Genesis, and therefore some will perhaps say that you are impious, or ungodly, or even an apostate. What are you going to do?[30]

This story vividly portrays the role that our worldview plays in how we interpret data which are presented to us. Fortunately, Buckland kept his integrity (as well as his faith), presented his findings at face value even though they refuted the idea that the flood was a global one, and helped the scientific world and the church take another step toward resolving the story. I recommend reading Greene's account of this seminal event in the science-faith dialogue.

I've already explained how radioisotope dating of elements other than carbon-14 corroborates the claims that earth is billions of years old, and devastates the claim that it's only thousands of years old. In chapter 6, I will elaborate on how the genetic record does the same. Many have questioned the logistics and feasibility of an ark that could contain representative pairs of all the species on earth, along with all the food they would need, and that those animals could be cared for by a team of only eight people.

---

29. Montgomery, *Rocks Don't Lie*, 13.
30. Greene, "Genesis and Geology Revisited," 139–40.

They are equally skeptical that those animals radiated from the final resting place of the ark and distributed themselves across the entire face of earth, and that some of those animals would have had to travel thousands of miles without leaving any fossil evidence of their journey: for example, there's no evidence of fossils of kangaroos outside of Australia, or of polar bears outside of the polar regions.

Diluvian proponents also provide no reasonable explanation for the precise layering of fossils in an order which happens to be consistent with evolutionary theory: ancient and primitive species in the lowest layers, with higher layers featuring animals of increasing complexity. One explanation some will give is that flood waters would also cause dead bodies to layer out. But that explanation isn't thought through carefully enough. The bodies that would layer out first would be those that are large and dense, while those that would layer out last would be the lightest ones, including single-celled animals. We see precisely the opposite in the geological record: small and simple animals in the lowest layers and large complex animals in the higher layers. Moreover, a flood would mix the bodies of fish with land-based animals, and would mix up less advanced animal forms with more advanced animal forms. The fact remains that we have multiple specimens of fossilized bones for thousands of different types of animals, and also of mankind, but there isn't a single case where we find fossilized remains of humans mixed together with fossilized remains of dinosaurs or trilobites. If they had coexisted and they all died together in a global flood, then you should find all three mixed up together. Likewise, we have lots of fossilized footprints for humans and for dinosaurs, but never the two together. This is just not consistent with the Genesis flood story. Evolutionary theory, on the other hand, predicts these will be found in very distinct layers, and that is exactly what we find. I could go on, but the point is clear.

Just the same, some find it necessary to ignore certain scientific findings, and to distort others, in an attempt to fit it all into a certain theological framework. For no other reason than to align it with a story written by Bronze Age authors. This needs to stop.

I'm not going to recapitulate that whole flood debate, complete with the pros and cons on both sides and whether or not the evidence supports the Diluvian story. The stated goal of the present book is to highlight the tendency of theists to bend science to fit their theology. And this debate around the flood in Genesis chs. 6–9 has proven to be another classic example of exactly that; perhaps one that epitomizes the faith-versus-science dialogue over the past couple centuries. I came across one particularly amusing illustration of this kind of self-delusion while reading Montgomery's book that is worth repeating, though.

Johann Scheuchzer, a professor of mathematics in early seventeenth-century Zurich, Switzerland, had an acute personal interest in natural history, including the fossil evidence for Noah's flood. He collected extensive fossil evidence of plants and animals in sedimentary rock and wrote several books on this subject. These included fossils of marine life found high up in the Alps, which he insisted were deposited there by the flood. He also acquired a large specimen which he interpreted to be the skeleton of a human victim of the flood and named it *Homo diluvii testis* (Latin for "man who testifies to the flood"). He published descriptions of this find in British, French, and German journals, as well as a short book. In these, he not only emphasized that this victim served as a witness and testimony to the historicity of the flood, but also the skeleton's unusually large size affirmed to him the biblical passage that "there were giants in the earth in those days":[31] in his mind, this was one of the Nephilim! Scheuchzer dedicated his life to educating the public on the historicity of the flood while fending off criticism from his fellow scientists. In the end, though, his story didn't fare well. As Montgomery wrote: "Ironically for a talented naturalist, Scheuchzer's faith that the geologic record told the story of Noah's flood led him to the colossal blunder he is still lampooned for today. As [Georges] Cuvier pointed out, Scheuchzer's flood victim was a giant amphibian"[32] (see figure 5).

Figure 5. Scheuchzer's "Flood victim."

Epic of Gilgamesh

It goes without saying that the OT is important to the Christian faith. It comprises the larger amount of biblical text and the greater number of books (thirty-nine in the Protestant Bible and forty-six in the Catholic one, versus twenty-seven in the NT of both versions), and it covers a far greater period of theistic history (arguably four thousand years versus one hundred years, respectively). It sets the stage upon which the life and ministry of Jesus Christ is cast. Not surprisingly, then, Christians will be incredibly cautious around reinterpretations of the OT, and defensive against new

---

31. Gen 6:4 KJV.
32. Montgomery, *Rocks Don't Lie*, 83–84; also see Brown, "Noah's Flood," 131–32.

data which demands any such reinterpretation. For this reason, the archaeology of the Middle East has long been a hotbed of controversy within the Christian community, and a battleground between biblical literalists, Christian Liberals, and non-theists.

Archaeologists have long been finding artefacts from ancient civilizations with inscriptions in a cryptic language which mystified them. Many readers will be able to picture hieroglyphics in Egyptian and Mayan temples. With painstaking effort, these have been deciphered and translated, opening up whole new insights into those ancient civilizations.

Innumerable clay tablets were found in ancient Sumerian and Babylonian ruins, located in what is now modern Iraq. Many of these simply documented financial transactions and legal disputes and proceedings: this was indeed a highly organized and sophisticated society. A large number of other clay tablets, however, comprised copy after copy of religious stories and myths. As the translation of the latter proceeded, the Christian community was knocked back on its heels. One of these stories—the *Epic of Gilgamesh*, the oldest known human story that we have in text form—made reference to an anecdote with incredible parallels to the Genesis account of Noah's flood.

The eleventh tablet of the *Gilgamesh Epic* and Tablet III of the *Atrahasis epic* tell a mythic story of when one of the Babylonian gods had decided to flood the whole world in order to exterminate mankind, apparently because they were making too much noise and keeping the gods awake at night (I kid you not!). A righteous man named Ut-napishtim was warned about this impending flood and instructed to build a boat to escape the destruction. The tablets describe him building that boat, caulking it with bitumen, loading it with his family and with animals, then riding out many days and nights of rain which flooded earth and killed all living things. Once the storm abated, Ut-napishtim sent out three birds—a dove, a swallow and finally a raven—to search for dry ground. Eventually the flood waters receded and the boat rested on a mountainside (Mount Nizir). The boat's human and animal occupants disembarked, and Ut-napishtim offered up a burnt sacrifice to the deities, who were in turn "well pleased" with this offering. The parallels between this Babylonian myth and the Genesis account are astounding.

Did one borrow from the other?

Some might claim that the Hebrew version of the flood myth is the "true" or original version. However, the Sumerian civilization itself goes back to the year 4000–4500 BC, whereas the Hebrew civilization goes back to only ~2000 BC. The oldest fragments of OT texts we have date to ~600 BC, whereas the oldest clay tablets date back to ~2000 BC.

So on what basis can a claim be made that the biblical version of the story is the original one, other than simply wanting to believe that to be the case?

Others might say that the Sumerians incorporated the Hebrew story into their culture. But why would they do so?[33] The Sumerian/Babylonian civilizations ruled the

33. Enns, *Evolution of Adam*, 47; Cahill, *Gifts of the Jews*.

known world at that time, and had been doing so for thousands of years: why would they adopt this origins story coming from some small nomadic tribe of shepherds in one of their conquered regions? Even those Hebrew writings themselves say the nomads were never more than wanderers or slaves until ~2000 BC (around the disputed time of the exodus) or ~1000 BC (approximately when David is said to have ruled the nation of Israel)?

Another claim might be made that the Hebrew and Babylonian versions of the flood story are so similar because both are describing the same event, but from two different perspectives. That is indeed a possibility if the flood event did happen, and those who accept the latter as fact can explain it in this way. But a majority of scholars—both Christian and atheist—conclude that the flood was not historical, simply because the expected evidence of this global flood doesn't exist, as I've already explained in the preceding section. This issue of the historicity of the Noah flood is a hugely contentious one, and one which has divided the Christian community for several hundred years. Much of that division has also centered on the interpretation of scientific data: the geological formations, fossils in the rocks, ice cores, sedimentary layers, and many other lines of evidence. Countless books have been written on those data, and I'm not going to recapitulate them here.

But I will point out that according to biblical literalists, the flood would have occurred around the year 2350 BC, yet historical records from China and Egypt long precede that date and skip through that era without any mention of a massive worldwide flood that would have wiped out all civilization. The Babylonian clay tablets do contain the Gilgamesh story, including the anecdote of a massive flood, complete with Ut-napishtim and his boat full of animals. The Babylonians ruled the ANE for thousands of years, and influenced the mind-set of the world at that time. Their stories and myths would have permeated the culture of the regions ruled by the Babylonians. They would have been reenacted in the plays and dramas performed by actors and dancers at annual celebrations. Songs which celebrated the characters and events in those stories would have been composed and performed by musicians. Any children who received an education would hear those stories, read about them, and memorize them. Out of that backdrop, we find a couple—Abram and his wife Sarai—embarking on a journey to start a whole new existence in a far off land, complete with the promised hope of a new nation.[34] The early literature which the new nation of Israel produced (particularly Genesis) bears a striking resemblance to the literature of the Babylonian civilization from which their forefathers came. That Babylonian literature would also have been found in the major libraries of the world at that time, including the Egyptian libraries in which Moses was educated[35] and the Babylonian libraries to which the post-exhilic priests would have had access (it's still debated whether Moses or those priests produced the Pentateuch). Is it really so hard, then, to consider the

34. Gen 17.
35. Acts 7:22.

possibility that the Hebrew literature was a reinterpretation of the Babylonian stories, or at least influenced by them?

Meteorology

Weather is such a fundamental component for every facet of our modern day-to-day experience. Think how much more this would be the case for a premodern society. With simply an open fire pit to serve as a source of warmth and light at night, and no reservoirs or irrigation system to guarantee water for crops, the ancient farmer would be completely and utterly dependent upon the vagaries of the weather for survival.

They would have little or no understanding whatsoever of the factors that determine the weather that the next day brings. No advance warning that an incoming cold front would drop temperatures and bring torrential rain. No way to know that the drought that had lingered over their land for the past month would extend for yet another five weeks. That vitally important aspect of their existence was completely out of their control.

However, they almost certainly had a belief in some kind of deity, since this seems to be a fundamental characteristic of *Homo sapiens*. And it would be easy to think that deities could influence the weather, let alone have primary responsibility for determining it. The sun god. The rain god. The god of thunder. The winds being the breaths of gods. Perhaps it could be possible to appeal to the god(s) for their help in this area. Offer sacrifices and perform a dance or some other ritual to bring on the spring rains.

The Hebrew mind-set was no different. The book of Job quotes YHWH himself in describing his own direct role in every aspect of the weather. The point I'm making here is not whether he controls the weather (many theists would say that, ultimately, he does). Instead, the key issue in the next two paragraphs is *how* he describes himself as doing so. Rather than referring to atmospheric parameters like temperature, pressure and moisture content, or to mechanisms like the clouds and the sky doing what they do naturally, the biblical text describes God's very hands-on approach to shaping every detail in the weather report, and orchestrating every storm.

YHWH controls the rain by hanging the clouds in the sky (37:15–16) and loading them with moisture (37:11) and cutting channels for the rain and paths for the thunderstorms (38:25). During the thunderstorms, he vocalizes the thunder (37:4–5) and unleashes and scatters the lightning (37:3, 11), dispersing it from his storehouse of lightning (38:24–25). If you hate the static shock you get when you shuffle across a carpet and touch a door handle, imagine trying to navigate your way around in that storehouse.

YHWH also has storehouses for snow and other storehouses for hail (38:22), and he tells the snow and rain to fall (38:6), breathes out ice (37:10), births the ice and frost (38:29), and fathers drops of dew (38:28). Notice the different verbs used there—"birthing" and "fathering"—probably because ice and frost just appear very slowly, growing gradually in some mysterious way.

Today, most Christians have no problem attributing all these meteorological events to solar energy being injected into the planet's atmosphere and surface, and then being redistributed by the winds and the water cycle. We treat the weather more like a machine now, one that just keeps running all on its own and can be predicted by computer models.

Yet old habits die hard: we do still appeal sometimes to God to directly control the weather for us. Imagine the dilemma God faces when, on a hot day in August, he hears a Christian farmer praying for rain because there hasn't been a drop of precipitation in five weeks and the fields are dry as a bone, while at the same time a Christian bride and groom are praying for a clear sky so that their wedding day won't be ruined. Or in a different example which always makes me wince, imagine the thoughts of the people picking through the rubble of their demolished houses while listening to their neighbour exclaiming to the Channel Four TV News crew about how "we saw that tornado coming right at us but God spared us—he moved it right out of the way—it just went right around us and kept on going and our house wasn't even touched!"

People Outside of the Line of Adam

The reader should not think that it has only been in the recent past that the Christian church resisted against the idea that the Adam of Genesis was not the first human to have existed. In fact, that challenge to the literal interpretation of Genesis even long preceded the discovery of human or humanlike fossils in the nineteenth century. David Livingstone provides an excellent exploration of the pre-Adamite theory,[36] another book which I highly recommend to readers.

In it, Livingstone first summarizes the various lines of evidence from Egyptian and Chinese records of historical timelines much longer than the ones derived from the genealogies given in the book of Genesis. Taken at face value (and why would one do otherwise, unless one were simply trying to rescue and protect an inerrant and infallible view of Scripture), these suggested that people not only preexisted the Hebrew Adam of Genesis, but were recording history in lands as distant as the Orient: it would have taken a long time for humans migrating out of the Mesopotamian basin, if the YEC view is to be taken, to reach those distant lands and set up such extensive empires.

Livingstone also summarizes the superstitious beliefs among first-century Romans in a race of humanlike creatures with grotesque body shapes called the Plinians (so named because they were described by Pliny the Elder). Most readers will be familiar with the one-eyed Cyclops, but there were many others:

---

36. Livingstone, *Adam's Ancestors*.

> Amyctyrae (beings with protruding upper or lower lip), Anthropophagi (cannibals who drank from skulls), Artibatirae (those walking on all fours), Blemmyae (creatures with faces on their chests), Cyclopes (one-eyed beings), Cynocephali (dog-headed tribes), Martikhora (human-headed, four-legged peoples), Sciritae (noseless, flat-faced races), Ethiopians, and Pygmies.[37]

Of course, the point here is not whether these races ever did exist, but that even in the first-century Greco-Roman world, people believed in such creatures, challenging the Hebrew idea of common ancestry from one human, Adam. Clearly, they would say, there were other non-Adamite tribes.

As Europeans began to cross the continents and oceans and explore distant lands, they took these superstitions and beliefs with them. Their maps of the known world were decorated with warnings, always written at the borders of their maps, of the presence of such strange beings as well as of other monsters and dangers. Those exploring Europeans began to encounter humanlike beings who were similar to them, and yet different. The latter had very different colored skin. They often had very distinctive facial features and somewhat different body forms (the most extreme of these would have been the pygmy tribes of Africa). The explorers would have added to this list of differences their observations of the "strange" ways in which these people dressed (or didn't), their "strange" diets (eating insects; drinking blood; cannibalism), and their "strange" customs (tribal dances; sacrifices to gods)

Their encounters with such foreigners raised questions in the minds of the European explorers. Where did they come from? Why were the ancestors of these aboriginals not destroyed in the flood? In their European minds, Adam was Caucasian (just look at any of the paintings from that era), so clearly these beings had some kind of extra-biblical origin. Were they human? Did they have souls? Did they too have original sin, and therefore need redemption? If they were not human, but were instead only beasts of the field, could they be harnessed to do work just like horses and oxen?

These kinds of questions were raised about the indigenous people of North and South America, Africa, Polynesia, Australia and many other regions. One sketch from Livingstone is representative for all of these discussions:

> In the half-century or so after Columbus's venture, the debate on the nature and status of the American peoples was engaged and nowhere so conspicuously as in the papal junta at Valladolid, Spain, in 1550, when the doyen of Spanish Aristotelian scholarship, the humanist Juan Ginés de Sepúlveda, and the former Dominican vicar of Guatemala and now bishop of Chiapa, Barolemé de Las Casas, vigorously disputed the subject of how the American Indians should be treated by Europeans. The papal legate had been dispatched from Rome to Spain to determine once and for all whether the Indians shared the *imago Dei* or were a distinctly other species, whether they were fundamentally

---

37. Ibid., 11.

bestial and fitted only for slavery or sufficiently advanced that they should not be considered barbarians.[38]

History bears sad testimony to how this question was so often answered.

This theological bigotry continues even today. In articles, comments, and rebuttals posted on the internet and in social media on the topic of anthropological discoveries, some insist categorically that hominids such as Neanderthals and *Australopithecus* could not have had a soul, and absolutely did not possess the *imago Dei*. I'll elaborate on this point later on pages 203–214. My intent here is not to defend such rights on behalf of those other hominids. Instead, I only want to point out how the theological bigotry seen in the sixteenth century, and which we might criticize today, reemerges in the modern context.

Livingstone dwells at length on a landmark event in the dialogue between the church and science, given its groundbreaking impact.[39] In 1655, Isaac La Peyrère, a Calvinist of Portugese Jewish origin from Bordeaux, published a book—*PraeAdamitae* (translated in English as *Pre-Adamites*)—which met with much opposition by the Catholic Church. In it, he described the discovery that Greenland was completely geographically isolated from America and Asia, and yet was populated by a race of non-Caucasian humans (whom he described in very denigrating terms). After sifting through the observations and speculations, La Peyrère came to the central conclusion of the book: that human beings existed before the biblical Adam, and that the origin of humans was polygenic not monogenic (multiple beginnings, not one). Livingstone quotes from the book of Mungo Ponton, a fellow of the Royal Society of Edinburgh: "There may have been a Negro Adam, a Mongolian Adam, and perhaps two or three more besides the Caucasian Adam."[40] La Peyrère's idea helped explain paradoxical passages in the Bible, such as Cain's fear of being murdered, Cain's wife, and Cain building a city (why build a city if the only people on earth were oneself and one's parents?).

Livingstone provides a very interesting description of how this concept of pre-Adamite peoples was unpacked and debated for five hundred years afterward, and the strange and/or unacceptable ideas (by my standards) which arose from within it and the opposition raised against it. The latter include:

- the gap theory: the idea that there were in fact two separate creation events, one occurring fifteen billion years ago to accommodate geological and astronomical measurements of the age of the earth and of the universe as well as the fossil record, followed by a great Divine destruction of all living creatures and then a new re-creation including Adam and Eve in the garden of Eden;

---

38. Ibid., 19.
39. Also see Brown, "Noah's Ark," 121–24.
40. Livingstone, *Adam's Ancestors*, 135.

- Hebrew as the very first original human language. This was the adamant belief of many theologians at that time. "Saint Augustine and most of the early church fathers were convinced that primeval Hebrew was the primary and perfect language . . . all existing languages contained traces of corrupted Hebrew vocabulary."[41] Livingstone also recounts how "at the tail end of the seventeenth century . . . a certain John Webb advanced the novel hypothesis that . . . the Chinese language is the purest version of Adamic Hebrew, and only the Chinese, having lived for millennia without suffering foreign invasions, preserved it in its original purity."[42] Pre-Adamism would conclude that human languages were also polygenic, just like the human groups who invented the languages.

- Adam and Eve were Caucasian (just look at any painting of a biblical story from the seventeenth century);

- the Christian Identity or Kingdom Identity movements, which postulate that the Anglo-Saxons of Britain and North America are of Israelite descent (through Ephraim, one of the ten lost tribes of Israel);

- scriptural justification for the subjugation of whole races of people into slavery, because of their "non-Adamic" and therefore "nonhuman" status;

- racial (white) supremacy, and anti-Semitism;

- the fall in the garden is a metaphorical recounting of a sexual liaison between Eve and her pre-Adamite (black) gardener (others proposed the paramour to have been Chinese or Mongolian), and that God's judgment was in fact for her/their failure to keep the racial bloodlines pure.

After reading Livingstone's book, I was struck by the irony engendered by La Peyrère's pre-Adamite theory. It was initially proposed in order to explain how Scripture could be accepted without taking it literally. But I now see it being used by YECs to substantiate its authority and infallibility in the face of scientific discoveries of other hominids such as Neanderthals, *Australopithecus* and *Homo naledi*. It was originally formed as a tool of biblical scepticism, but is now used by biblical fundamentalists. As Livingstone put it:

> The very idea that initially represented a secularizing trend in scientific endeavour by challenging the detailed accuracy of scriptural history and by liberating anthropological investigation from Mosaic strictures, has ended up being valorised as a reconciling tactic among conservative believers. . . . Beginning as a theory to recapture the centrality of the Jewish experience in scriptural economy and global history, it has ended up as a weapon of flagrant anti-Semitism. La Peyrère cast Adam as the father of the Jews, with the pre-adamites as the gentile nations . . . to sweep all nations into the benefits enjoyed

41. Ibid., 49.
42. Ibid., 10.

by the chosen people—the adamic Jews. . . . Latter-day white supremacists presume the adamic line to be Caucasian and find justification in their reading of Genesis for locating Jewish origins in some act of interracial sexual union with corrupt pre-adamites. . . . The genealogy of pre-adamism traces out a path from communalism to elitism, from humanitarianism to bigotry.[43]

Atheists Do It, Too?

People of faith aren't the only ones who are guilty of allowing their worldview to color how they view scientific observations. As I said in chapter 1, everyone has one or more worldviews, and scientists are no exception. They too have been known to insist on a particular interpretation, or to avoid certain interpretations, because of perceived conflict with their worldview.

Many readers will think that it's always been theists who have opposed Big Bang theory. But actually, atheists were the very first to resist it, and for theological reasons no less!

Prior to the proposal of this theory, atheists preferred to think of the universe as eternal, having no beginning whatsoever. It had just always been. This was the easiest way to do away with the concept of a creation event (and thus, of a Creator).

Then astrophysicists began playing in their minds with the equations that define the expanding universe as we know it, and ran them in reverse. They found they could describe the universe getting smaller and denser until it reached a point of unfathomable density and temperature. Beyond that point, the equations no longer work. They become nonsensical or non-solvable when the value of time is set to zero. In fact, Stephen Hawking and James Hartle, in 1983, said it was impossible to get any closer than 10^{-43} seconds after the Big Bang.

From those mental gymnastics, they proposed the idea that the universe began as a singularity—an infinitely hot and dense object—which then exploded, producing what we now call our universe. Certain colleagues didn't like the idea purely for theological reasons: it meant effectively that the universe had a beginning, which played right into the hands of creationists, and they couldn't allow that. It was the hugely influential astrophysicist Fred Hoyle who is credited for being the first to refer to this theory mockingly as the "Big Bang model," and the moniker stuck. Eventually, resistance toward the idea of the universe having a beginning waned, but it still hasn't gone away entirely.

Perhaps there is no better example of atheistic scientists feverishly resisting an idea largely because of its theological implications than the antagonism they show to Intelligent Design (ID; I'll define and discuss this in more detail later on pp. 269–271). I'm not here advocating or defending ID. I'm simply calling attention to their typically

---

43. Ibid., 221.

very emotional response to it. All scientific disciplines other than biology are comfortable with invoking intelligence to explain the discovery of something containing information.

The SETI project has pointed radio telescopes out into space to scan for signals: if they were ever to pick up a complex signal—such as the sequence of prime numbers portrayed in the Hollywood movie *Contact*—they would immediately seek out whether it could be attributed to an alien intelligence.

When archaeologists find paintings of deer on the wall of a cave, or carefully shaped flints on the ground, they immediately attribute that to an ancient civilization.

When computer programmers find malicious code that they didn't write in some of their software, they immediately suggest it was intentionally put there by a malevolent coder.

I could go on with many other examples from diverse scientific disciplines of researchers invoking intelligence when they observe a phenomenon which leads to increased information. This is in large part because every time we've observed an increase in information in some system in the past and we discover exactly where that information came from, we find it inevitably points back to some form of intelligence. In every branch of science across the board. Except for one: the field of biology. Biologists resist viciously any idea that a designer is behind the complex coding found within our cells. We have few examples of genetic mutations giving rise to a significant increase in information or a more complex gene sequence. Many examples of large evolutionary steps via gene mutations that we've been able to document comprise the *reduction* of information: the inactivation of a gene or the functional neutralization of its gene product. And yet atheists will insist that all the amazing complexity of life is the result of a long repeated series of genetic mutations leading to a fantastically increased level of information. One has to admit it's a bit of a stretch.

Or a very few atheists will put up with panspermia, the idea that life didn't appear on earth spontaneously or unaided, but was instead seeded by some kind of alien life-form(s). This was the premise in the Hollywood movie *Prometheus*. Proponents of this idea don't seem to recognize that this doesn't solve the mystery of the origin of life and all its complexity. All it does is move the goalposts further down the field: if life on earth was engineered by aliens, then how did those aliens come to evolve? We have no evidence whatsoever for any such aliens, and in order to be able to navigate interstellar distances to seed life, they would have to be at least nearly infinitely knowledgeable and powerful: to us, they would appear as gods. But try referring to them in any kind of theistic terms and you will be dismissed as ridiculous, unsophisticated and naïve. Richard Dawkins is famous for defining biology as "the study of things having the appearance of design."[44] Talk about winning the war by defining the rules of engagement!

---

44. Dawkins, *God Delusion*, 1.

Theists are often accused of having a God-of-the-gaps mentality. But I find frequently that atheists exhibit a naturalism-of-the-gaps mentality. They don't know how science will explain a certain phenomenon, or develop a certain technology, but they express an unshakeable certainty—an undying faith—that in time, science will do so.

Finally, in chapter 5, as I summarize what we've learned about the origin of mankind through the science of anthropology, there will be many other examples of how those scientists, many of them atheists, allowed their preconceptions and biases to color how they interpreted any new data which came to light. In many cases, there will be groups of scientists who hold up the same piece(s) of data and come to startlingly opposite conclusions about it.

Conclusion

Once again, the main reason I wrote this chapter and the one before it was to show how theists have either persisted in a certain theological belief despite persuasive scientific evidence against it, or eventually changed their thinking 180 degrees when reason prevailed. It's important to lay that foundation for this book, because our generation is now in the middle of yet another turning point in its thinking, this time vis-à-vis the theory of evolution. More importantly, there's a tsunami on the horizon about which most theists are only just becoming aware: genetics has become a powerful tool in science, and its findings go straight to the heart of many key theological tenets.

So now we come to the primary purpose for this book. In the next few chapters, I'll try to help the average reader understand the scientific tools which generate these new data (ch. 4), then summarize the actual data which will force us to rethink our theology (chs. 5 and 6), and then highlight the specific theological beliefs which now need to be reexamined (ch. 8).

4

A Basic Understanding of the Science

JUDEO-CHRISTIAN THEOLOGY HAS HAD to redefine itself repeatedly over the past several millennia in response to many major scientific advances and cultural changes. The Copernican Revolution, outlined in chapter 2, completely overthrew not just our understanding of the structure of the cosmos, but more importantly our place in it. We went from being the very center of the universe, and the focal point of everything that is important in the great grand scheme of things, to being simply "a pale blue dot" revolving around an insignificant star in the corner of some galaxy, one of billions of other galaxies scattered helter-skelter across a massive universe. Likewise, chapter 3 in this book identified many other revisions to theistic explanations for observations in nature when science began to peel back the veil of ignorance.

The main purpose for those two chapters was not to reeducate the reader. I highly doubt there were any readers who still thought that earth is flat, or that YHWH has storehouses full of lightning bolts. Instead, the main point behind those chapters was to drive home repeatedly that the church once held ideas that were firmly based on Scripture and resisted for a while against the "onslaught" of science, but it has since grown completely comfortable with relegating those ideas to the trash heap of ancient, premodern thinking. No one gets worked up anymore when some heretical anatomist claims that women have the same number of ribs as men do, or a sacrilegious astronomer insists that the constellations are actually stars separated by millions of light years rather than some kind of deity or heavenly creature hanging from a dome just beyond our reach.

The scientific and intellectual advances of the nineteenth century have had a brutal impact on creationist beliefs. Many geological discoveries in that century began to make it increasingly difficult to accept that the earth is only six thousand years old, but was in fact many millions of years old. Other geologists began peeling back the sedimentary layers in the earth and found evidence that refuted the story of Noah's flood. Darwin published *Origin of Species* in 1859,[1] which explained how all species on earth arose by random mutation and natural selection. In 1856, the first Neanderthal was presented to the scientific world as a predecessor of humans, and soon many other "predecessors" began to be found. The rediscoveries of Babylonian and Egyptian civilizations and their texts and artefacts began to give us whole new insights into the pre-Hebrew, ANE world and its zeitgeist. In 1870, George Smith translated the *Epic of Gilgamesh,* and its amazing similarity to the Genesis account knocked the theological world off its feet, making many question whether the historically more recent Genesis account was merely a plagiarism. Archaeologists began making major discoveries which discredited the stories of Israel's invasion of Canaan. Out of this came a form of biblical criticism which completely changed our view of the Old Testament:

> Old Testament scholar Julius Wellhausen (1844–1918) . . . proposed a theory about the authorship of the Pentateuch that, although both strongly contested and widely accepted, has had an unparalleled effect on how the Pentateuch is viewed—and Old Testament scholarship has not been the same since. The bottom line is that for Wellhausen and many other biblical scholars before and since, the Pentateuch as we know it (an important qualification) was not completed until the postexilic period.[2]

Altogether, these findings seriously undermined the foundations of Christian belief, and provoked a theological knee-jerk reaction: Fundamentalism. Absolute insistence on a literal interpretation of the Genesis accounts despite evidence to the contrary.

Certainly there is still a segment of Christian society which resists Big Bang cosmology and Darwin's theory of evolution. YEC and flood geology have certainly dwindled from being the mainstream paradigm that dominated for several millennia to representing now a fraction (albeit, a very vocal one) of the Christian community. Those that perceive that the YEC view is growing in strength should take a global perspective on this question rather than simply considering what's going on in the United States. I'm also beginning to see this as a generational thing: increasingly, the children of Fundamentalists and literalists are coming out of high schools, colleges and universities with a full acceptance of the immense age of the universe, and the common descent of living organisms. They don't understand why this is such a big deal to the people of their parent's generation. They shrug their shoulders at the suggestion that the biblical flood might have been only a local phenomenon. Impassioned defense of YEC views are disappearing as the older generation dies off and the next generation replaces them. In that sense, the situation bears a vague resemblance to the biblical story of the Israelites marching through the desert for forty years, with a long trail of dead bodies marking their journey from captivity in Egypt to freedom in Canaan.

1. Darwin, *Origin of Species*.
2. Enns, *Evolution of Adam*, 5.

But for those Israelites, arriving at Canaan also meant engagement in a major battle against a battle-hardened army of giants. In the same way, today's new generation of believers face a new and much bigger threat to their beliefs. In the present century, jaw-dropping advances in the fields of paleontology and genetics have necessitated a whole new framework for the origin of mankind, with tremendous theological implications. That has been the raison d'être for my writing this book. But before overwhelming the reader with those new findings, I need to equip the reader with a basic understanding of the science and the techniques used to obtain those data. Given the scale of the changes to theology, it is important to describe and justify the science. "Extraordinary claims require extraordinary evidence."[3]

The purpose of this chapter is solely to give the average reader who has little or no formal training in science a working knowledge of paleontology and genetics, and of various concepts and terminologies that will be essential to the next few chapters. Some readers might find this chapter overly simplistic; others might find it unnecessarily scientific and tedious. In either case, those readers might be willing to try to wade past this chapter and skip straight through to chapters 5 and 6 without this background information. If so, please feel free to do so: you can always come back if you feel you may have misjudged yourself.

But before any readers either proceed with this basic introduction to anthropology and genetics or skip to chapter 5, it's absolutely essential that all readers who wish to be well-informed on this subject of human origins make sure they properly understand one simple concept: what it does *not* mean when a scientist refers to evolution.

Evolution: What It Is Not

Most readers today understand the term evolution used in a biological context: how groups of living organisms gradually change from one form to another. Such readers will also know that this change is often caused by genetic mutations combined with some sort of natural filter that selects out one form from another, often on the basis that those forms are either better or lesser fit for survival or reproduction because of the genetic change. And that the natural filter could pertain to how the organisms interact with their environment, their predators, their prey, or even their mates with whom they reproduce.

Although many readers think they have a good idea of what evolution is all about, they often hold one or more misperceptions which prevent them from accepting it as an excellent explanation of nature around us.

Misperception #1: It should be possible to arrange all the species into a long sequence that ends with humans. Actually, this is not true, because evolution is branching in nature, not linear. The latter statement may sound simple and unimportant, but

---

3. Popularized by Carl Sagan, but Laplace and David Hume made very similar statements.

much of the confusion, ignorance, debate and rancor which creationists bring to the table rest on this core misunderstanding.

Yes, evolution theory proposes that one kind of animal evolves into another, which in turn evolves into another, and so on: this is what I mean by the linear nature of evolution. But it's important to realize that many evolutionary changes lead to branching events where two or more new lines arise from one common ancestor. Those new lines then proceed to evolve, in either linear or further branching fashion, completely independently.

Figure 6. Classic evolution logo. Public domain.

This is the biggest stumbling block for creationists who refer to missing links, especially the link(s) they expect should be found between apes and humans. As will become clear over the next few chapters, scientists don't believe that humans evolved from apes or monkeys: instead, they propose that humans and apes both evolved from a common ancestor. Unfortunately, this fundamental misunderstanding is only reinforced by the extremely popular logo for human evolution (see figure 6), used by scientists and lay-people alike from both theistic and atheistic perspectives. So are museum exhibits which line up a series of hominid skulls or skeletons side-by-side in a way that suggests a progressive change from one to the other.

A BASIC UNDERSTANDING OF THE SCIENCE

Figure 7. Evolution of letters. Each letter arises through a series of small changes producing intermediate forms (represented by the lighter grey shapes). Notice the "missing links" between certain letters, which instead arose from a distant common ancestor.

To better illustrate this point, I often use an analogy involving a hypothetical evolution of alphabetical letters (see figure 7). Imagine the letters of the English alphabet representing twenty-six different species of living organisms. The letter *o* looks very similar to the letter *c* (just a small piece of its body removed) and to the letter *a* (just a short line added). The letters *b*, *d*, *p*, and *q* also look very much like the letters *o* and *a*: one just simply has to add a slightly longer line to the *o*, or extend the line on *a* a little more. The same thing could be said for the letter *u*, which is simply the letter *a* lacking a small piece, the letter *n* which is simply *u* turned upside-down, and the letter *m* which is simply *n* with another added bit (or you could say that *n* went through a partial duplication event).

75

One could do the same thing for all the letters of the alphabet: showing how all of them can be derived by a very slight modification of another letter, whether that be the addition or deletion of a small piece, a minor lengthening of a bit, a rotation or mirror inversion, a partial duplication, and so on.

This may not seem to be an earth-shattering revelation. However, the absolutely, fundamentally important thing to appreciate is that this hypothetical evolution of letters can only occur in a branching fashion, and not in a linear fashion. There's no easy way to get directly, in a strictly linear fashion, to all the letters of the alphabet without any branching events. Because of the branching nature of this analogy, one can see that the letters *w* and *s* have a common ancestor: the letter *v*. We don't have to look for an intermediate between *w* and *s*: there is no such thing as a missing link between them. They've diverged. Likewise for the letters *m* and *t*.

This branching characteristic of evolution naturally led to a metaphor of a tree—the Tree of Life—during the early 1800s. This term may resonate with some creationists and grate against others because of the biblical references to a Tree of Life (second and third chapters of Genesis, as well as chapters 2 and 22 of Revelation), but many other theists have no problem with it. In fact, Jean-Baptiste Lamarck, arguably a theist, produced the first tree of animal life. Edward Hitchcock, former pastor and later president of Amherst College who ardently sought to reconcile faith and science, published the first tree of paleontological life.

Given the branching nature of the Tree of Life, opponents to Darwinism should no longer ask "why don't we see monkeys turning into humans anymore?" or "where are the missing links?" Given enough time, monkeys will continue to evolve (as will humans) but will almost certainly not go through a human phase. In fact, the descendants of the monkeys will likely look quite different from anything we've ever seen on earth. Likewise, there should no longer be an insistence that by now we should have found the missing link between apes and humans, or between chimps and humans. We no longer believe those links ever existed.

Misperception #2: There should be an abundance of transitional forms between one species and another. This is another concept which anti-evolutionists who point accusingly to the absence of missing links need to get clear in their minds.

The branching nature of evolution does indeed predict the existence of transitional forms. For example, whales and hippopotamuses are believed to have evolved from a common ancestor, a mammal that lived on land and breathed through nostrils at the front of its head. It's thought that as the two species diverged, one maintained the terrestrial form (hippopotamuses), while the other took on a more aquatic form (whales), which required their nostrils to migrate to the tops of their heads. As such, one might expect to find transitional forms with various intermediate positions for their air holes, and will be surprised to learn that we don't have any such intermediate forms. The same conundrum can arise for the many other examples of transition from one species to another.

The answer to this question is that it's all in the numbers. Generally speaking, a given species which is well adapted to its environment will live for a relatively long period of time (perhaps measured in millions of years) and reproduce in great abundance, possibly producing billions or trillions of structurally similar descendants. When they encounter some kind of selection pressure which is strong enough, they can die out relatively quickly and a new form which is better adapted to that new environment can then multiply in great abundance, again producing billions or trillions of descendants. The transition period, however, will generally be short, and a relatively much smaller number of those intermediate forms will have been formed.

Complicating this rarity of transitional forms (relative to the parent and daughter species) is the fact that fossilization is a rare and brief event, requiring very stringent conditions which arise only sporadically (explained in more detail on p. 185). For this reason, only an exceedingly small fraction of all the organisms which have ever lived have become fossilized. Thus, it shouldn't be at all surprising that it will be much more likely for the fossil record to exhibit representatives of the fully differentiated forms (which may have numbered in the billions or trillions) than the much, much fewer intermediate forms. In fact, the latter often don't even appear because the transition was too quick and the appropriate conditions for fossilization never materialized during that relatively short transitional period.

Some readers might see this as simply a dismissive hand-waving exercise. Perhaps a couple modern-day examples would help to clarify its relevance and legitimacy.

Mankind for millennia have needed to do complex calculations to make buildings and boats, transact business, conduct warfare, and many other things. For thousands of years we made our calculations using physical devices such as beads on strings, the abacus, tables of numbers, and complex clock-like machines full of interlocking gears. But then we invented electronic devices which could calculate much more quickly and efficiently. At first, these computers relied on large vacuum tubes and bulky wiring which used massive amounts of power, generated a lot of heat and filled up the space of whole rooms. But a few decades later we developed circuit boards and solid-state electronics which could do the same job as those vacuum tubes but which could be scaled down to nearly microscopic sizes and use a fraction of the energy. Just a few decades later, those electronic circuits are found in every computer and in almost every modern device. Today, you won't be able to find any devices still running on vacuum tubes (unless you go to an antique store). If some futuristic alien were to dig through the archaeological ruins of a nuclear-devastated earth, all over the globe they would find countless relics of those physical calculation devices (beads on strings; abacuses; tables of numbers; slide rules) and of electronic devices (computers; calculators; range finders), but they probably wouldn't find any devices which used vacuum tubes. The latter were a transitional form which didn't last very long and weren't mass produced to the extent that the other kinds of devices have been.

We see the same phenomenon when we look at data storage. For millennia we did this by creating markings on some kind of physical medium (stylus-and-clay; pencil-and-paper; chalk-and-slate), but now almost everything is recorded in digital form on electronic chips (computer hard-drives; thumb-drives; camera/phone memory cards). For a few decades, though, during the transition, electronic storage of read-write memory (information which could be added to, edited and modified) was done using materials that could be easily magnetized and demagnetized: these included reel-to-reel tape recorders, cassette tapes, VCR tapes, and floppy disks. Once again, if one were to dig through the garbage dumps of cities around the world, you would find countless examples of the physical writings and of the electronic forms of memory, but you would rarely find examples of the magnetic recordings (other than the magnetic strip in credit cards, which are used as a permanent form of memory, not an editable one). Technology which used a magnetic form of read-write memory was just not used widely enough (in a global sense) nor long enough.

One cannot at all deny that vacuum tubes and magnetic data storage were instrumental in the transition from one way of doing things in the past to another way of doing those things today. But the archaeological evidence a thousand years from now will likely provide little or no support for the proposed existence of those transitional forms. The artefacts dug up by those future archaeologists will show only the fully differentiated ancient and modern forms without any transitional ones.

Misperception #3: All species must undergo change over long periods of time. This common misperception reveals itself in comments about extremely ancient fossils that look exactly like animals alive today. For example, the fossil of a turtle which was dated to be 120 million years old[4] and which looks exactly like turtles today. The question that is often asked is: "Why haven't they changed? Evolution requires changes over time, yet this looks exactly like living animals today?"

In fact, evolution theory doesn't require organisms to change. If an organism is well adapted to its environment, there is little or no selection pressure for it to change. If there are also no circumstances which lead to genetic drift (such as some kind of reproductive isolation), it is entirely possible for the organism to continue for millions of years with little or no change.

In the examples I gave above for misperception #2, the fact remains that many people around the world still use the abacus to do calculations and still record writings on paper, even though the more "evolved" electronic forms of calculating and recording are also in wide-spread use around the world. Those old-school ways of doing things still work perfectly well and are in that sense perfectly well adapted for continued use today. In fact, we'll probably still be counting with our fingers and writing things down on paper a thousand years from now. We don't have to stop using those strategies just because there are newer and "better" ways of doing things.

---

4. Cadena and Parham, "Oldest Known Marine Turtle."

Misperception #4: All changes in a species improve their survival and/or reproductive success. Not all genetic changes emerge through some kind of active selection filter. Sometimes a change occurs which has no impact on reproductive success, neither positive nor negative, but nonetheless becomes fixed within the genome. This is especially the case when the organism becomes reproductively isolated. For example, when the lake they live in becomes permanently separated from other bodies of water as a result of changes in the geological plates, or shifting weather patterns create a permanent desert between two halves of a large grassy savannah or a rain forest, just to name a few examples. The population of a given species is thus divided and gene pool mixing between the two is prevented. Over time, the one population may take on certain characteristics while the other population can take on other characteristics. Eventually, this can give rise to two or more separate species.

So it is a misperception that all genetic changes are a product of increased survival and/or better reproductive success. It isn't always necessary to explain a particular evolutionary change in those terms. Some changes just happen.

It's a Family Affair

Humans are very good at organizing things into categories.

For those readers who are reasonably computer-friendly, think about how your computer's files can be arranged into different directories, each of which can have sub-directories, which in turn can also have sub-directories, and so on.

For those who are seasoned shoppers, think about how Walmart will have all the food products separate from all the pet supplies, clothing, automotive parts, and so on. Then, within the food area, you will have the aisle for canned goods and another aisle for baked goods or meats, and within the meats aisle will be found the section for bacon separate from the section for fish or frozen beef, and so on.

We humans like to organize, categorize, catalogue and list or name things. We see this already in the very opening chapters of the Bible, with Adam naming all the animals that YHWH had just created.[5]

Carl Linnaeus has been credited with setting up a system of organizing all the different forms of life on earth, a system we call taxonomy. This includes a variety of categories and subcategories and sub-subcategories. Once we get near the bottom of this categorizing system, we find organisms grouped into whole families (some of which can be separated into subfamilies and then into tribes), which in turn will be broken down into different genera (the singular of which is genus), and each of those in turn into different species.

Perhaps a couple examples might be helpful.

---

5. Gen 2:19–20.

The animals we call the common domesticated house cat (which includes breeds or phenotypes such as Siamese, Calico, Persian, and Angora) are all one genus (*Felis*). This genus is distinct from other genera such as *Panthera* (lion, jaguar, leopard, and tiger), *Leopardus* (mountain cat and ocelot), *Lynx* (lynx and bobcat), *Puma* (cougar and jaguarundi) and many others.

Likewise for dogs and wolves, which are two different members of the family *Canidae*, which is broken down into different genera and species which include foxes, jackals, and coyotes. But not the hyena or thylacine (marsupial wolf): these are both members of whole different families.

These terms family, genus and species are important ones and will come up frequently in this book.

The Scientific Method versus a Theological Approach

Many theists have the idea in their heads that science and scientists are intent upon destroying faith and belief systems.

This is simply not the case. Scientists also have belief systems, although they often don't recognize the latter or won't admit to them. But what scientists do is gather facts in a pursuit of knowledge, and in the process they may dismantle an understanding which is built on belief rather than cold, hard fact. That dismantling isn't directed specifically against religious beliefs. It's also directed against beliefs that one race is superior to another. That earth is flat. That sasquatches live in the Himalayas. That immunizations and fluoride in our drinking water are bad things. That eating your placenta after giving birth is healthy. If the facts don't support the belief, or even speak against it, scientists will speak out against the belief.

The scientific method is a process which non-scientists use all the time, even daily. Usually it begins with a question ("Are there any fish in that lake?"), which itself may spring out from an observation ("Something just splashed the water in the middle of the lake. I wonder if a fish did that?")

The next step is to perform an experiment (for example, throwing a piece of cookie on the water), which usually involves having made a prediction ("If I throw a piece of cookie on the water, the fish might go for it") or a hypothesis ("An aquatic animal is present within the body of water, and can be enticed to the surface by distributing portions of baked hydrocarbons upon the aqueous layer").

Observations are made, which in turn lead to a conclusion or a modification of the initial hypothesis. If a fish does indeed gobble the bait, the conclusion is obvious, and may lead to further questions and experimentation ("I wonder how I can catch it?"). If nothing happens, the hypothesis can be revised ("maybe the splash was a stone thrown by my brother behind me," or "maybe the fish isn't hungry, but I can scare it into moving with a stone").

This, of course, is a very simple example, but science is rarely this straightforward. Sometimes there are many explanations for a given observation, and many different kinds of experiments need to be done in order to convince the scientist.

Often, scientists are trying to make sense of events which happened in the distant past. Much of what they conclude is based on consequences and inference to the simplest explanation of those consequences. Again, all of these are things everyone does every day.

While driving down the highway, you will see tire skid marks on the pavement, crumpled guard rails to the side, garbage in the ditch, deep tire tracks leading directly toward a broken fence and a scorched patch of grass surrounded by glass and fragments of a car. All of these consequences provide glimpses into actual events that occurred on that highway, some perhaps even many years ago.

We find a tree in the middle of a forest and infer the obvious and simplest explanation: it grew from a seed dropped by one of the surrounding trees. Sure, it's entirely possible that it came to be here by someone coming along and planting a two-year-old seedling that was germinated thousands of miles away. But that latter explanation is quite unlikely, much less so than the one claiming the tree to be simply the product of a randomly scattered seed.

So in science and in everyday life, we collect evidence to put together a likely story. The more lines of evidence we have, the more likely that the story is accurate.

Returning to the highway, we examine the fragments of car and make further observations which tell us about what kind of car it was. We find smudges of a different color of paint on several of the fragments, perhaps even find a part of a different kind of car, and infer that another car was involved in the collision. Perhaps find a license plate or some paper debris which tells us the name and address of one of the drivers. We use that information to collect more information from newspapers to determine the date, and the circumstances. The more lines of evidence we collect the more accurate and likely is the explanation we come up with.

Sometimes we make assumptions about the data at hand, and a little piece of additional information suddenly puts a whole new and different perspective on them. A colleague pointed out a humorous illustration of this. If I were to say, "this morning I got out of bed and shot an elephant in my pajamas," you would interpret it a certain way. But if I followed that up with, "why he was wearing my pajamas I'll never know," your first interpretation would immediately evaporate and you'd now be working with an entirely different image.

Returning to the tree in the forest, we are disturbed to find that there are no other trees of that particular species in the entire forest: in fact, the tree is indigenous to a different part of the world. And then we notice a plastic band around the base of the tree, and on that plastic band the name of a tree farm based in a different country. We talk to a local forestry warden and learn that twenty-eight years ago a family from that foreign country died while camping in this part of the forest, and grieving relatives returned to the site a year later. Now, having obtained a more extensive and detailed data set, the simplest explanation for the existence of that tree is beginning to look like the one that had previously been quite unlikely (that a seedling from another country was intentionally planted in this very spot, rather than a random seed germinating here).

An important quality of a good scientist is the need to continue collecting more evidence and testing predictions from the model to get the best inference, and to be ready to discard a current theory when the data begin to stack up against it.

Many lay people see our inherent skepticism as a weakness or a flaw. But we scientists see it as strength. I recently listened to a paleontologist interviewed about her discovery, together with an expert in biomaterials and microscopy, of collagen fibers and red blood cells inside fossilized dinosaur bones.[6] One part of the interview which really caught my attention was her response to data which conflicted with what she expected or "wanted" to find:

> When he told me this, I was extremely skeptical. . . . Being a paleontologist I "know" you can't find soft tissues in fossils that are seventy-five million years old. I said, "No, no, that can't be. They must be bacteria or pollen or something completely different." . . . We went through various steps to try and eliminate those. . . . We tried [this] and we tried [that] and we tried [the other thing]. . . . I was incredibly skeptical when he first told me what he thought these things were. . . . I really didn't believe him at all. I was really skeptical and I actually went through a long process of trying to prove him wrong. But in the end I lost and he won.[7]

This vignette describes the dynamics that often take place within the mind of a scientist, between different research groups, and even in the scientific community in general. We don't just flip the coin or peek inside the box once and then go with that single outcome. We repeat the experiment(s) several times, design other experiments which will further support the hypothesis and/or which rule out alternative explanations. And we actually encourage attempts made by other scientists to disprove a given hypothesis.

This contrasts markedly with the attitude I often see in theists when someone asks a question that casts doubt on a particular theological point: dismissal, condemnation, sometimes even ostracism. Or sometimes the response is to put limits on how far the investigation can go. I once asked a group what they thought about the new findings from genetics that were hard to reconcile with a historic Adam and Eve, and one person said: "I don't care what you do with that as long as you leave me with Jesus dying on the cross for my sins." I winced. Not because I disagree with that theology or that aspect of Christian history. But instead because the sentiment expressed was

---

6. Bertazzo et al., "Fibres and Cellular Structures."
7. McDonald, "Dinosaur Fossils."

simply: "you can interpret it however you want, as long as you interpret it this way." Why not be open to letting the data speak for themselves? If it's true, it's true, and the data will inevitably get us to that endpoint, or at the very least not prevent us from holding to that endpoint. But don't constrain the search to that endpoint even before the investigation has begun.

I once read a statement from a staunchly YEC believer who emphasized and repeated that science must give way to "a plain reading of Scripture," that evolutionary theory must therefore be rejected and that no matter how much evidence that kind of science collects, it must be seen as misleading and bad science.

This is the kind of smothering perspective I'm trying to confront in this book.

Basic Anthropology

Louis H. Sullivan, an influential American architect, is noted for having said, "Form ever follows function." Another famous American architect, Frank Lloyd Wright, revised that expression: "Form and function should be one, joined in a spiritual union."

What they were saying was that the purpose for which an object is meant to be used determines how that object should be designed. Changes to function should likewise lead to corresponding structural changes in that object.

This applies not only to buildings and other man-made objects, but also to biological objects and the process of evolution.

Hominids are believed to have evolved from tree-dwelling ancestors. Tree-dwelling primates today have powerful arms, at least as strong as their legs but sometimes more so, carrying the full weight of their body and helping them climb trees and swing through the branches. They usually also have tails to help them grasp branches and/or serve as a counter-balance when they leap from limb to limb. Our ancestors were also omnivores—meaning they ate anything from fruits, nuts and vegetation to other animals—and therefore their jaws feature pointed canine teeth for piercing, broad sharp incisors for cutting, and flattened molars for crushing. They had small brains (because big ones take millennia to develop, and require an exceptional amount of metabolic energy to maintain), and therefore they needed only small skulls. As hominids continued to evolve, culminating in *Homo sapiens* today, all of this changed. They left the trees, transitioned from walking on all four limbs (quadrupeds) to walking upright on legs (bipeds) in search of a more varied diet, and developed ever larger brains.

Consistent with Sullivan's and Wright's statements about form and function, all of these functional changes, and many others, required massive changes to the relative shapes and sizes of their arms and legs, as well as adjustments to their spine, pelvis, jaw bones and skull. Like a good mathematical equation, this relationship works in the reverse direction as well: when anthropologists examine a bone closely, they can look back in time and infer a great deal about the individual that produced them, and therefore the species from which that individual derived.

The skull gives clues about the intelligence of the hominid. This is not only because overall brain size correlates with intelligence, but also because castings made from the interior space of the skull can also reveal evidence of folding in the brain's surface, which is an indication of neural development and intelligence. Moreover, the attachment of the skull to the spine gives glimpses into the manner in which the animal walked (quadrupedal versus bipedal).

The jawbone and its teeth can say much about the diet of the animal: the relative proportion of sharp incisors versus flat molars for example, or the strength of the attachment sites of the chewing muscles. The age can be inferred from the teeth: the presence of "milk teeth" suggests a very immature individual, while excessive wearing indicates a very elderly one. Some even inferred from the attachment sites for the muscles of the tongue whether the individual was capable of speech or not.

The shape of the pelvis tells us whether it was a male or a female (because the latter have wider hips to facilitate childbirth), as well as the degree to which it was a quadruped or biped.

Of course, the leg bones provide an excellent indicator of the height of the animal, but also tell us whether the animal stood erect (as humans do) or hunched over (like apes).

The shape of the feet can also tell whether the animal was bipedal (more flat-footed) or whether it tended to grasp tree branches and swing through the trees (rounded, curving toes).

The general appearance of the bones can suggest its general health, or a disease with which it struggled or possibly even died from. When those bones are accompanied by tools and other artefacts, anthropologists can begin to paint a picture of the sociology, intelligence and behavior of those hominids.

A skeleton which is carefully buried betrays a belief in the afterlife and therefore also the capacity for abstract thought, especially if the burial is accompanied by jewelry, items of food or religious artefacts; this in turn strongly suggests language.

Also, many fossilized bones are found embedded in rock, the original bones having been buried under layers of silt, sand, gravel and other crushed material which then hardens into various forms of sedimentary rock. Although this makes it incredibly difficult to expose the bones without damaging them—often requiring months or even years of careful and meticulous chipping and scraping away of the sedimentary rock off the underlying mineralized bone—it also allows the researchers to determine the age of the bones without damaging them directly by dating other materials in the adherent sedimentary rocky material which would have become embedded at the same time, or by knowledge of when the sedimentation occurred (for example, burial in an ash layer from a known volcanic eruption).

On the other hand, if the bones are found loosely in mud or sand, one can't conclude anything about the age from that surrounding material: that fossilized specimen could have become enveloped in the mud along with other objects of widely varying ages. Their only recourse in this case is to perform radioisotope dating of the bone material itself, which involves complete destruction of the portions which are used to do the dating; this difficult decision is exacerbated by the fact that the certainty of the measurement decreases as the amount of material sampled (and destroyed) decreases.

It has now become possible to do computerized tomography (CT) scans of the bones embedded in the rock, much like doctors performing CT scans of your brain inside your skull, and then use a 3D-printer to make a replica of the bone even while it remains embedded within the rock

There is so much that can be inferred from old dry bones. It is now even becoming possible to extract DNA from these bones (I'll elaborate on this astounding development in the second half of this chapter), which in turn reveals genetic relationships between them, and their placement within the Tree of Life.

Altogether, a tremendous amount of information can be gleaned from simply finding a skull, or a jaw bone or a femur. Newspaper headlines which trumpet the discovery of a new hominid often mention only one or another of these key items. Currently, we already have an incredible number of hominid remains, including many which are fairly complete and even articulated to varying degrees (meaning the bones are found more or less in their relative positions within the body when they were connected by ligaments and muscles at the time of death, rather than being scattered and randomly distributed).

As anthropologists continue to find more and more of these specimens, they can develop an averaged picture of what the species looked like, and define certain anatomical measurements and relationships that can be used to identify a newly found bone as belonging to this species or that one.

It is even possible now to rebuild the faces and physiques of the specimens. Anatomists will take a cast of a relatively complete skull and build up layers of clay on top of it to mimic the layers of skin and muscle. This is possible because of careful study of how those different kinds of living tissue adhere to an underlying skeletal structure, and mold around edges of bone. This skill is now frequently used in forensic criminal investigations to obtain a facial likeness of a victim they've found in the hopes of soliciting help from others in identifying the victim. That technique is now equally being used to get a glimpse of our earliest ancestors and evolutionary relatives. The exact same principles can be used for an entire skeleton. The attachment sites for the muscle to the bone will give tell-tale signs of whether the muscle was large and powerful or not and thereby help in the rebuilding process.

In this way, anthropologists are convinced they now have a reasonably detailed representation of the branch of the Tree of Life from which the twig labelled *Homo sapiens* pokes out. But as much as we have learned about ourselves from dry bones unearthed from rock, we have now found a whole new way to travel back in time and observe the evolution of mankind: genetics.

The Study of Bones Can Be Deceiving

Despite the gains which have been made through the study of dusty old bones, there is good reason to be cautious. Biologists have for centuries been categorizing living organisms into groups according to their morphology, and making claims that one group descended from another based on gradual changes in form. The inference makes perfect sense. But it looks like it may have led us down the wrong path at least a few times in the past.

Take, for example, *Thylacinus cynocephalus*, the marsupial wolf which had lived in Australia, Tasmania and New Guinea, but went extinct approximately a century ago. In the shape of its body, it looks every bit like a wolf. With one notable exception: it has a pouch in which it carried its young. Hence the prefix marsupial, which refers to animals having pouches (kangaroo; opossum; Tasmanian devil): *Thylacinus cynocephalus* is literally dog-headed-pouched-one. There would otherwise have been every reason to put this animal in the wolf/dog lineage. Except that subsequent genetic testing (to be described below) showed the marsupial wolf to be closely related to other marsupials, especially those found in Australia more so than those from South America.[8] This, then, is an example of convergent evolution: a process whereby unrelated animals in similar environments and subject to similar pressures evolve similar forms and behaviors. In this case, two separate lines of wolf-like animals evolved among marsupials and a third one evolved among placental mammals. Without the help of genetics, we might have inferred that these were all more directly related.

There are two lessons to be taken from this. One needs to maintain a bit of skepticism about family trees or Trees of Life built up according to bodily forms, especially those built up solely upon the finding of skeletal remains. More importantly, the new science of genetics is a powerful technique, offering a form of time travel into the distant past. As will become clear in the next few sections, it is in fact a far more discriminating tool than paleontology (the study of fossil bones) because it can identify individual mutations, while a detectable change in bodily form requires dozens if not hundreds of mutations.

DNA and Genetics

Nearly everything that defines who and what we are as organisms is determined by the genetic code.

It isn't just the differences in the coded message itself which distinguishes one from another, but the way that the coding is decoded also makes a huge difference. For example, every cell in your body has the same genes, but a nerve cell only decodes one part of the entire collection of genes (which we call the genome), and does so in a certain order during its lifetime, while an immune cell may decode a different part of

---

8. Pääbo, *Neanderthal Man*, 45; Miller et al., "Mitochondrial Genome Sequence," 213–20.

the genome at a different time. The same can be said of all the other cell types in your body. It can also be said about your body as a whole: you decode very different sets of genes during each of the major stages in life: fetal development, growth, puberty, menopause, and senescence.

We also inherit traits and characteristics from certain cell parts that made up our mother's egg, other than the nucleus and the genes contained within it. This is a field of science referred to as epigenetics, a subject that I won't get into in this book.

But ultimately, one can say that the genetic coding is hugely influential in determining our very being.

So what exactly is this genetic code?

Books have been written which describe the structure of a gene in tremendous and minute detail: we have indeed learned quite a bit in less than a century of picking this question apart. For the reader I'm targeting in this book, though, it's enough to simply say that the genetic code consists of a sequence of four chemicals or bases: cytosine, guanine, adenine, and thymine, which we otherwise refer to using their first letters (C, G, A, and T, respectively). Every gene in our body can be represented by a long string of these letters which appear to be randomly ordered in seemingly non-repeating sequences.

The entire collection of genes in humans, our genome, consists of just over three billion of these letters. Nearly every cell in your body has a copy of that genome (the exceptions being your red blood cells, which don't have a nucleus, and your eggs or sperm cells, which have only half a copy of the genome). If those three billion bases were kept together as one single molecule, it would stretch out for three meters in length. For this reason, cells divide it up into smaller segments, which we call chromosomes (each of us has twenty-three pairs of chromosomes). They then twist each segment into a coil, then coil up the coils, and pack everything up tightly so that they can fit neatly inside a small compartment inside the cell: the nucleus. Imagine all the work that has to be done whenever the cell needs to unpack all of that in order to decode any given gene, something that is constantly happening in nearly every cell. It's absolutely mind-boggling, even for the scientists who work in this field every day. That process of decoding the gene is otherwise usually referred to as expression, and I will use that term through the remainder of the book.

To help you put into perspective what I've just described to you, think about the human genome as a library of books, and each chromosome as a shelf-unit full of books.

When it comes time for the cell to express a given gene, the cell pulls that book off the shelf (uncoils the appropriate segment of DNA) and opens it up to the relevant page, and begins to read the gene. The latter is merely the sequence of letters C, G, A and T, broken out into separate chapters, paragraphs and sentences (I'll explain a little later how those literary devices in turn can represent functionally relevant parts of genes).

When it comes time for the cell to reproduce—to divide into two daughter cells—it must pull out every one of those books off the shelves in the whole library and carefully copy every sentence letter by letter. It does this by unravelling all the segments of DNA, and matching every base in the original with a corresponding base in the copy. This would otherwise be a daunting task prone to accumulation of all kinds of errors, except that life has come up with an ingenious way to make the process not only simpler, but also self-correcting.

The four bases are not connected to each other directly. Instead, each base is pinned individually and separately onto a central axis, similar to laundry hanging on a clothesline. The clothesline itself is a long molecule made up of individual sugar molecules—millions of them, in fact—attached end-to-end like children holding hands in a game of Red Rover. The sugar is deoxyribose, the D in DNA (deoxyribonucleic acid). The dimensions of this string of sugars with bases hanging off of it, and the physical and chemical properties of those bases themselves, are exactly right such that if two lengths of DNA are stretched out parallel to each other, it is possible to exactly match the bases on the first clothesline with a certain complementary base on the second line. More specifically, a C pinned to one clothesline will match up only with a G on the other line, but not with another C or an A or a T. Likewise, a T on the one line will match up perfectly and only with an A on the other. The same can be said of the G (only matches to C) and the T (only matches to A). In the end, the second stretch of DNA becomes a matching opposite of the first. To use another analogy, the second is like a photographic negative of the first (although, I'm not sure in this day and age of digital photography whether the reader will understand what a photographic negative is).

There are two main reasons behind this precise matching of C only with G, and T only with A. One has to do with the physical dimensions of the bases. The second has to do with the chemical structure of the bases that allows for lines of attraction—referred to by chemists as hydrogen bonding—to form between the bases. The C and G base pair have just the right chemical structure such that three hydrogen bonds form between them, while A and T have a different overall chemical structure such that they form two different hydrogen bonds.

The cell keeps both lengths of DNA paired up together in this way (and then coiled and supercoiled when it's packed away in the nucleus) to help in error-detection and duplication. Whenever a mutation occurs—one or more of the bases is altered—the change can be immediately recognized and fixed. Also, when it comes time to divide and reproduce, the cell "simply" pulls those two lines apart and rebuilds a new complementary partner for both of the separated strands.

This is an extremely brief introduction to basic genetics, and leaves out so very many fascinating details. This doesn't even mention how the information in those lengths of DNA is then translated into making proteins, which are the ultimate reason we have DNA and chromosomes. Contrary to what is sometimes claimed about DNA, genes are not blueprints. They're more like recipes than blueprints. All they do is contain the information needed to make simple building blocks, as well as synchronization of time clocks, and placement of various beacons and flags. The cells then simply begin producing building blocks (proteins and some genetic materials), which in turn build other building blocks (other proteins, sugars, lipids, and so on) and start placing flags and beacons and triggering various stop-watches and starting-guns. The product of all this mindless action is a living organism!

This summary of cell biology is admittedly overly simplistic, but it is absolutely essential for laying down a foundation for understanding the new lines of evidence from genetic studies that now challenge some key theological concepts.

Early Tools of Genetics: Chromosomal Staining and Banding

First-century Romans began to explore how the appearance of objects was changed when they were viewed through melted sand, which we now call glass. But it wasn't until the thirteenth century that one of their descendants turned that into a lens which could help its wearer see better (many give credit to Salvino D'Armate for this, but others dispute this). Anton van Leeuwenhoek, a Dutch scientist is often credited with making the first microscope in the seventeenth century.

Scientists now had a way to look at extremely small things, and did so like rabid explorers on an entirely new continent or planet. But despite their discovery of whole new forms of life and answers to some of biology's riddles, when they looked at cells, the fundamental building block of life, they often saw only bags of water. In fact, it was like looking at crushed ice cubes in a glass of water: you could barely make out certain edges in the interior of the cell. Eventually they found out that certain stains could be used to highlight various features of the cells, essentially coloring the different ice cubes in the glass of water in different ways to make them stand out from each other and from the background bag of water. One of the earliest pioneers of this new technique, Ramón y Cajal, gave us a classic quote of irony: "I expressed the surprise which I experienced upon seeing with my own eyes the wonderful revelatory powers of the chrome-silver reaction . . . and the absence of any excitement in the scientific world aroused by its discovery."[9]

As cell staining techniques developed, organelles and other features of these microscopic machines were revealed. Not only could the nucleus now be delineated, but one of these techniques highlighted certain structures which were eventually identified as the chromosomes: the individual strands of DNA which make up the organism. Biologists began to realize that different organisms have different numbers of chromosomes, and that the chromosomes often differed in size. They also noticed that certain sections of the chromosomes might stain more darkly or more lightly than

---

9. Cajal, *Recuerdos de mi vida*, 76.

others such that the chromosomes looked like multi-banded caterpillars. The bands differed considerably with respect to thickness, intensity and location, producing unique banding profiles which can help distinguish one chromosome from another. In one sense, they look very much like the UPC product codes on anything one buys in a store, and which are used to identify that unique product in the store's computer database (figure 8). An image of chromosomes stained in this way is referred to as a karyotype: this will become important and useful in pages 135–136 when we begin to look at the similarities between human genomes and their differences with the genomes of other animals.

Using these new nuclear stains, biologists could now observe the cell as it went through the process of dividing into two daughter cells, and in the process they watched the pairs of chromosomes line up, duplicate, and then get pulled to opposite sides as the doors between the two daughter cells slammed shut.

Figure 8. UPC bar code.

But they hadn't yet made the connection between these chromosomes and the roles they played within the cells, including the inheritance of genetic traits. Centuries later, once that connection had been made, this technique of chromosome staining would play an important role in an intriguing detective story, which I'll tell in chapter 6.

Genetic Flags Which Mark Out the Path of Evolution

When police are looking for a fugitive-of-the-law, they'll put out a description which includes distinguishing features. They won't mention that the suspect has ten fingers, but they might mention that he has only nine fingers (if that were the fact).

Likewise, when a lost item is found, one generally confirms it by looking for its anomalies. "How do you know he has your bike?" "Well, just look, Mom. There's the scratch on the front fork when I hit dad's car, and look how the seat cover's ripped from when I . . ."

In the same way, scientists can identify distinguishing features and anomalies in DNA which help them to identify it as unique to a certain species, or to characterize how related two different samples of DNA are. These can range from mutations of a single base pair to major changes in long stretches of DNA: several of these genetic markers are described in detail in the next few sections.

Mitochondrial DNA

Many readers will understand that DNA is found in the nucleus of the cell and encodes the information which defines who we are. But this is like saying "books are found in large public libraries and summarize an immense amount of human knowledge": the statement is true, but it's far from the whole truth. To really understand how cells work and, more importantly, how this new science of genetics relates to certain aspects of theology, we need to be more precise. Books are also found on coffee tables of nearly every house and nearly every doctor's or dentist's waiting room: these provide a form of entertaining reading that you generally won't find in a public library. Books are also found in glove compartments of cars, and provide important user information on how to maintain the car and its registry: information you probably won't find in a public library or doctor's office. They're found under pillows of beds around the world, and those books allow incredibly detailed insights into the private thought-life and history of an individual in the form of regular entries which often begin with "Dear diary." Books can be found in many, many other places, and all of these can serve many other purposes distinct from the books found in public libraries.

The same is true of DNA. The cell is more than simply a floppy bag with a nucleus in the center. It has many compartments in which the cell keeps specific materials, and in which it performs some very specific functions. In this sense, it's similar to an average house in which you have laundry rooms for one purpose, bathrooms for another purpose, kitchens for yet another, and so on. Many of these compartments are referred to as organelles, or "little organs," drawing an analogy between the cell and its organelles and a whole body with its different organs.

There's a large mountain of evidence, not to be reviewed in this book, which tells us that these organelles arose billions of years ago during the course of cellular evolution: one cell type invading another cell type, in the same way that bacteria and viruses invade the cells in your body today. Long before there were large multicellular plants and animals, the world was populated by single-celled ones: bacteria. Each was a species unto itself, and carried out all the necessary functions that cells need to do: find food, eat, grow, reproduce, avoid danger, and so on. This meant that each cell had to carry an incredible amount of genetic material to code for the proteins required to carry out any one of those functions at any given time. At the same time, these bacteria were competing with each other for the limited resources—food molecules—available in their environment. In many cases, they had to develop strategies to kill off their main competitors, which added to the list of genetic information that needed to be carried and to the list of jobs that needed to be done.

At some point in time, the bacteria developed a whole new trick that suddenly made life easier, made them incredibly more powerful, and catapulted them into an entirely higher category of life. Rather than competing against each other as individuals, they learned how to cooperate and collaborate.

Bacteria had long before this point developed the ability to invade other types of bacteria and then either kill them immediately or else live inside them like a parasite, slowly draining the resources from the host while they reproduced until the host eventually died of exhaustion; or they actively hacked their way out like the extraterrestrial in Ridley Scott's popular movie series *Alien* writhing out of the chest of the human it had infected/impregnated. After which they had to then find a new host and start all over again.

Eventually, one type of bacteria found a way to invade another, but then drew up a mutually beneficial contract with its host. The invader would concentrate on performing certain specialized functions that would benefit the host, while the host took care of nearly all the other essential functions of life (collecting food molecules, taking out the garbage, evading predators, and so on). This meant that the invader could get rid of a lot of bulky DNA—the genes that were necessary to carry out those other essential functions of life—such that the two organisms could now get by with only a fraction of the DNA that the two would have needed if they lived separately. It also meant that the invader could spend all of its time doing the specialized function, and thus do it that much better, or get a lot more of it done in a given amount of time because it wasn't having to worry about the other essential functions of life. In this sense, it's like a Renaissance nobleman's palace in which one had servants to take care of all the housecleaning, cooks to simply make meals for everybody, one person to take care of all the repairs to the building and make more rooms and buildings, a resident artist to compose timeless concertos on the piano, a resident astronomer to develop the heliocentric theory, and an accountant to pay for all the bills.

In one example of this mutual agreement between bugs, a bacterium which had the capability to carry out photosynthesis (converting sun light into useful sugar molecules) invaded a host, but then turned itself over solely to photosynthesis, sharing the sugar molecules with its host, who in turn took care of the humdrum tasks of housekeeping. The host and invader coordinated their cellular activities such that every time the host bacterium reproduced, so did the invader. In that way, all the daughter cells benefitted from this cooperative relationship. In time, after much modifications including losing much of its DNA, the invading bacterium became merely an organelle of the host: more specifically an organelle we call the chloroplast. This new type of cell eventually evolved into the multicellular plant life we see today.

In another case, a different kind of invading bacterium turned its attention solely to generating high-energy molecules, particularly one called adenosine triphosphate (ATP), which both the invader and the host used to power all the functions of life. Again, the invader eventually gave up most of the DNA it shared in common with the host, and even transferred some of its DNA to the host's nucleus, hanging on mostly to those genes it needed to make the high energy molecules. In time, the much reduced invader became simply another organelle, one which we call the mitochondrion. This new type of streamlined cell could produce massive amounts of energy in short periods of time, paving the way for newer evolutionary developments which required massive amounts of energy.

So far in this section, I've simply been trying to introduce the readers to the idea of cells having organelles which came from invading bacteria that gave up their own DNA in a mutually beneficial cooperative relationship with the host. Now we come to the main purpose of this section: to talk about mitochondrial DNA and how it has become a powerful argument against the concept of a historical Adam and Eve.

Fast forward billions of years to the modern day, in which we have plants and animals made up of trillions of cells. In those organisms which reproduce sexually, the mother and the father of the organism contribute half of the normal amount of DNA which is then mixed, shuffled, sliced and diced and otherwise recombined in a myriad of different ways within the nucleus such that an entirely new individual is produced, one having a whole different set of characteristics than did each of its parents. In some cases, it might be possible to identify a certain trait that came from either the mother or the father, such as the shape of the nose, or the tendency to have a certain disease. But in most cases, each of the progeny (children) of those two parents can look and function entirely differently from each other and from the parents. For this reason, geneticists who want to trace the inheritance of a certain gene down through the family tree find that after a few generations, the inheritance pattern becomes too complicated and it usually becomes nearly impossible to retrace the lineage (although there are special cases in which it can be traced back a great many generations). If one were to somehow obtain a sample of DNA from a person who lived many hundreds of years ago—for example, someone found frozen in the Arctic tundra—and then analyze samples from thousands of people still living today in an attempt to find out who might be related to that unfortunate frozen corpse, it might be impossible to positively identify a descendant.

Unless one looked specifically at the mitochondrial DNA.

During reproduction of modern day cells, the father's sperm cell injects only its DNA content into the egg cell of the mother. Whatever remains of the sperm cell is then discarded and digested. The maternal and paternal DNA are shuffled and recombined, and then the daughter cells get equal copies of both.

The daughter cells also inherit all the cellular machinery needed to thrive and reproduce themselves, but they inherit those only from the mother. More specifically, the mitochondria, including the mitochondrial DNA contained within them, comes *only* from the mother: there is no paternal mitochondrial DNA to recombine and shuffle with. In other words, excluding the occasional mutation which can sometimes occur at any point during the lifetime of the cells, the daughter cells have an exact copy of the mother's mitochondrial DNA, which in turn is an exact copy of the grandmother's mitochondrial DNA, and that of the great-grandmother's, all the way down the maternal lineage for millions of years. Except when genetic mutations (mistakes) arise.

Mutations do sometimes occur, but are rare. The DNA copy/repair machine is amazingly powerful and efficient. But not perfect. And geneticists have determined the background rate at which those mutations occur, and can use that as a molecular clock. In other words, if the background rate were one mutation in four thousand reproductive cycles, and the organisms on average reproduce once every fifteen years, and geneticists discover five mutations in a sample of mitochondrial DNA, then they can estimate the age of the sample: (5 mutations x 4,000 cycles/mutation x 15 years per cycle), or 300,000 years old.

Not only can the geneticists use that information to determine the age of one sample, but they can also determine the relationship between many samples. Imagine they analyze the mitochondrial DNA from several hundred women still living today in different parts of the world, and find eight differences (mutations) among them, which they label using letters from the alphabet from A to H. More specifically, they notice that these women have different combinations of those eight mutations (see figure 9). Some have A, B and G, but others have A, B, D and H, while yet others have A, C and F. This may seem perplexing at first, but after carefully thinking through the combinations, one can actually arrange them into a family tree (figure 9). This example is greatly simplified: in

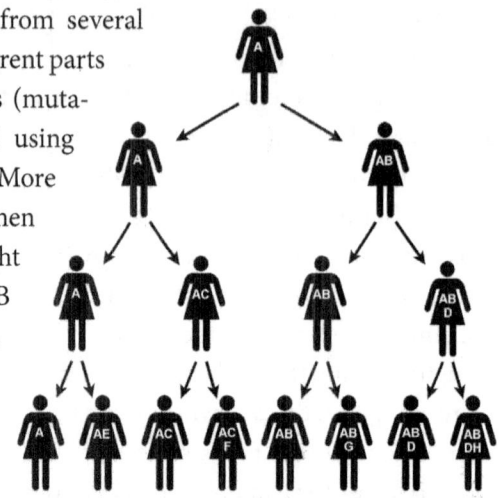

Figure 9. Maternal inheritance of mitochondrial DNA patterns.

real life, the numbers of mutations are much larger, and the number of combinations incredibly complicated, such that one needs computers to analyze the data. But suffice to say here, for the purpose of this book, mitochondrial DNA can provide powerful insights in the study of fossil samples.

But it isn't just maternal lineages which can be inferred from the mitochondrial DNA of samples. We have a similar phenomenon going on when we consider the Y-chromosome of males. Almost all genes expressed in the nucleus can be obtained from either the mother or the father (in some cases, certain genes from both parents can be expressed). However, the genes on the Y-chromosome are obtained *only* from the father. There is no shuffling or recombination possible for these genes, because the mother does not have a Y-chromosome to contribute. So all those cells which later derive from that fertilization event have an exact copy of their father's Y-chromosome, which in turn is an exact copy of the grandfather's Y-chromosome, and so on. So in the exact same way, it can be possible to trace the paternal relationships between samples backwards in time to a common ancestral father.

This strategy is particularly useful in determining the relationship between two samples of DNA, or in tracing the relative lineage in a large number of samples. In chapter 6, we will see exactly how these strategies have been applied to thousands of human DNA samples obtained from around the world, and been used to trace the original Adam and Eve. The answer crumbled the foundations from under the theological concept of humans arising from one man and one woman living approximately six thousand years ago.

Single Nucleotide Polymorphisms (SNPs)

Genes are fairly robust entities. Cells have developed a whole host of repair mechanisms to take care of various forms of damage to the precious genetic family heirloom. However, sometimes a mistake gets through. Some of those mutations lead to major deleterious changes which eventually result in the removal of that particular aberration from the gene pool. However, some mutations are much less harmful or even neutral, some possibly even beneficial. In those cases, the slightly altered gene sequence—different at that one particular position where the mutation occurred—is passed on to the daughter cells and eventually into the gene pool. It is now a different form of that gene, and for that reason the gene can now be described as being polymorphic: literally, "multiple forms." More precisely, we refer to the two (or more) different forms of this gene as being single nucleotide polymorphisms, or SNPs. These slightly different gene sequences can now be easily detected using modern genetic tools. Several generations later, it may be possible to take DNA samples from a large number of individuals and determine which ones descended from the individual that first acquired that mutation, because they possess the form of the gene which is different in that one position.

The main reason that I call attention to SNPs is that they are incredibly useful in science and medicine. Geneticists will use SNPs to ascertain whether individuals belong to one lineage or another. The ability to do so can be highly useful to healthcare professionals in following the inheritance of a particular disease, such as cystic fibrosis, hemophilia, sickle-cell disease, or Down's syndrome. Evolutionary biologists can use them in dissecting evolutionary changes within a population.

Pseudogenes

Within our genomes, there are many occurrences of gene sequences which have all the hallmarks of a fully functional protein-coding gene. They may have the usual signals (called codons) that trigger the starting and stopping of decoding. They may have a number of signals, called promoter regions, which enhance gene decoding (expression) under certain conditions. More importantly, though, they also have long sequences which could easily encode the sequence of amino acids in a protein: in many cases, these have been matched to gene sequences found elsewhere in the genome of that species or even in the genome of another unrelated one. And yet this gene does not code for a protein because of some particular problem within it that interferes with normal expression. There are a myriad of possible ways in which it can lose functionality, so I won't list them here. This gene—which looks like a gene but doesn't work at all like one—we call a pseudogene.

Humans have thousands of pseudogenes. In fact, one survey suggested that the number of pseudogenes in our genome exceeds the number of functional genes.[10] I'll give a few examples of these in chapter 6.

The fact that pseudogenes share so much coding in common with functional genes suggests that it once was a functional gene (when it existed in some ancestor). The similarities and differences within the pseudogenes, compared to their fully functional counterparts, contain clues of the gene's history. Geneticists can tease out some of that history of the gene's evolution. They can also use the pseudogene as a marker in comparing DNA samples.

Pseudogenes are not under any kind of negative selection pressure, since they're not involved in the ongoing life of the organism (by definition, they don't code for functional proteins). This has important implications within evolution theory: they can continue to accumulate mutations without consequence until eventually a new protein with a new function results. Geneticists "speculate that a considerable fraction of the human genome has evolved from pseudogenes."[11]

Redundant or Unnecessary Genes

In some cases, a species finds itself in a situation in which it doesn't need a certain copy of a gene anymore. This may be because extra copies of that particular gene have been added to the genome, and therefore all but one are redundant. Or it may be that something changes in the environment or in the behavior of the species such that it just doesn't use or require the protein(s) for which it codes.

I've already described one example which covers both of these situations: one bacterium invading another one and taking up residence inside its host. The invader possesses genes for a life of swimming in a pond, evading predators, capturing prey, detecting danger, exuding enzymes to digest plants and absorb complex chemicals. The host also has genes for these very same functions. As such, the invader no longer needs its own copies of those genes for two reasons. First, it is no longer swimming in the pond but is instead subject to the whims of its host, going wherever the latter decides to go. Second, the host can do all the work of exuding enzymes and digesting complex chemicals: all the invader has to do is absorb the energy molecules and building blocks from the host itself. For this reason, there is no problem if the invader loses any or all of those redundant genes, or allows them to mutate into other genes.

There are many other examples of an organism losing the need for particular genes. The olm, or proteus fish, which has lived for millennia in complete darkness in subterranean caves, no longer needs to be able to see (indeed, there is nothing for it

---

10. Torrents et al., "Genome-wide Survey," 2559–67.
11. Ibid., 2565.

to see): these retain only the degenerated remains of functionless eyes. Humans have the genes to create a tail at the bottom of our spines. This becomes expressed during the development of the fetus, but the tail which is formed is then normally lost before birth; in exceptionally rare cases, a baby has been born with a tail as much as twelve centimeters long,[12] which can be later removed by surgery.[13] We humans also have another gene which in egg-laying animals produces the major protein in egg yolk (vitellin), but those genes have been turned off.

In situations such as these—when a given gene is either no longer needed for its original function, or it is a duplicate gene—there is no longer a critical need to maintain it in pristine working order. Mutations and other types of DNA damage which occur all the time no longer need to be carefully repaired. In time, those genes can disappear from the genome without jeopardizing the organism's existence.

Sometimes, though, something much more interesting can happen. The organism can redeploy the proteins encoded by those genes toward other newer jobs, or mutations can be allowed to accumulate in that gene until it encodes a new protein with a different set of useful properties.

For example, this seems to have happened when the ability to see in color developed. Rhodopsin is a protein which changes shape when a photon of light strikes it. Cells have therefore built up a whole suite of proteins around this molecule such that the cell or organism can now see. It's beyond the scope of this book to explain how the process of sight works, but the relevant part of this story is this: when we look within our genome for the gene which encodes rhodopsin, we actually find many of them. But they're not all identical: each encodes a slightly different rhodopsin which responds to a different range of light wavelengths (otherwise known as colors). When we look in the genomes of other animals such as dogs, we find some of these same genes, but not all of them. And as we pursue this question in yet simpler organisms, we find they have only the one rhodopsin gene. One explanation for this is that during the course of evolution, after an organism acquired the gene to express rhodopsin and therefore the ability to see, there was a gene duplication event. It didn't need both rhodopsin genes, so it didn't matter if a mutation occurred in one or the other gene. However, certain mutations changed the properties of the rhodopsin protein such that it could now detect a different set of wavelengths: perhaps see more into the blue part of the spectrum. By being able to now detect the relative brightness of light in one part of the spectrum differently from the brightness in a different part of the spectrum, the organism could now discern more details about what it was looking at. That conveys an incredible advantage on the organism. And if it were to happen again—another gene duplication followed by a mutation to produce a rhodopsin that worked better in the red part of the spectrum—that organism would now see in three colors. On a blog which I maintain, I have included a photograph which powerfully depicts how a very

12. Dao, "Human Tails," 449–53.
13. Dubrow, "Detailing Human Tail," 340–44.

small genetic mutation can convey an amazing advantage to an organism and thereby catapult the organisms which inherit the change into a whole new level of competitive superiority.[14] It shows tree leaves and berries as viewed by two organisms that can see in the green spectrum: both can distinguish the berries from the surrounding leaves due to the very different shapes of those two items. However, the one organism also acquired the ability to see red colors (through a duplication of a "green gene," followed by a subtle mutation). It's undeniably much easier when the berries are colored differently from the leaves. But even more, now the organism blessed with that mutation can discern how ripe the fruit is, as the latter transitions from green to red: it can selectively eat the ripe ones and avoid wasting time on the unripe ones. This amazing new ability—which confers profoundly powerful, advantageous abilities to the host organism—requires only a very minor, otherwise insignificant mutational change in an extra copy of the rhodopsin gene such that the wavelength of light that excites the molecule is subtly shifted to a different color. The same thing can happen when a subtle mutational change in the protein used to detect smells (odorant receptors) or in the enzyme used to metabolize a certain kind of chemical food source opens up the world to a whole new range of things to work with.

Proviral Sequences

Viruses are a unique type of organism, different from all other species. Unlike all the latter, they have no cell bodies or organelles. In essence, they are nothing more than genetic material wrapped up in a protective coat. Even sperm cells are more sophisticated than these. In one sense, viruses are nothing more than a complex chemical: a particle.

And yet they're incredibly dangerous and destructive. One reason for this is that they turn our own cells against us. Their strategy is simple but effective: insert the genetic material into our own cells and then subvert our cells into becoming factories to produce more virus particles.

One group of viruses do this by injecting their RNA genome into our cells and immediately producing an enzyme (called reverse transcriptase) which turns that into DNA. This is the opposite of what normally happens. The usual sequence of events is that double-stranded DNA is used to create a single-stranded RNA molecule (which in turn is used to direct the synthesis of a protein); here the single-stranded RNA molecule is used to create a double-stranded stretch of DNA. For this reason, we refer to these as retroviruses.

A retrovirus, then, cuts a little hole somewhere in our own genome and splices itself in between the two cut ends, at which point we call it a "provirus." From then on, our own cells treat it as if it is part of our own body: we transcribe its genes into

14. Janssen, "Little Things."

proteins (a variety of coat proteins, as well as some enzymes to coordinate the building of other viral particles), and duplicate its genes every time the cell divides into two daughter cells.

Some of those inserted bits of information—called proviral gene sequences—remain in our DNA and can even be passed down to our offspring if the proviral insertion occurs in an egg- or sperm-producing cell. A substantial portion of our genome represents proviral sequences such as these.[15] Given that these sequences originate from completely different organisms than ourselves, geneticists can design tools which easily find them in samples of our tissues and make them stand out as unique.

Non-coding Sequences and "Junk DNA"

Generally speaking, genes code for specific protein(s). It once had been thought that every gene codes for one and only one protein, but we now know that isn't the case. We've since learned that some genes can be read (transcription) and/or interpreted (translation) in many different ways, depending on a number of different conditions, and in so doing produce different proteins. Others don't code for any protein at all. Pääbo Svante, a world-leading researcher in human genetics and the one who helped reveal to us the Neanderthal and Denisovan genomes, has said that we understand the function of only 10 percent of the nuclear genome.[16]

These non-coding regions were at first a complete mystery to us. Since the only function for DNA which had been recognized till that time is to encode proteins, we concluded that these must be useless pieces of genetic material: so-called "junk DNA." Theists took great offence to this term, because of its theological implications: "God doesn't make junk," they would say. But scientists have long abandoned the term because of what we've learned about the many functions of these previously mysterious sections of DNA.

For example, genes have associated with them a number of nucleotide sequences which don't directly determine the amino acid sequence, but instead regulate the transcription of the gene. These include promoters which turn on or enhance transcription, and others which do the opposite (down-regulate transcription).

Other non-coding sequences are found within the genes themselves. That is, many genes are split up into separate segments which are expressed (referred to as exons) interspersed by intervening non-coding segments (called introns). The introns serve in part to control whether or not certain exons are included in the final translated protein. The exact how's and why's of this differential expression aren't important for the purpose of this book.

---

15. Lander et al., "Initial Sequencing," 860–921.
16. Pääbo, *Neanderthal Man*.

Yet other non-coding regions are found at the ends of the chromosomes, referred to as telomeres. These help ensure the health of the chromosome: certain cancers and the process of aging are associated with greatly shortened or damaged telomeres.

Finally, there is a central region in the chromosome which serves as handles to move the chromosome around during cell division: these are referred to as centromeres.

A very large number of other sections of DNA also don't code for proteins, but produce very short lengths of genetic signals which we call microRNAs. We're only beginning to learn that these microRNAs, as well as some of the pseudogenes referred to above, play an important role in regulating how/when the other protein-encoding segments are expressed.

These discoveries that "junk DNA" actually isn't junk led to a massive, multinational collaboration called the ENCODE Project Consortium, ENCODE being the acronym for "*En*cyclopedia *of DNA E*lements." The ENCODE project stretched out over several years and involved over 400 international scientists studying noncoding DNA in the human genome. They produced many groundbreaking papers, and learned that the vast majority of the human genome shows biochemical function: "These data enabled us to assign biochemical functions for 80 percent of the genome, in particular outside of the well-studied protein-coding regions."[17]

Modern Gene-Cloning and Gene-Sequencing Techniques

The past few decades have seen mind-boggling advances in the field of genetics. It was only in the middle of the last century—just over fifty years ago—that we were first given a model of the structure of the DNA molecule. There's a whole intriguing story behind this profound scientific breakthrough: scientists around the world proposing all kinds of theories and models, and then James Watson and Francis Crick stealing a glance at the unpublished work of Rosalind Franklin, without her knowing, before madly scrambling to their own laboratory to put it all together into the picture we have now.

A couple decades later, we were just beginning to understand how this amazing molecule is replicated enzymatically, and making our first attempts in test-tubes using DNA and enzymes freshly isolated from living cells. We began picking genes apart one base at a time in order to puzzle together the first gene sequences.

In 1984, the field had developed sufficiently that we committed ourselves to the audacious goal of characterizing the complete human genome, and this goal was finally achieved in 2003.

---

17. ENCODE Project Consortium, "Integrated Encyclopedia," 57–74.

Now, only a decade later, we can sequence the genome of any organism of choice in a matter of weeks using massively parallel, cutting-edge techniques, and have already done so for hundreds of species.

What's more, we've begun to apply these techniques to DNA samples from animals long dead. Not merely having been dead for a few weeks. Not even dead for a few years. But dead for hundreds of thousands of years! In 1988, researchers sequenced two fragments of mitochondrial DNA from a one hundred and forty-year-old museum specimen of an extinct relative of the zebra (a quagga).[18] In 1985, nuclear DNA from a mummified Egyptian child who died 2400 years ago was cloned.[19] In 2013, the full genome of a species of horse that roamed the Middle Pleistocene permafrost of North America more than seven hundred thousand years ago was published.[20] We have also sequenced, substantial portions of other long extinct species, like the mastodons,[21] mammoths,[22] Neanderthals, and Denisovans (to be elaborated upon in chapter 6), and there is even discussion now of the looming possibility of resurrecting those dead species by reintroducing the gene sequences into living cells.

The developments are breath-taking, and the pace of discovery and achievement continues to accelerate exponentially. And so do the ethical implications: heated discussions about resurrection of extinct species; of treating disease by gene therapy; the development of genetically modified foods; genetic selection of fetuses. And so do the implications to theology. In fact, that is the primary purpose of this book, and the subject of chapters 7 and 8.

The Investigative Power of DNA

As I stated earlier in this chapter, much of what defines us as individuals is found in the genetic code within our DNA. This isn't only true of humans, but also of every species to have ever lived on earth. Many readers will already know it is possible to take the DNA from a given organism and use it to make multiple exact copies of that organism. Almost two decades ago, newspapers around the world trumpeted the fact that this had been done for Dolly the sheep. DNA was taken from one female sheep, injected into a cell obtained from a second female sheep, and this new fertilized egg was implanted in the womb of a third female sheep: Dolly had three mothers and no father! Since then, dozens of other types of animals have been cloned. It has even been done with human DNA simply to prove it is possible, but those cells were not allowed to develop to maturity, given the uncharted waters of legal and ethical concerns surrounding this.

18. Pääbo and Wilson, "Polymerase Chain Reaction," 387–88.
19. Pääbo, "Molecular Cloning," 644–45.
20. Orlando et al., "Recalibrating *Equus*," 74–78.
21. Yang et al., "Phylogenetic Resolution," 1190–94.
22. Hagelberg et al., "Ancient Mammoth Bones," 333–34.

Careful analysis of the DNA reveals minor differences in the genetic sequence of one individual compared to another, and that some of these differences are either responsible for a given disease (for example, the mutations which cause cystic fibrosis), or at least indicate a heightened susceptibility to certain diseases (like hypertension, cancer, and Huntington's Disease, for example). The confidence in this technique, and the ease with which genomes can now be sequenced, has resulted in laboratories springing up on the landscape which will provide you with a written copy of your own genome and predictions from that of the likelihood of you coming down with various diseases, presumably to help you plan for your future health care. Medical scientists are also trying to develop ways to introduce segments of DNA in an attempt to correct the mutations and thus treat disease, a procedure referred to as gene therapy.

DNA is like a UPC bar code (see fig. 8). It can be used to identify the tiniest of remains of a victim of a tragic accident (such as those found in the rubble of the Twin Towers collapse in 2011), or identify a criminal who was careless enough to leave a hair at the scene of a crime. It can be used to confirm/determine the paternity of a child, or the genetic relationship of ancient bones with one hominid species or another (a strategy which will be highlighted in the next chapter). It can track worldwide migrations of animals, and of people (again, next chapter). Knowing the average rate of spontaneous mutations which occur in DNA, it is possible to measure the progressive accumulation of mutations in two different species and thereby determine how long ago they diverged on the Tree of Life. In this way, the DNA can act like a molecular clock, complementing measurements made using radio-isotope dating techniques (such as carbon-14).

New investigative uses for DNA are being announced at an ever increasing rate. Some of these have put large question marks around certain beliefs we once held firmly about the origin of humanity. The church will need to respond to these: we can't simply put our heads in the sand like ostriches and hope that the danger will eventually go away.

As the church comes to grips with these new developments, the words of Cardinal Bellarmine several centuries ago become particularly meaningful: "One would then have to proceed with great care in explaining the scriptures that appear contrary."[23]

Conclusion

We now have many amazing tools for discovery. We've developed a rigorous strategy—the Scientific Method—to gather data, propose hypotheses and devise tests intended to knock down the hypothesis if it is actually false. We've also developed a skeptical attitude which is prone to question everything and ready to reject an idea if it doesn't stand up to scrutiny. Together, these have allowed us to accumulate a

---

23. Brooke and Cantor, *Reconstructing Nature*, 114.

vast encyclopedia of knowledge. Admittedly, from time to time, certain pages in that encyclopedia have undergone changes, but always change toward truth.

This body of knowledge is now so vast and is still growing exponentially that scientists have had to specialize into ever more highly focused areas of research. There is just too much to know: we've moved far beyond the day when the local scholar knew everything there was to know at that time about astronomy, chemistry, physics, mathematics, medicine, and philosophy combined. Now, scientists become experts on a very narrow range of scientific investigation, and within their respective areas of specialty, they have a highly detailed knowledge about nearly everything that touches on that subject, having spent decades dwelling and ruminating on it. This makes one wonder how nonexperts can feel justified in simply dismissing the information which scientists bring to the table, especially when there is a general consensus among scientists from various different but overlapping disciplines.

Ever since the ancient Greeks described Plinians to us, we humans have toyed in our imagination with images of our primitive ancestors. The latter have become a familiar part of our psyche. As a child, I grew up watching *The Flintstones*, a cartoon series surrounding the adventures of two Stone Age families of cave people living with dinosaurs (quite the anachronism, now that I think about it). I still remember the first time I read Arthur C. Clarke's novel *2001: A Space Odyssey*, which opens up with prehistoric people first learning how to make tools, putting them on the fast-track in their evolution. The concept of prehistoric humans is now old hat: one would think that we were used to it. And we have since learned so much about our origins, particularly from two areas of scientific specialization which have brought a completely new perspective and level of certainty on that question: anthropology and genetics. The next two chapters will summarize the amazing discoveries which those two disciplines have given us in only the past one or two centuries, millennia after the writers of Scripture tried to make sense of the world.

5

Origin of Humanity
The Paleontological Evidence

ON JUNE 30, 1860, bishop of Oxford Samuel Wilberforce and the scientist T. H. Huxley faced off in a historic debate over Darwin's new book *Origin of Species* at a meeting of the British Association of Oxford. It was a long and heated debate, one with which any serious student of the dialogue between faith and science must acquaint themselves. One particular quote from this debate has often been repeated, although some question whether the quote was accurately recorded. At one point in this debate, the bishop is said to have asked Huxley, "Is it on your grandfather's or grandmother's side that you claim descent from the apes?" to which Huxley snorted, "I would rather be descended from an ape than a bishop!"

Ever since the delivery of those two *ad hominem* jabs, there have been armchair quarterbacks dissecting whether either antagonist was true to the name *Homo sapiens* (Latin, for "wise human") when making those wisecracks.

This exchange epitomizes the stance of the church and science on the question of the origin of humanity. The church has held for millennia that humans are completely distinct from the other animals: created in God's image and absolutely unique. Scientists were now claiming something very different: something that would snatch the proverbial clay from God's hands.

But on what evidence could they build such a provocative theory? Much has been written on the breadth and scope of the fossil evidence which answers this question. But two books which I highly recommend to a general audience are both written by John Reader.[1] He presents this information in a highly readable fashion which is easily within reach of the nonexpert, and includes many stunning photographs (he himself is an expert photographer). Although he is not a credentialed academic in this area, he holds an honorary research fellowship in the Department of Anthropology at University College London, is a fellow of the Royal Geographical Society, and his

---

1. Reader, *Hunt for Earliest Man*; Reader, *In Search of Human Origins*.

books contain forewords and endorsements from leading experts in the field which validate the reliability of his writing.

Setting the Stage

Imagine a child growing up in a home thinking for years that she was an only child. Her parents never spoke of there having ever been any other children in her family, and there was no evidence anywhere to indicate the possibility. She grew up for years with this idea that there had only ever been her and her two parents.

Then one day in her early teen years, while exploring through the attic, she comes across a box with a photo album, and in it pictures of her parents holding two babies, and sitting beside them on the couch is a beaming little boy wearing a fireman's helmet and holding a "Welcome" sign with her own name and a boy's name on it. Also in the box is that fireman's helmet and a few other toys for a young boy, as well as some baby clothing, all for a little baby boy.

Imagine the questions that now flood the young girl's mind!

Up until the time that Darwin proposed his theory of gradual evolution of one species into others, there was no evidence that the human species had brothers or sisters of any kind. No bones. No abandoned cities. No literature or artwork. Nothing on which to base the idea that there had been other beings like us around at some time in the past. In fact, the Western world had the biblical texts which told us explicitly for millennia that we were only-children, created by God.

But then we started exploring in the attic.

Neanderthal Man

In 1856, laborers employed by a limestone quarry were clearing a cave in the deep and narrow ravine known as the Neander valley in Germany, through which the Düssel River flows. The entrance to the cave was ten meters up a steep cliff and itself only a meter high, but opened up into a large cavern. Buried in the mud of the floor of that cave the workers found a skull cap and some limb bones of what looked to them to be some poor unfortunate human. The remains were eventually brought to a local expert—Professor Shaaffhousen—who recognized that these were in fact not human. He noted that the limb bones were exceptionally thick, with very pronounced muscle attachments denoting an extremely powerful individual. Also peculiarly different from human skulls were the very prominent eyebrow ridges, which were characteristic of the facial conformation of apes. He presented "Neanderthal Man"—named after the valley in which they were found—at a meeting of the Lower Rhine Medical and Natural History Society in Bonn, three years before the publication of Darwin's *Origin of Species*. There was mixed response from his listening audience, but not much happened until he published the find in the scientific journal *Natural History Review*, and that in turn was translated to English by George Busk, then professor of anatomy at the Royal College of Surgeons, who added some commentary of his own interpretation of the new data, emphasizing the Neanderthal skull's overall resemblance to that of the gorilla and chimp. The scientific community grappled over the details of this report and plaster casts of the bones which had been found, and divided itself between two interpretations: the bones were either from an ancient individual belonging to an early stage of human evolution linking man to an apelike ancestor, or from a modern human with gross pathological deformities. Some concluded that Neanderthal Man was merely a fully human hermit who suffered from a bad case of rickets. Another interpreted these bones to be from an ailing deserter of a Cossack army that had camped in the vicinity of the cave half a century before who had hidden in the cave and died there (it was pointed out that the bow-legged appearance was a result of a lifetime spent in a horse's saddle, and the imagery of a Cossack army deserter helped the German audience explain the powerful and yet unintelligent appearance of the individual).

Unfortunately, nothing in the cave's mud in which the bones were found shed any light on how ancient (or not) this find was.

In actual fact, this was not the first time that Neanderthal bones had been found, and certainly not the last. Neanderthal skulls were found in the area of Belgium in 1829, and another skull (including the entire face, upper jaw, and most of the teeth) during construction of military fortifications at Gibraltar in 1848, although none of these were presented to the scientific community until after the one first described by Hermann Shaaffhousen, which is therefore designated the type specimen (the one against which subsequent finds must be compared). Two skeletons were found in a cave near Spy in Belgium in 1887. Another skeleton near La Chapelle-aux-Saints in France in 1908, several more from La Ferrassie in 1909, and from La Quina in 1911. All of these were strikingly similar to the original Neanderthal specimen, especially with respect to the prominent eyebrow ridges and thick, powerful skeleton.

Clearly they all represented a race of humanlike beings that had populated Europe from Belgium to Gibraltar. Yet there were still some scientists resisting the idea that Neanderthals should be placed within the family of humans, some for religious reasons, and others because they found it offensive that humans and Neanderthals would be related. To them at that time, and to many people still today, Neanderthal Man was a shambling, frowning brute of low intelligence.

Grafton Elliott Smith, distinguished professor of anatomy at the University of London and a very influential investigator in the search to understand human evolution, was one who felt that humans did not evolve from Neanderthals. In his book *The Evolution of Man*, he gave this description:

> A clear-cut picture of the uncouth and repellent Neanderthal Man. His short, thick-set and coarsely built body was carried in a half-stooping slouch upon short, powerful and half-flexed legs of peculiarly ungraceful form. His thick

neck sloped forward from the broad shoulders to support the massive flattened head, which protruded forward, so as to form an unbroken curve of neck and back, in place of the alternation of curves which is one of the graces of the truly erect *Homo sapiens*. The heavy overhanging eyebrow-ridges and retreating forehead, the great coarse face with its large eye-sockets, broad nose, and receding chin, combined to complete the picture of unattractiveness, which it is more probable than not was still further emphasized by a shaggy covering of hair over most of the body. The arms were relatively short, and the exceptionally large hands lacked the delicacy and the nicely balanced co-operation of thumb and fingers which is regarded as the most distinctive of human characteristics. The contemplation of all these features emphasizes the reality of the fact that Neanderthal Man belongs to some other species than *Homo sapiens*.[2]

Since then, we have collected Neanderthal bones from hundreds of individuals (not all are complete skeletons), and from those pieced together an understanding of where they came from, how far they spread out and how they lived.[3] It seems they diverged from our ancestral line approximately four hundred thousand years ago, made their first appearance as a distinct species two or three hundred thousand years ago, reached a population peak of seventy thousand, and went extinct roughly forty thousand years ago, possibly at our hands. It's unclear whether we viewed them as competitors and killed them off intentionally, or whether we just introduced diseases to them for which they didn't have any immunity, much as the European explorers did to the indigenous people of South America. Some say that our domestication of dogs made us a formidable hunting team against which the Neanderthals couldn't compete.[4] Their extinction also coincides with the start of a very cold period in Europe, which might mean that they simply couldn't cope as well with climate change as we did.

From the beginning their bones suggested that they stood upright and were fully bipedal, and a later study of their feet confirmed this. The average male Neanderthal stood just a little shorter (164–168 centimeters high) than the average human male today (177 centimeters), and the females approximately ten centimeters shorter yet. Despite their shorter height, they had much more powerful bodies—as indicated by their larger, thicker bones and attachment sites for muscles—and a much wider, barrel-shaped rib cage. Their brains were also larger than ours, with an average capacity of 1600 cm^3 (compared to the modern human average of 1400 cm^3).

They were almost certainly quite hairy, given that they thrived even in the northern parts of Europe and Asia, and a genetic study of Neanderthal DNA suggested that their hair color could include red and blonde and that they had light skin tone. It's been said that they looked much like we do now, albeit apparently with a smaller chin

2. Reader, *Hunt for Earliest Man*, 34; also see Reader, *In Search of Human Origins*.
3. Stringer and Andrews, "Genetic and Fossil Evidence," 1263–68.
4. Shipman, *Invaders*.

and bigger nose. In fact, as John Reader put it: "Given a bath, a collar and tie, [Neanderthal Man] would have passed unnoticed in a New York subway."[5] As a result, we ourselves have out-grown our need to distance ourselves from them by referring to their characteristics in such derogatory terms.

There is evidence that we mated with them at some point: as will be explained in the next chapter of this book, genetic studies tell us that as much as 4 percent of the genome from modern-day non-African humans derives from Neanderthals (we picked that up when *Homo sapiens* left Africa and encountered them while migrating into Europe and Asia, but Neanderthals never migrated into Africa where they would have encountered members of our lineage who had stayed there).

We've also found abundant evidence that they made stone tools, harnessed the power of fire, created their own superglue to attach stone blades to spears, painted images on cave walls, made jewelry, buried their dead (which suggests a belief in an afterlife), and other lines of evidence of a highly developed society.

But Neanderthals are not the only well-documented representative of fossil man.

Homo erectus

Eugene Dubois, entered medical school in 1877 "a devout Dutch Catholic" and emerged "an ambitious, determined and intractable believer in evolution."[6] After a few years toiling within the academic setting and getting into some hot water with the professor of the department for publishing a paper without the name of the latter appended (a duty generally paid out of political respect), he "decided that he would much rather look for fossils of early man than become a professor of anatomy. In 1887 he resigned his lectureship, leaving behind only the rumor that he had promised to return with the 'Missing Link.'"[7]

He signed on with the Medical Corps of the Dutch East Indian Army and arranged to be assigned to Java to oversee convict labor and to search for fossils of human ancestors, and soon found a fragment of a primate jaw bone with one canine tooth still in place. Although quite human in appearance, he nonetheless concluded it to be "of another and probably lower type" because in his opinion the attachment site for the digastric muscle was not conducive to normal functioning of the tongue for articulate speech. The next year, he found a tooth and a skullcap, and tentatively concluded these to belong to *Anthropopithecus* (troglodytes), the scientific term for "man-like ape" (and the term used at that time to refer to chimpanzees). Yet another year later, he found a fossil thighbone which was clearly not apelike and very much like that of humans, showing clear signs that the individual stood and walked upright. The thighbone was found some distance from the skullcap: unfortunately, Dubois's

---

5. Reader, *Hunt for Earliest Man*, 36.
6. Ibid., 41.
7. Ibid.

record-keeping was quite sloppy and his assertions of the distance separating them changed frequently. Dubois nonetheless was convinced that both came from the same individual. By the end of that same year, they had found a few more thighbone fragments.

As he wrote up a paper summarizing his finds, Dubois struggled with what to conclude. At this point in the story, John Reader makes a very interesting statement, one which is very relevant to the goal of my own book: "In truth, the evidence was too scanty, and the current state of knowledge too slight, for any specific attribution at all. But an absence of evidence encourages speculation; and Dubois was free to reach the conclusion which became a point of faith (and argument) dominating the rest of his life. He decided the bones had belonged to an apelike man. In other words, he reversed the earlier attribution and *Anthropopithecus* became *Pithecanthropus*."[8]

Once again, as was the case for the announcements of Neanderthal Man, the response from the scientific community was mixed (and it goes without saying that Dubois also attracted a great deal of acrimony from the religious community). Everyone did indeed celebrate the discoveries themselves, since the finds greatly increased the data available on the question of mankind's origin. However, many questioned his interpretation of the data: more specifically the relationship between this new species and mankind. Dubois maintained that it was an ancestor of *Homo sapiens*: that we evolved from it. However, together with other similar fossils found in Peking, the scientific community was increasingly coming to the present-day conclusion that these all represented a separate branch on the hominid tree—one of our cousins—rather than a distinct intermediate ancestor between man and ape. Eventually, the name of this species was changed to *Homo erectus* (sometimes referred to as *Homo ergaster*).

Dubois fought bitterly against this to the day of his death. John Reader quotes from an obituary written by Sir Arthur Keith to celebrate Dubois's life, a quote which also struck me as being acutely relevant to the goal of my own book: "He was an idealist, his ideas being so firmly held that his mind tended to bend facts rather than alter his ideas to fit them."[9]

Australopithecus

In 1924, Raymond Dart, professor of anatomy at the Witwatersrand University in Johannesburg, South Africa, obtained a complete skull with intact jawbone from a limestone quarry near a place called Taung in South Africa. This skull lacked a great eyebrow ridge and its jaw did not jut forward, as is the case for apes. The first molars were emerging, but it had a full set of milk teeth indicating this was a juvenile, not an adult. The cranial capacity and the in-folding of the brain was intermediate between apes and humans (although it was far too small to have been a recent ancestor of

---

8. Ibid., 46–47. In other words, "man-like ape" became "ape-like man."
9. Ibid., 54.

humans). The central positioning of the hole in the base of the skull suggested to him that this creature walked upright. Ultimately, he named the find *Australopithecus africanus*—"southern ape of Africa."

Dart presented his find to the popular press before proceeding through proper peer review by his scientific colleagues, and did so in a very sensationalistic manner. "The creature could appreciate color, weight and form, he claimed: it knew the significance of sounds and had already passed important milestones along the road towards the acquisition of articulate speech."[10] Dramatic headlines appeared in newspapers all around the world. For years, Dart promoted his find through the lay media and did not follow up with requests from the scientific community for more substantive evidence. In fact, he apparently never even attempted to find other fossilized remains to further support his claims. John Reader's conclusion to this story is that Dart merely rode out this one-trick pony into ignoble obscurity, and didn't make any further contribution to our understanding of the evolution of mankind.

However, other investigators were looking, and another fossil was discovered by quarry laborers clearing a limestone cave in South Africa (Sterkfontein), this one comprising pieces of skull and badly crushed face as well as four upper teeth. This discovery was valuable in its own right, but was not enough to persuade everyone that Raymond Dart had been right in making his claims. There may have been disagreement as to whether this animal was more related to the gorilla or to the chimpanzee, but there was general agreement that *Australopithecus* was not an ancestor of mankind.

Since then, a great deal of diverse and yet supporting data have been collected, and there is now agreement that *Australopithecus* is a species close to the apes that was habitually erect and bipedal with man-like dentition and relatively small brains. In other words: "good candidates for the ancestry of mankind."[11]

But the contribution to this area of research that is most widely recognized by today's nonexpert audience has been made by the Leakey family, who shifted the focal point of the story from South Africa to a different part of that continent.

Leakey's Lucy

Louis Leakey carried out most of his work in the Olduvai Gorge in Tanzania, and surrounding regions, beginning in the mid-1920s, with the help of his wife and son and the collaboration of hundreds of other researchers, grad students and employees. Over the course of a few decades they've found an abundance of fossilized remains of many different animals and hominids, as well as mountains of stone tools, mostly hand-axes.

Much of their hominid finds were quite human in appearance—others felt they were very much Australopithicene in nature—and indeed quite old, seeming to

10. Ibid., 87.
11. Ibid., 134.

predate all other existing finds at the time, as indicated by geological determinations, including the presence of certain signature animal fossils. However, Leakey did not want to conclude that they were a species separate from *Homo sapiens*, simply because they were found among so many stone tools. He assumed that only mankind were toolmakers, and for this reason struggled against a conclusion of an ancestor of mankind that made tools: in other words, he allowed his preconceptions to color his interpretations of the data. They even tried to explain away this discrepancy by claiming any bones found there which were clearly Australopithicene in nature were victims of the tool-making species, and the latter in turn didn't leave any fossil evidence of their own existence. For this reason alone, he created a new genus to accommodate the phenomenon—*Zinjanthropus boisei*, or "Boise's East African man" (Charles Boise being a major financial backer of Leakey's early work). The rest of the scientific community was not convinced that this warranted a separate species: to them, the bones were clearly *Australopithecus*.

Through meticulous record-keeping of the precise locations where items were found, they were even able to show areas on the ground having an exceptionally high density of tools as well as the kinds of animal bones associated with the largest amount of marrow and edible muscle (for example, limb bones). These specimen-rich areas were surrounded by narrow bands of areas devoid of either, which in turn were surrounded by wide areas of ground featuring smaller bones with less edible parts (for example, knuckles, ankles, hands) and tool fragments scattered helter skelter. Their interpretation of the data was that this peculiar distribution reflected permanent settlements enclosed by walls made of brush and tree branches, much as certain African tribes still do in that part of the world today. The richest parts of the animals they'd killed would be taken inside the enclosure and consumed (accounting for the high density of those kinds of bones in those areas of the ground), while the less useful parts of the animals were left outside of the enclosure and ripped apart by scavengers. The narrow bands of ground which contained few bones or tools at all indicated the location of the brush/wood enclosure which had long ago decayed or burned away. As such, this shows that this species, two million years ago, had developed some kind of culture, a social structure, and a concept of home and communal living.

As time went on and digging continued at the Olduvai site which they referred to as FLK, bones from a different hominid were discovered:

> Among the hand bones, anatomist John Napier found evidence of at least two hands (one juvenile and one adult), an opposable thumb and the physical capacity to manufacture the Oldowan tools found on the living floor. From the foot bones another anatomist, Michael Day, reconstructed an almost complete adult left foot; it was entirely human, with no sign of the ape's divergent big toe and every indication that its owner had stood erect and walked with a bipedal and free-striding gait—a view which a third anatomist, Peter Davis, confirmed in an independent study of the tibia and fibula from FLK. In the

mandible, Leakey found the front teeth relatively large and the cheek teeth relatively small—quite different from the australopithicines, he said, and quite appropriate for a new and distinct type of early hominid. From the parietals and other fragments, yet another anatomist, Phillip Tobias, reconstructed a skull and estimated its cranial capacity as 680 cubic centimeters; nicely beyond the australopithicine average and approaching the *Homo* range. So the scanty remains from FLK NN appeared to represent a hominid with a relatively large brain, thin humanlike skull bones, *Homo*-like dentition, manipulative hands and the ability to make stone tools—evidence which, together with the age of the FLK NN living floor, most persuasively suggested that the new fossils must represent man's earliest ancestor.[12]

Leaky felt justified in naming this a separate species: *Homo habilis*, "handy man" (because of all the stone tools), our true ancestor. Altogether, then, he claimed they had evidence at FLK for three separate species: *Zinjanthropus boisei*, *Homo habilis* and *Homo erectus*. They were convinced that *Australopithecus* was not an ancestor of humans, but a side-branch.

Others weren't so sure. Could the three species described by the Leakeys instead represent a continuum of evolution of one species? "Was *habilis* the most advanced *Australopithecus* or the lowliest *Homo*?" they asked. There was no way to answer this question using anthropological techniques and anatomical data alone.

The interesting thing is that both these competing views are based on the same evidence, interpreted differently. As John Reader put it: "The arguments are united by familiar undertones—each reflects preconception as much as interpretation."[13] "For the time being, the ambiguous nature of fossil evidence obliges paleoanthropologists to pursue the truth mainly by hypothesis and speculation. And in a science powered by individual ambitions and so susceptible to preconceived beliefs, interpretations are bound to differ whenever the evidence is sufficiently ambiguous."[14]

As time went on and they extended their exploration to the ravines and valleys of the Hadar River in the Afar region of north eastern Ethiopia, they found many fossils including one that most nonexpert readers may recognize: "Lucy." This one looked like the three million–year-old *Australopithecus* specimens described earlier, and yet they looked a bit different. Eventually they decided this was yet another hominid subspecies: *Australopithecus afarensis*, with an age of 3.7 million years.

Standing back and looking at all the data we have on *Australopithecus*, it seems they evolved in eastern Africa around four million years ago and went extinct approximately two million years ago. During that time, they spread throughout the continent, and different subgroups of them developed into a large number of subspecies which appeared and disappeared in different places at different times: *A. afarensis*,

12. Ibid., 187.
13. Ibid., 194.
14. Ibid., 226.

A. africanus, A. robustus, A. boisei, A. anamensis, A. bahrelghazali, A. deyiremeda, A. garhi, and *A. sediba*. In general, they had a much smaller brain than our own (much less than half our size), which leads many to think that they would have appeared no more intelligent than modern apes. However, they used stone tools[15] (then again, modern apes have been observed using stones to crack open nuts and sticks to harpoon termites, while chimps use sticks and branches for hunting or as a form of defense). They were clearly bipedal, and relatively smaller in height than ourselves: 120–140 centimeters tall.

Ardipithecus

In the early 1990s, a research team from the famed Leakey dynasty—this one headed by Timothy D. White—found various parts of approximately half of a skeleton including the skull, mandible, teeth, as well as the long bones of the arm and leg in a river valley in Ethiopia.[16] Radiometric dating of the volcanic ash in which these bones were embedded placed them between 4.35 and 4.45 million years old (although some put this at approximately 3.9 million years).[17] The teeth suggested it was a frugivore (ate fruit) and omnivore (ate meat and plant material). The structure of their limbs was consistent with this kind of varied diet: the big toe of the foot was still adapted for grasping tree branches, but less so than modern chimpanzees, and showed changes consistent with early bipedalism. It had a quite small brain: 300–350 cubic centimeters, slightly smaller than that of a modern bonobo or chimp, and much smaller than *Australopithecus*, let alone modern humans.

These were at first described as a subspecies of *Australopithecus*, but then quickly renamed as a separate species: *Ardipithecus ramidus*, "Ardi" and "ramid" deriving from the Afar language of the region in which it was found, and representing the words "floor/baseline/ground" and "root," respectively (as already mentioned above, *pithecus* comes from the Greek for "ape").

Since that initial discovery, many other *Ardipithecus* fossils have been found by other paleontologists exploring the Afar region of Ethiopia. A few years ago, teeth and other bits and pieces of the skeleton were found which closely resemble *Ardipithecus*, but were dated to roughly 5.6 million years ago. It's still disputed (in large part because the supporting evidence is still scanty) whether this represents a subspecies of *Ardipithecus*, an ancestor of *Ardipithecus*, or even a separate species. At this point, it's been named *Ardipithecus kadabba* (*kadabba* is an Afar word which refers to the father of a family; hence the full name could be translated as "progenitor of the ground ape").[18]

---

15. McPherron et al., "Evidence for Stone-Tool-Assisted Consumption," 857–60.
16. White et al., "*Australopithecus ramidus*," 306–12.
17. Kappelman and Fleagle, "Age of Early Hominids," 558–59.
18. Gibbons, "New Kind of Ancestor," 36–40; Haile-Selassie et al., "Late Miocene Teeth," 1503–5.

Homo floresiensis ("The Hobbit")

In 2001, Mike Morwood (a New Zealand-born Australian palaeoanthropologist) went to Indonesia and paired up with Raden Soejono (from the Indonesian National Centre for Archaeology), with the goal of finding the first modern humans who had travelled from mainland Asia to Australia. At that time, it was thought that *Homo sapiens* was the only hominid that had travelled that far east out of Africa: that we left Africa, encountered Neanderthals in central Asia and Europe, the latter eventually died out (either naturally, or at our hands), and we continued expanding across that part of the globe. *Homo erectus* was already known to have made it to China (Java man), but the expansion into Australia was still unclear.

For two years, all they found were bones of pygmy forms of stegodons (now-extinct primitive elephants), Komodo dragons, rats and giant stork. Stone tools were also found, which encouraged them on.

Then, in 2003, they unearthed a very small hominid skeleton, which they first thought was that of a human child. But as the layers of sediment were carefully removed, they noticed the molars in the jaw had fully erupted, and the skull had distinct eye brow ridges, both signs indicating that this was a mature adult. However, the height of this individual would have been only one meter. Carbon-dating of materials co-buried with it suggested an age of approximately eighteen thousand years. The hips indicated it to be female.

Many of the details were hard to reconcile. The extremely small brain size of this adult (only 400 cubic centimeters) was similar to the very premodern human *Australopithecus* (Lucy), which lived approximately 2.5 to 3 million years ago (although, again, this skeleton was dated to less than 20,000 years). Also, its "broad flared hip-bones, short collarbone and forwardly positioned shoulder joint all resembled the pre-human group known as australopithecines."[19]

On the other hand, the more human features made them wonder whether this was in fact a descendant of *Homo erectus*, which had lived in Java, a little to the north and on the mainland, until about 150,000 years ago. They wondered whether this descendant, and her relatives, survived the Ice Age on the island of Flores, undergoing insular dwarfism in the process. The latter is a process whereby large mammals find themselves on small isolated islands with very limited food resources and absolutely no predators and evolve into much smaller versions of their parent line (larger size is a very metabolically costly form of defense against predation). This might also explain the presence of the dwarf stegodons.

In the end, the discovery team proposed this to be a new species, one they named *Homo floresiensis*.[20] Morwood occasionally referred to it in jest as "the hobbit": unfortunately, the latter was the name the public media quickly latched onto.

---

19. Stringer, "Small Remains," 427–29.
20. Brown et al., "New Small-Bodied Hominin," 1055–61.

One interesting and unfortunate detail in this story, one which epitomizes the passion of this field of science, involves the arrival on the scene of the head of Indonesia's national palaeoanthropology institute in late 2004. In the spirit of collaboration and open inquiry, the discovery team happily showed him the bones. The visiting guest promptly packed the bones into a suitcase and marched out the door: he had decided that they belonged in his own lab! There, his group carried out some studies on the bones, including trying to create casts, but in the process damaged it (the lower jaw was broken, and the skull damaged). Eventually, the bones were returned to the original discovery team.

In 2009, Morwood published much more detailed analysis of this skeleton as well as many other bones which had since been found, including a second jawbone and part of limb bones of up to eight other individuals.[21] The feet suggested these were avid climbers (which one of the researchers attributed to a defense against the Komodo dragons which also populated the island, and which can't climb trees).

There is still debate as to whether this represents a new species of *Homo* or not: one expert associated with the original team made comparisons with modern microcephalics using 3D-computerized tomography and concluded it to be a distinct species,[22] while another expert made comparisons against modern microcephalics using MRI and concluded it to be a fully modern *Homo* with microcephaly.[23]

At this point, the scientific community in general is divided over whether this was a dwarfed archaic human (for example, *Homo erectus*, which had already been found a little bit north in Java), or a fully modern human with some disease which severely distorted its skeletal features.

Denisovans

In 2008, Russian scientists from the Institute of Archaeology and Ethnology of Novosibirsk were investigating a cave in south western Siberia named after an eighteenth-century Russian hermit named simply "Denis." There they found a finger bone of a hominid, together with a bracelet and other artifacts in a soil layer that was dated to around forty thousand years ago. The bone is a very tiny one, barely bigger than a cubic centimeter.

Previously, such a small find would barely merit any kind of attention whatsoever. Such an insignificant bone from the hand would reveal very little, if anything about the individual that owned it using standard anthropological techniques. Nonetheless, on the basis of this meager find, they announced that they had discovered a young female from an entirely new species, which they named Denisovans,[24] and

21. Morwood and Jungers,"Conclusions," 640–48.
22. Falk et al., "Brain Shape," 2513–18.
23. Vannucci et al., "Craniometric Ratios," 14043–48.
24. Krause et al., "Complete Mitochondrial DNA Genome," 894–97.

later added that she had brown skin and brown hair.[25] Moreover, they claimed that the Denisovans originated from an ancestral migration that left Africa around eight hundred thousand years ago, (which in turn gave rise to Neanderthals approximately six hundred thousand years ago; that migration was followed later by two other out-of-Africa migrations that gave rise to *Homo erectus* and modern humans).

How could they possibly conclude so much detail from such a small knuckle bone?

By extracting DNA from that tiny bone and applying cutting-edge genetic sequencing techniques to it. I won't say anything more about those genetic techniques here: that's the subject of chapter 6. However, together with that powerful new technique and the additional finding of two teeth, a toe bone and some other artifacts left behind by the Denisovans, we have been able to learn a great deal about this new species. It ranged throughout Asia: from Papua New Guinea to Spain, and as northward as Siberia. They used stone tools and decorated themselves with jewelry (such as the bracelet which was originally found). They interbred with Neanderthals and anatomically modern humans, and left their mark on humans (in the form of diagnostic remnants in our own DNA).

Homo naledi

In 2015, another new candidate hominid species was announced[26] after a large collection of bones was found in the Rising Star cave system near Johannesburg, South Africa. A few of the bones had first been discovered by spelunkers (people who crawl through caves for fun), who brought back pictures to a fellow caver and geologist, who in turn alerted Dr. Lee Berger, a professor at the University of the Witwatersrand in Johannesburg.[27] Extensive searches of the cave in 2013 and 2014 recovered more than 1500 fossil fragments. This find, all within one little cave, is the largest hominid fossil find in Africa (far bigger than the finds in Tanzania and Ethiopia). In fact, other than the tremendous discoveries in *"the Pit of Bones"* in Spain and the Zhoukoudian *Homo erectus* finds in China, this is the largest collection of bones found yet.

Altogether, it looks like the bones comprise at least fifteen individuals, including what looks like eight children (three infants, three juveniles, two older children) and five adults (including two young adults or older adolescents and one very old adult with extremely worn-down teeth). As such, this find gives us great insight into the life history of this species: the changes which occurred as they grew up from infants into adulthood and then into old age, sexual dimorphism, and the range of variability in the shape of the various individual bones. The cave and other passages around

---

25. Meyer et al., "High-Coverage Genome Sequence," 222–26.
26. Berger et al., "*Homo naledi*," e09560.
27. Shreeve, "This Face Changes the Human Story."

it haven't been thoroughly explored yet, so it's possible that more bones may yet be found.

The team leader for this discovery, Dr. Lee Berger (a professor at the University of the Witwatersrand in Johannesburg), announced this to be a new species in the hominid line, and named it *Homo naledi*. *Naledi* is the word for "star" in Sesotho, one of the official languages of the area in which this was found.

As has often been the case when newly discovered fossilized bones are presented and interpreted, this find stirred up considerable debate. One of the points of contention was whether this should be labelled a new species: some experts say it was too early to make such a claim.

The original description of the find goes through each of the bones (over 1500 of them!) and points out many similarities between certain features and those of other hominid species, but also very meticulously points out a great many differences. In general, this creature provides an unprecedented and provocative blend of australopith-like and humanlike features.

The adult males stood approximately 1.5 meters tall.

The size of their brain is much smaller than that of modern humans, approximately 560 cubic centimeters for the males and 465 cubic centimeters for the females: approximately the size of a gorilla's brain and a bit larger than that of a chimpanzee. The teeth are small and simple (which is interpreted as being primitive).

The hands are quite modern in shape, and appear to be very well suited to making and using tools, although no tools were found with the bones. Just the same, the fingers are slightly curved, as is the case for apes which spend much of their time in trees.

On the other hand, their legs are proportionately long, and the attachment sites for muscles as well as the shape of the feet and ankles clearly indicate that *Homo naledi* walked upright.

Altogether, the bones suggest that *Homo naledi* belonged to our distant past—but as a relative, rather than an ancestor—around the time that *Australopithecus* and *Homo* were diverging down their separate evolutionary paths approximately three to four million years ago. Unfortunately, the age of the bones is not yet known. They were not found co-buried with the bones of other signature animal species or stone tools, which would have helped to place them in time. The bones were found loosely in the mud of the cave bottom, rather than being embedded in rock, so it isn't possible to radiometrically date that surrounding material. It is possible to date the bones themselves radiometrically, but that requires them to be destroyed (pulverized and effectively incinerated), so that technique won't be used until the bones have been carefully studied and three-dimensional casts made.

Another contentious issue is how the bones came to end up at the bottom of this cave.

There's no evidence that they were carried there by flowing water or predators.

To get to this cave, one has to crawl down through a long winding passage to a widened area, then squeeze through another long passage known as "Superman's crawl" (certain sections are less than twenty-five centimeters high, so to pass through one must extend one hand and arm forward, and keep the other one tight against one's body). This crawl opens up to a very a large cavern, from which one has to climb up about fifteen meters to another very tight squeeze called "Dragon's Back" and proceed approximately five meters horizontally and then twelve meters vertically down to the chamber where the bones were found.[28] A diagram of this demanding and claustrophobically frightening path is given in National Geographic,[29] and is worth viewing to get a sense of how unusual it is that these specimens should be found there and how hard it was to get them. In fact, it is so difficult to traverse the passages that the anthropological team had to advertise on Facebook for amateur cavers to retrieve the bones rather than get them themselves.

Since the bones don't indicate traumatic injuries from the precipitous drop, it seems the individuals crawled there themselves while alive or their bodies were carried there by their relatives. Also, it's unclear how or why *Homo naledi* would have crawled so far into a deep, tight dark cave together with several infants and juveniles and at least one old person in tow (it's possible that there once had been another entrance which has since collapsed or filled in, but there's no evidence to substantiate this claim). In fact, the way that the bones are scattered at different depths in the cave's soil suggests they arrived there at different periods of time. Berger has concluded that the bodies were placed there as part of some kind of ritual burial practice, which raises the possibility that the species had a sense of the afterlife. However, many experts and other critics vigorously challenge that proposition. The team leaders also suggest that the long and tortuous nature of the cave in complete darkness suggests this species used fire torches to guide them through the cave. But again, this is only speculation.

This anthropological find has also raised big questions about the current ideas around the sequence of events which occurred during human evolution. It had previously been thought that a number of adaptive changes happened around the same time in our evolution: increased brain size, larger body size and longer legs, tool use, and dietary changes which resulted in smaller teeth. The belief that all these changes were approximately concurrent kept anthropologists engaged in endless chicken-and-egg debates: was increased brain size a result of a higher-quality diet, or did it produce the dietary change by empowering the species with tool-making skills; was the dietary change a result of becoming bipedal, or the other way around; and so on. But the finding of *Homo naledi* changed that, because it provided a species which had a very small brain and small primitive teeth, yet walked upright and had hands that would have been well-adapted for toolmaking. The fact that the legs and feet suggested bipedalism

---

28. Dirks et al., "Geological and Taphonomic Context," e09561.
29. Shreeve, "This Face Changes the Human Story."

while the shoulders and hands suggested it still spent a fair amount of time in trees added a puzzling aspect to this new species.

It's exciting to realize that the Dinaledi Cave has not yet been completely explored: who knows what remains yet to be found (play-on-word not intended).

Ancient Homo

We are a very gregarious species (preferring to live in societies), and seem to be driven to keep moving to the horizon. Because of those two factors, together with the relatively recent period of time since our ancestors lived (measured in tens and hundreds of thousands of years, rather than millions of years), anthropologists have been finding large collections of bones from our direct ancestors all across the globe.

For example, one of the biggest of these collections was found in 1997 at a series of limestone caves in Atapuerca in northern Spain when a railway line was being cut through a hill. In fact, one of these caves has since been named the *Sima de los Huesos*, Spanish for "pit of bones." These contain the bones, including many fully articulated skeletons, of hundreds of hominins that have been dated in various ways to several hundred thousand years ago. They also contain countless numbers of bones of the animals they killed and butchered (and the predators which stalked them), as well as the stone tools they used and many cultural artefacts. Many other such sites have been found all across Africa, Europe and Asia, albeit not as large.

Careful analysis of the bones and cultural artefacts tells an interesting story of several lines of *Homo* cohabiting, interbreeding, engaging in economic trade, developing various technologies (metallurgy) and arts (musical instruments; sculptures; jewelry; wall paintings and engravings). Their apparent ability for such abstract thinking, as well as the anatomy of their jaws, including the attachment sites for muscles for their tongues, suggest they were capable of speech.

Even more interesting, from the point of view of this book, we see evidence that these ancient people, hundreds of thousands of years ago, had a sense of the afterlife. Their bodies are carefully buried, sometimes with jewelry or tools. There is even some controversial data to suggest flowers being planted at a Neanderthal grave site.[30] One very recent find in Siberia describes *Homo sapiens* being buried with Venus-like figurines.[31] Very few people today who are familiar with the data, including Christians of all theological stripes, will deny that these people not only kept chasing the distant physical horizon all around them, but also had their eyes on the boundary between this life and the afterlife (although some Christians will dispute the dating of those discoveries).

Altogether, the general picture we get from the massive amount of data begins with *Homo ergaster* and/or *Homo antecessor* (there is dispute, as always seems to be

30. Leroi-Gourhan, "Shanidar," 79–88.
31. Raghavan et al., "Upper Palaeolithic Siberian Genome," 87–91.

in this field, as to whether these are distinct species or not). These gave rise to *Homo heidelbergensis*, also known as *Homo rhodesiensis*, which in turn gave rise to Neanderthals, Denisovans and ourselves. It needs to be emphasized to the reader that this picture is not based on merely a few partial skeletons found in a few places, but the bones of literally many hundreds of individuals, together with their tools and other artefacts, scattered all across the globe. A proverbial mountain of evidence.

Laetoli Footprints

In the classic tale *Robinson Crusoe*, by Daniel Dafoe, the main character finds himself shipwrecked for twenty-eight years on an uninhabited equatorial island. Despite the desolate isolation, he assimilates himself quite well and learns how to live comfortably and free of concern. Until the day he finds a human footprint on the beach. This encounter, one of the most famous episodes in English literature, marks a dramatic change in Crusoe's outlook and demeanor: someone else has been here!

That account is fictional, but a version of it has played out in real life in the world of paleontology.

In 1976, Mary Leakey was exploring the Laetoli site in the Olduvai Gorge in Tanzania, and found a series of footprints in the rock surface. These are remarkably similar to the footprints of modern humans, complete with impressions for the toes and the ball of the foot, and very much not apelike. What's surprising about these footprints is that the rock in which they've become fossilized has been dated to approximately 3.6 million years ago,[32] and that the footprints clearly indicate the three individuals who made them to have been fully bipedal, since there are no knuckle impressions. This is the paleontological equivalent of the video many parents now make of their baby's first steps!

It's believed that the rock which recorded this prehistoric hike had previously been a volcanic ash layer which was essentially turned into cement by gentle rains and then baked under a hot sun, but only after our venturesome trio passed through the area and left their impressions on the prehistoric "sidewalk."

Footprints may at first seem to be uninformative, but a great deal can be deduced from them. We can easily detect that the three individuals were fully bipedal, and get a sense of their heights from the length of the stride. Computer simulations are being used to estimate their walking speed and the nature of their striding gait. Of course, the naked footprints give us a crude idea of the shape of their feet. We also get a picture of their social bonding: a tight little group of three individuals walking cautiously across an open muddy space, constantly looking around them for prey or predators: one of the individuals is carefully stepping into the footprints of the other.

---

32. Raichlen et al., "Laetoli Footprints," e9769.

The only bipedal hominid we know of in that place and at that time with those general anatomical features is *Australopithecus*, and this is the sensible conclusion made by most paleontologists.

Others, however, who view the data through a very rigid presuppositional YEC perspective (many of whom are not trained experts in this field), elect to conclude that these footprints were made by modern humans (*Homo sapiens*). That in turn allows them to claim that humans walked with the dinosaurs, which then justifies their dismissal of evolution theory. They're free to make such a choice, but it's completely inconsistent with the dating of that rock layer, as well as a large body of other fossil evidence.

"But isn't this all based on a hoax?"

Before leaving this summary of the anthropological evidence for the origin of mankind, it is important to discuss an interesting chapter in the scientific exploration to answer that question. Many theists will have heard that this exploration was fraught with at least one hoax in which atheistically biased and motivated scientists "faked the data." And that revelation alone will be enough for those theists to jettison the entire set of data.

Yes, there was a hoax: Piltdown Man. But it was also uncovered by scientists, many of them also atheistic and also biased and motivated (in this case, though, biased and motivated to find truth), and it was quickly condemned by those same motivated scientists. For the sake of informing the reader, here's the story:

By the late 1800s and early 1900s, the scientific community had only two candidates for early man: Neanderthals and *Homo erectus* (otherwise known as *Australopithecus*). However, these are both quite distinct anatomically from humans, and so there was great interest in finding more transitional forms; of course, creationists were stirring up the hornet's nest with their demands for the "missing link." The thinking at this time was that the origin of mankind was a very recent event and that Neanderthals and *Homo erectus*, and any other candidates yet to be found, must be direct ancestors of mankind: discrete steps in a linear sequence that culminated with *Homo sapiens* (as opposed to being branches of the *Homo* tree).

Charles Dawson, initially training to be a lawyer but ultimately letting his hobby of collecting fossils become his profession, investigated a gravel pit adjacent to Piltdown Common in Sussex, where farm laborers had discovered unusual brown flints. There he acquired many fossilized fragments of a human skull and one part of the lower jaw which he was able to fit together like a jigsaw puzzle. These were all found in a layer of soil which also contained bones of other prehistoric animals (fragments of an elephant tooth, beaver teeth, a worn mastodon tooth) and flint fragments, and were in that way dated to the early Pleistocene (~2.5 million years ago) or even Pliocene era (many millions of years ago).

Apparently there was some room for error in putting the skull-and-jawbone puzzle back together. The paleontologist originally connected to the find put it together in secrecy, without any help (or even awareness) of his colleagues at the British Museum; the latter distanced themselves from this reconstruction later on in the unfolding drama. Eventually, he unveiled his discovery: a hominid with the jaw of an ape, the skull of a man and a cranial capacity (1070 cubic centimeters) which was intermediate between the two.

But Arthur Keith, a competent anatomist disagreed, pointing out that the chin and front teeth were re-formed too much like that of chimpanzees and that they had "fitted a chimpanzee palate and jaw on a skull that could not possibly carry them ... with the result that the upper joints of the backbone were so close to the palate that there was no room for windpipe or gullet and Piltdown man would have been unable either to eat or breathe."[33] He obtained casts of the fragments and reassembled them himself using principles from his training in anatomy to produce one in which "the chin and front teeth were entirely human and the cranial capacity was 1500 cubic centimeters—greater than the average for modern man."

The scientific community vigorously debated the two very different models for quite some time. Once again, John Reader summarizes the situation in a way that is so relevant to the purpose of my own book: "Thus Piltdown Man contributed to the intellectual climate of the 1920s and 1930s when some significant discoveries were scorned because they did not conform with accepted beliefs, while others, less accurately founded, were welcomed because they conformed only too well."[34] (the beliefs being scientific in nature, of course, not theistic). To make matters worse, a new dating technique—fluorine absorption—came onto the scene which seemed to indicate that these fragments were not so ancient after all.

For decades, questions about this apelike jaw and humanlike skull were clouded by doubt, debate, confusion, accusation and animosity.

And then, an important development took place: Joseph Weiner and Professor le Gros Clark at Oxford entertained the unnerving possibility that Piltdown Man was a hoax. All the evidence was carefully reexamined from this new perspective, and a common theme emerged. Circumstantial evidence pertaining to the finds from different sites had coincidentally identical details between them. There was the fact that new finds always lacked those crucial parts of the entire structure of the head which would definitively indicate the size of the brain case and the attachment of the jaw to the skull. There was unusual wearing of certain teeth which now looked peculiarly artificially caused. And worst of all: "Further fluorine testing revealed that the jaw was not just recent, but not long dead ... the remains were all stained to match the

---

33. Reader, *Hunt for Earliest Man*, 66.
34. Ibid., 73.

Piltdown deposit, so too were the mammalian fossils with which they were associated. The hoax had been ingeniously planned, carefully carried out and completely unsuspected."[35]

The actual perpetrator(s) of this hoax have not (cannot?) be definitively determined. Once again, John Reader closes this vignette in the scientific pursuit of mankind's origin with a particularly poignant statement: "When preconception is so clearly defined, so easily reproduced, so enthusiastically welcomed, and so long accommodated as in the case of Piltdown Man, science reveals a disturbing predisposition towards belief before investigation."[36]

Half a century later, we have another episode which had equally negative and long-lasting repercussions in the faith-science dialogue. A tooth discovered in Nebraska was all-too-quickly announced as coming from a hominid, the first North American apelike species: hence, it was named scientifically as *Hesperopithecus haroldcookii* (the first part of the name being translated "ape of the Western world," and the latter referring to its discoverer), and more colloquially as Nebraska Man.[37] However, the scientific community was skeptical and quick to examine the claim more closely, finding it instead to have originated from an extinct species of peccary (pig), resulting only a few years later in a retraction of the claim.[38]

Anti-evolutionists have been quick to pounce on both episodes and frequently use them to discredit the entire theory of human evolution from a common ancestor. They should recognize, however, that it was the pro-evolution scientific community itself, not anti-evolutionists, who exposed the intentional Piltdown hoax as well as the over-enthusiastic and premature Nebraska Man mistake. This is painting with a very broad brush. There have been all kinds of Ponzi schemes, yet no one would say that the entire investment world is built on a lie. Athletes who test positive for performance-enhancing substances, yet we continue to celebrate sports. Religious cults that deceived many out of their money or even their lives. Singing sensations that lip-synched their way through concerts. And yet we continue to promote all these human activities, despite those anomalies because they are exactly that. Exceptions to the rule, carried out by misguided people.

Turning the Pages of the Family Photo Album

The publication of Darwin's *Origin of Species* in 1859 started the world wondering whether we might have ancestors, although no one at that time was aware of any candidates. The truth is, some skulls that would later be identified as Neanderthal had already been found in 1829 and in 1846 but were not recognized for what they

35. Ibid., 80–81.
36. Ibid., 81.
37. Osborn, "*Hesperopithecus*," 463–5.
38. Gregory, "*Hesperopithecus*," 579–81.

are until Hermann Shaaffhousen made his presentation around the very same time as Darwin's announcement.

A few decades later (1891), Eugene Dubois presented his find of *Pithecanthropus* (now called *Homo erectus*), followed shortly after that (1924) with Raymond Dart's presentation of *Australopithecus africanus*. That decade also marked the beginning of a tremendous exploration and series of discoveries by the Leakey family (Louis and Mary, their son Richard and daughter-in-law Maeve), which ultimately netted thousands of finds and introduced the world to several new fossil hominids.

Collectively, the work of all these explorers told other paleontologists where to look and how to find the specimens. As a result, our collection of fossilized human and premodern human bones grew exponentially. Joining "Lucy" (*Australopithecus*) and Java Man (*Homo erectus*) in the photo album were "Ardi" (*Ardipithecus*) in 1994, "the hobbit" (*Homo floresiensis*) in 2004, Denisovans in 2008, and *Homo naledi* in 2015.

Too often, the discovery of a new set of bones would be hailed as a new species, to be met with support or skepticism from the rest of the scientific community. But eventually the collection was large enough that one could take a step back and see resemblances between certain bones, and it was possible to merge them into groups of highly related individuals.

At the same time, we learned how to date the fossilized finds by taking careful note of the rock or soil layer in which they were embedded. Determining the age of that layer using various forms of radionuclide dating techniques, or detecting signature components such as volcanic ash or lava from an eruption which in turn had been dated by various techniques, or noting the other animal species which were co-buried with it and knowing when those in turn had lived and/or gone extinct. All these techniques and others have allowed us to place the hominid finds into a sequence. We can now track the changes over the past several million years.

"But what about all the missing links?"

There are still some who throw up the canard: "there are too many missing links." Often they are just parroting what they've heard someone else say. They simply haven't done enough homework for themselves: I encourage them to do some reading and internet searches and investigate how much information we now have. The situation is very analogous to looking through a family photo album. The latter will document the growth of the family, and the maturing of the individuals within that family, but almost never does so by presenting a picture taken every day during the entire lifetimes of the family members. Instead, there will be groups of photos taken within a short period of time, followed by long gaps: fifty pictures taken around the Christmas / New Year's holiday, then nothing until the summer vacation seven months later during which another fifty are added, and then twenty more added a couple months later during someone's birthday. There may even be gaps spanning a year or two. Sometimes there can be a dramatic change in the appearance of an individual: perhaps they went through a dramatic growth spurt, or experienced a disease or accident which markedly changes their appearance. Most of the photos will feature certain members of the specific family, but every now and then there will be a sudden appearance of new and other individuals: cousins, best friends, teachers or bosses. But nobody would deny that these photos, which hopscotch through time, with the occasional appearance and disappearance of certain peripheral individuals, clearly tell the story of this one family.

In exactly the same way, the fossil evidence documents our relationship within an extended hominid family. Today, we recognize quite a number of different species and subspecies in that extended family, although we're still trying to puzzle out the exact relationships between different individuals, much in the same way that one tries to dig up their family tree by going back through a variety of historical records. But as different pieces of the puzzle fall into place, it becomes easier to see where other pieces fit.

Disputes over the naming and placement of different fossil finds still continue within the scientific community, but this should not be construed by opponents of evolution theory to mean that the whole story is suspect or untrue. The situation is no different than two grandparents turning the pages in the family album and arguing whether a given picture depicts "Bobby at summer camp when he was eleven years old" versus "Bobby at thirteen when the family took a trip to the Grand Canyon" versus "Michael [Bobby's brother] on Uncle Sam's farm." The central story is still the same: there was a young Bobby who had at least one sibling Michael, and both had the same blond hair and other family resemblances.

As I explained above (pages 73–76), underlying this confusion over the "missing link" is the misperception that evolution is linear. It is not. In my analogy of the evolution of letters, all the letters can be linked up to one distant common ancestor—the letter *o*—but as the subtle modifications accumulate, certain letters diverge down different paths such that there is no missing link to be found between the letters *k* and *s*, or between *s* and *f*. Anthropologists have long ago moved away from the idea that humans evolved from apes (or even from chimpanzees, which are more evolutionarily closely related), and are even moving away from trying to find linear relationships between the hominid species they find. Instead, they're now describing new hominid finds as having varying constellations of traits. They will show how a given bone in a skeleton has certain characteristics reminiscent of one species (or are clearly distinct from other hominid species) but another bone in the same skeleton as resembling that of yet a different species. An excellent example of this can be found in the recent original report of *Homo naledi*.[39] This may sound suspicious or even scandalous to some who look for any opportunity to discredit the theory of evolution, but it is in principle no different than saying a child has her mother's nose but her father's eyes.

---

39. Berger et al., "*Homo naledi*," e09560.

Which is not to say that the novel skeleton in question is directly the product of the sexual union of two different hominid species; rather, it acknowledges that the information needed to produce different body forms are buried deep within our genetic code and are brought to the surface by varying environmental stresses. Stresses such as the need for climbing trees versus running on the ground (which affects the structure our limbs and spine), or the need to exploit alternate sources of food (leading to dental changes), for example. Other features don't seem to be related to a response to some environmental stress, but instead seem to be simply an inherited family trait, such as our eyebrow ridges.

Conclusion

We now have an abundance of fossil evidence for the evolution of mankind, beginning in Africa and spreading around the globe. Bones and tools and some cultural data for the very apelike australopithicenes, the most famous representative of which is "Lucy," arising approximately four million years ago. Another pile of similar data for *Homo habilis* and *Homo erectus* arising a couple million years ago. And there's tremendous evidence for several "cousins" who branched away from our roots. *Homo naledi*, who seems to bridge the gap between the apelike and humanlike species. *Homo neanderthalensis*, who spread throughout Europe and Asia, and whom we met again in the more recent past before they died out. As well as *Homo denisovans* and *Homo floresiensis*. It's interesting, and yet sad, to realize that although we once had so many different evolutionary siblings and cousins, we now are orphans and only-children on our particular branch of the Tree of Life.

It is true, though, that there is some ambiguity in how to interpret the details of that evolution, and considerable discussion and debate in the scientific community because of that. But although we may not have the precise details down yet, there is not enough ambiguity to suggest that it didn't happen. To deny this is to deceive oneself.

But now we've developed a whole new way to study the anthropological evidence and remove a great deal of that uncertainty: genetic sequencing of DNA extracted from the bones.

6

Origin of Humanity
The Genetic Evidence

THE THEOLOGICAL WORLD WAS caught quite off guard in the early nineteenth century when archaeologists found the Sumerian/Babylonian clay tablets describing a global flood. Detail after detail paralleled the Genesis version so closely that it quickly became impossible to not wonder whether one copied from the other (the evidence is quite in favor of the Babylonian version far predating the Hebrew version), or whether they're so similar because they describe a common event.

Now in the twenty-first century, a similar scenario is unfolding all over again. Scientists have developed the ability to read the genetic code and have begun doing so in various species. What they've found are amazing parallels in the sequences between the human and primate genomes such that it is becoming impossible to deny that one has been modified from the other. And scientists have found much more than simply the gross similarities in our gene sequences. Fine details are emerging which make the story absolutely compelling. Altogether, the science of genetics shatters certain interpretations of the Genesis account of the origin of humans.

Mitochondrial Eve and Y-chromosome Adam

"Knowledge is power," said Sir Francis Bacon. Knowing that mitochondrial DNA is passed down solely through the maternal line, without any of the shuffling and recombination that occurs between maternal and paternal chromosomes, one can virtually travel back in time and trace the lineages of individuals. The mitochondrial DNA sequences will be essentially identical between an individual and his or her mother, which in turn will be essentially identical to that of her mother, and so on from one generation to the next. The exception to this will be the steady accumulation of mutations, which usually occur at a relatively low rate. In this way comparisons between samples of mitochondrial DNA from many different individuals can be arranged in the form of a family tree, an exclusively maternal one, with those on the same branch having the fewest differences, and those on different branches having more differences.

Based on this understanding of mitochondrial DNA, one genetic study took samples from women all around the world and from many very different racial backgrounds, and created a family tree. What they found was that they could trace a common ancestor for all these women: the proverbial Eve. This so-called mitochondrial Eve lived approximately one to two hundred thousand years ago somewhere in Africa.[1]

A similar approach was used to look at the slow accumulation of mutations in certain pieces of the Y chromosome, which only males inherit (the Y chromosome is the one that has the genes that turns the developing fetus into a male).[2] Like the mitochondrial DNA being transmitted down an unbroken maternal genealogy, the Y chromosome DNA is transmitted down an unbroken male line of descent.

Both the mitochondrial DNA from the maternal lineage and the Y chromosome from the paternal lineage are passed down without any recombination between the genes from the two parents (which would blur the inheritance pattern). Just a slow and easily followed accumulation of spontaneous mutations. In this way, samples taken from men indicated that the proverbial Adam probably lived approximately 250,000 to 500,000 years ago, and also originated from Africa.

One might think that we should be able to trace our ancestry to only a single pair of individuals: an individual mother and individual father who lived together in the same place and time and gave birth to the ancestors that eventually led to us (of course, theists might be tempted to refer to these as the biblical Adam and Eve). But it doesn't work that way. It isn't as if some primal couple had one baby which eventually gave rise to humans, but also had another baby which eventually gave rise to Neanderthals or chimpanzees.

Instead, our species didn't arise from a single pair, but emerged from a large population out of a genetic bottleneck. At one point in time there were a very large number of our common ancestors. Using current technology, we don't know if there were thousands or millions. These common ancestors originally lived in Africa, but expanded out of Africa into the Middle East, Europe and Asia. As mutations and other genetic changes accumulated, they would have a wide range of gene variants, including a dizzying array of maternal lineages and paternal lineages, reflected in a large number of different mitochondrial and Y-chromosomal DNA patterns.

However, at some point it seems that we encountered some kind of natural disaster which nearly wiped out the species, leaving only a few thousand survivors. One leading scenario involves a gigantic volcanic eruption occurring roughly seventy thousand years ago in what is now the island of Sumatra in Indonesia. This vaporized nearly three thousand cubic kilometers of rock (nearly a thousand times larger

---

1. Cann et al., "Mitochondrial DNA," 31–36.
2. Mendez et al., "African American," 454–59.

than the eruption of Mount Vesuvius in 79 CE, which in turn was three times larger than that of Mount St. Helens in 1980), leaving behind what is now Lake Toba. This eruption covered much of the earth with a thick layer of ash which quickly killed off plant life and dimmed the skies for years. This all coincided with and exacerbated a global cooling that was already well under way. Needless to say, life on earth suddenly became very difficult, and many species weren't able to cope.

Humans too were nearly overcome, but a few thousand breeding pairs did survive (the overall population was likely larger than this, since there would have been other survivors that didn't contribute to the gene pool, such as old, weak and diseased individuals). Much of the globe was completely uninhabitable at that time, so the group(s) that survived would have been very localized: they would have been restricted to those much smaller regions which remained warm enough and had food resources abundant enough to continue to support them until the Ice Age retreated. And being so highly localized, they would have had a relatively limited number of paternal Y-chromosomal and maternal mitochondrial lineages. (There's no reason to think that those two distinct types of lineages must necessarily trace their distant origins to the same place and time, any more than one would expect that our grandfather's family must have the same geographical history as our grandmother's family.) All the other paternal and maternal lineages would have died off, but this now greatly reduced and genetically less diverse population would then eventually expand into the seven or eight billion humans which populate the earth today.

This is a difficult concept, and some readers might still be somewhat unsure of it or unclear about it. Let me use an analogy with which we're all familiar to make the point. The English language has evolved considerably over the past many hundreds of years. The vocabulary and speaking style that we use today are quite different from those which the people of the seventeenth century used (for example, the time of Sir Isaac Newton), which in turn was quite different from those of the sixth century (the time of King Arthur). Of course, the latter evolved out of the various languages being spoken in Europe for hundreds of years prior to that. There was no point in time at which the English language suddenly originated: it gradually evolved from something completely different and there is no point in history at which the elders of the population heard their grandchildren or even great-grandchildren speaking an unrecognizable language. Most readers should be able to recognize the similarities between the evolution of English and the evolution of humans. But let's now extend the analogy to bring in the idea of paternal and maternal lineages of DNA.

English-speaking peoples have always been moving, exploring and conducting trade with distant lands: in the process, we have been constantly spreading the gradual changes in our language which accumulate over time (analogous to mixing up the gene pool). Today, the English vocabulary is fairly uniform around the world: most speakers of English can engage in conversation with any other English speaker around the world with relatively little difficulty. However, there can be marked differences in how they pronounce that vocabulary: it's fairly easy to distinguish speakers as coming from Ireland, Scotland, England, or North America, and some listeners can even tell whether the latter come from New York, Boston, Georgia, Alabama, California, British Columbia, Northern Ontario or Newfoundland. We can take those diverse English accents to represent the various paternal and maternal lineages of human DNA: all using the same language, but having subtle differences which immediately set one speaker apart from all the others.

Now consider a hypothetical situation in which a man from a very rural part of Ireland marries a woman from the back waters of Alabama just before a global disaster which nearly wipes out the entire English-speaking part of the world: only a few small and geographically separated groups of individuals survive this disaster, including this couple I've just introduced. Their offspring will acquire some kind of interesting blend of accents and vocabulary which might be traced back through audio archives to two very different parts of the world, and with linguistic roots that go back to two very different points in history (the Alabaman accent, euphemisms, figures-of-speech, and speaking style have only developed over the past two centuries, while those of the Irish go back many more centuries). The global disaster will have wiped out many English accents forever, but elements of these two unique ones will persist.

In the same way, we can trace mitochondrial Eve and Y-chromosomal Adam to two different places and times through a relatively small group of ancestors comprised of several thousand.[3] Trying to look beyond that bottleneck, into the more distant past, it gets increasingly difficult to trace genetic lineages. Looking forward from that bottleneck, into the present, there are genetic lineages which are now forever lost and we will probably never retrieve any evidence of them.

A similar story seems to apply to Neanderthals, who also migrated out of Africa in a different wave and also expanded into Europe and Asia and also experienced natural disasters which did eventually wipe them out. The story of chimpanzees, on the other hand, is very different. They only ever lived in Africa, and did not roam the continent the way humans did, so they exhibit a far higher degree of genetic diversity than do humans.[4]

Humans Are Undoubtedly Related to the Chimps and Apes

At the time of the historic debate between Samuel Wilberforce and T. H. Huxley (see opening paragraph of chapter 5), the one which dragged Huxley's grandmother and grandfather into the argument about descent from apes, the only evidence supporting that theory was a very limited collection of fossil bones (exclusively Neanderthal). Not only has that collection since been massively extended and much more carefully studied and documented, but we now have an entirely different line of evidence which

---

3. Huff et al., "Mobile Elements," 2147–52.
4. Pääbo, *Neanderthal Man*, 75; Arnason, "Complete Mitochondrial DNA," 145–52.

is essentially a slam dunk in this dispute. Careful comparison of the genomes of humans, apes, chimps, orangutans, Neanderthals and Denisovans make it nearly impossible to escape the conclusion that we are all closely related.

I say "nearly" impossible because there is still one avenue left to avoid that conclusion: absolute and desperate denial of the mountain of evidence and blind faith in the tenet that "we just have to be different: we simply can't be like them."

And yet the evidence is there just the same. What exactly is this genetic evidence?

Genetic Plagiarism: Common Descent or Common Design?

Recall the analogy that I used to open up chapter 1. The one in which I portrayed myself marking student essays on American history and finding clear irrefutable evidence that one student had copied from another. In that situation, I might ponder the possibility of pretending it wasn't true, or trying to find excuses on behalf of the students. But deep inside I would know that the right thing to do would be to treat this for what it is: academic plagiarism. Even if it meant failing one or both of the students, and possibly launching an academic integrity complaint which itself could open up other negative ramifications for those students.

In the exact same way, an open-minded person just can't get around the fact that the human genome was certainly copied from the genome that also gave rise to other primates (chimpanzees, apes, gorillas, orangutans), as well as at least two very closely related genetic cousins, the Neanderthals and Denisovans. Why do I say that? Because they're riddled with the same "mistakes." In the next few paragraphs, I'm merely going to introduce these examples within a sentence or two to give the casual reader a sense of the breadth of the evidence for this genetic plagiarism. In the sections which follow, I'll explore each example in much greater detail for any interested and motivated readers.

Mistakes in the DNA—mutations—are random in nature. They can occur anywhere in the entire sequence, but we find many of them in the exact same places in the genomes of humans and great apes. They could have occurred in millions of other places, but many are in the same position in the various ape genes. It's like the fourth and fifth student essays both using the word "and" 1,683 times, but both misspelling it as "nad" only on page number thirty-seven, line number twenty-three. That's not random. That's copying.

There are genes which have part of their sequence cut off—much like a half-finished paragraph in a book—in the exact same places between the species.

Some genes which have become completely non-functional due to all kinds of damage and rearrangements (described in more detail below under "Pseudogenes"): again, these disruptions occur in the very same places in the human and ape DNA.

There are genetic scars in which one of our ancestors had a virus insert its viral DNA into the ancestor's genes, and then passed those viral genes on to all its off-spring, including you and me (described in more detail below under "Proviral sequences"). The interesting thing is that some of those genetic scars will be found in the exact same place in the human and chimp DNA, but be absent in the DNA of gorillas or orangutans (which are on a different branch of the evolutionary tree than humans and chimps).

Another fascinating piece of the story was found when the entire human "DNA book" was compared side-by-side with those from the great apes. While the metaphorical chapter layout for ours was mostly the same as theirs, ours seemed to be missing two whole chapters! As if we had just thrown away tons of valuable genetic information, which would be evolutionary suicide! But later we found those two chapters, spliced onto the end of one of the other chapters in a different part of the genome (described in more detail below under "Human Chromosome 2 Inversion").

On top of this new genetic evidence, and together with the large pile of radio-isotope-dated fossil bone and artefact evidence, there's also biological evidence that we evolved: certain aspects in the design of our bodies which don't confer any useful function, and which don't make sense if we were completely designed from start to finish as stand-alone units, but which instead make more sense if we are modifications of a previous design.

I mentioned above the formation of a tail-bone at the bottom of our spine.

We have muscles in our ears which in many animals allows them to turn their ears in the direction of certain sounds (just watch your cat or dog). Those are still fully functional in some monkeys, but in humans, orangutans and chimpanzees the muscles are present but are poorly developed (although it is possible to train and strengthen those muscles to make our ears twitch, as some unique individuals delight in showing people). Instead of turning our ears, we as a species have learned to turn our heads and train our more perceptive eyes, and use our brain's ability for depth perception, to identify potential threats.

The sparse hairs on our bodies, equipped with tiny muscles to raise them in response to cold or fear (goosebumps), which do nothing to keep us warm or appear more threatening, which is their function in animals.

The counterintuitive design of our eyes, our respiratory system, and some peculiarities in our cardiovascular system.

There's just too much evidence that humans evolved over millions of years from a common ancestor which also gave rise to other species of hominids (*Australopithecus, Homo erectus, Homo habilis*, Neanderthals, Denisovans) and primates. The alternative explanation is that God planted all that evidence intentionally. But as I already concluded above, that raises huge theological and moral problems. God becomes a deceiver, or allows a Deceiver to mislead us.

Some theists will say God didn't create us with those errors: that he originally made us perfect and that those errors appeared shortly after the flood. But that doesn't explain how such diverse types of mutations, which have always been random in nature, appeared in the very same places in those diverse genomes.

On top of that, another piece of genetic evidence is absolutely inconsistent with all of present-day humanity arising from one couple approximately six thousand years ago: the number of variants of any particular given gene. If all of mankind arose from one couple—Adam and Eve—then we can only have four different versions of any given gene, two from Adam and two others from Eve. And yet in many cases there are far more than four variants (described in more detail below under "Alleles").

This has been just a general overview and summary of the reasons why I feel one can only conclude that the human genome is a modified copy of the primate genome. In the next few sections, I'll delve more deeply into the technical details of all these points: some readers might find the remainder of this chapter too overwhelming with scientific concepts and may choose for this reason to skip this and proceed directly to chapter 7. But doing so requires the reader to accept the conclusion that humans evolved from an ancestor shared in common with the apes. If the reader is unable to accept that conclusion, then they will need to wade through the next set of technically demanding sections.

Shuffling the Cards

The genomes of complex organisms are large. The human genome comprises over three billion base pairs and would stretch out to three meters if it were kept in one piece (to put this in perspective, the size of an average human cell is 0.00001 to 0.0001 meters). To simplify the day-to-day handling of all this genetic material, rather than maintaining it all in the form of two long and flimsy molecules (one from the mother, the other from the father), cells have divided it up into many smaller more manageable units. Essentially, they have lined up the two strands of DNA side-by-side, matching up the genes in one strand with the corresponding genes in the other strand, then broken the two strands into smaller units called chromosomes (each individual strand of DNA within that chromosome is in turn called a chromatid).

In addition to the genes themselves, chromosomes possess a few key features which aid in their function.

One of these is a centromere (literally, "central body"). Its job is to keep together the two strands of DNA comprising any given chromosome, and to serve as handles when it comes time for the cell to divide: molecular strings (actin fibers) anchored at either pole of the dividing cell lasso the centromeres and pull them apart such that each daughter cell gets an exact half of the entire genetic package.

The chromosomes also have unique structures at each end of the DNA strands called telomeres (literally, "end bodies"). These protect the ends of the strands, much like the plastic ends on shoelaces, keeping the strands from unravelling during replication or being incorrectly recognized as broken DNA strands or viral DNA.

It goes without saying that different organisms can have different numbers of chromosomes, in part because their genomes can vary in size and because they may have had a different evolutionary journey. In chapter 4, I explained how chromosomes can be easily visualized and identified using special stains which produce unique banding patterns of the various genes they contain: their karyotypes. The centromeres and telomeres can also be visualized using special stains, and this aids in the identification of one pair versus another (for example, the centromere might be offset from the center differently).

We've known for many decades that there is an amazing similarity in the overall arrangement and structure of the human genome compared to those of apes, especially chimpanzees, and that both are arranged quite differently from most other animals.

For example, figures 10 and 11 show the karyotypes of mouse, human, and great ape chromosomes. Remember, karyotypes are obtained by staining the chromosomes to reveal unique banding patterns (pages 89–90), and are produced by differences in the degree to which different sections of nucleic acids take up the dye. Karyotypes don't tell us anything at all about the genes themselves within the chromosomes. They merely reflect unique ways in which the DNA is packaged, and distinguish one chromosome visually from another. Think of it like the colors of the covers of the books in your local library: the sequence of colors tells you nothing at all about what's written in the books, but it's possible to take photographs of the stacks of books and say, "This photo is obviously from aisle number twelve because of the sequence of colors of the book covers. I can tell because all these orange books are next to the blue ones, and then you have this section of yellow books at the bottom. This photo, on the other hand, is obviously from aisle forty-seven because . . . "

When one compares the human karyotype with that of other non-primate animals, they look very different. For example, figure 10 compares the human and mouse karyotypes. We humans have twenty-three pairs of chromosomes, with one pair comprising two distinct but related chromosomes: the X and Y chromosomes. Mice have only twenty-one chromosomes, and show a completely different banding pattern.

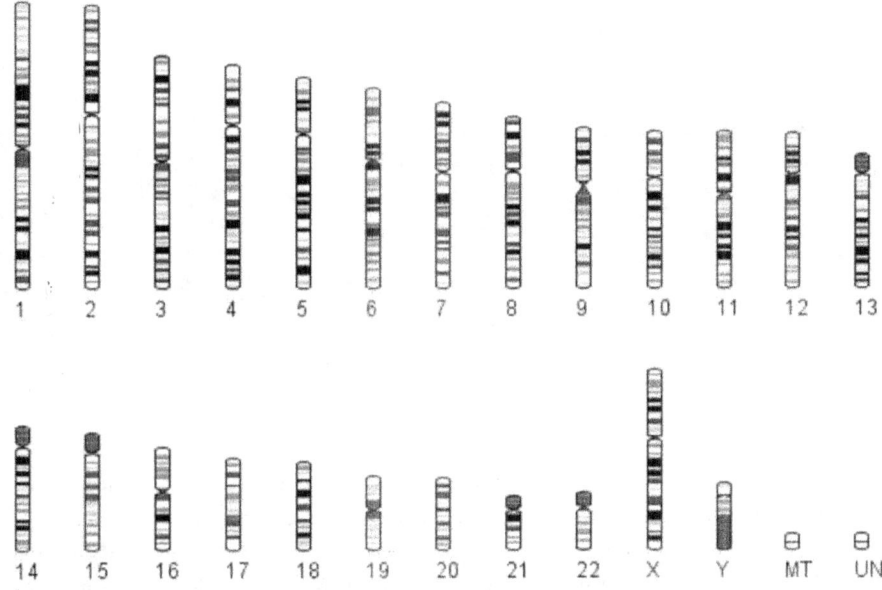

Figure 10. Karyotypes of the human (top) and mouse (bottom); notice different numbers of chromosomes and banding patterns for each. Permission granted under the GNU General Public License, version 2.

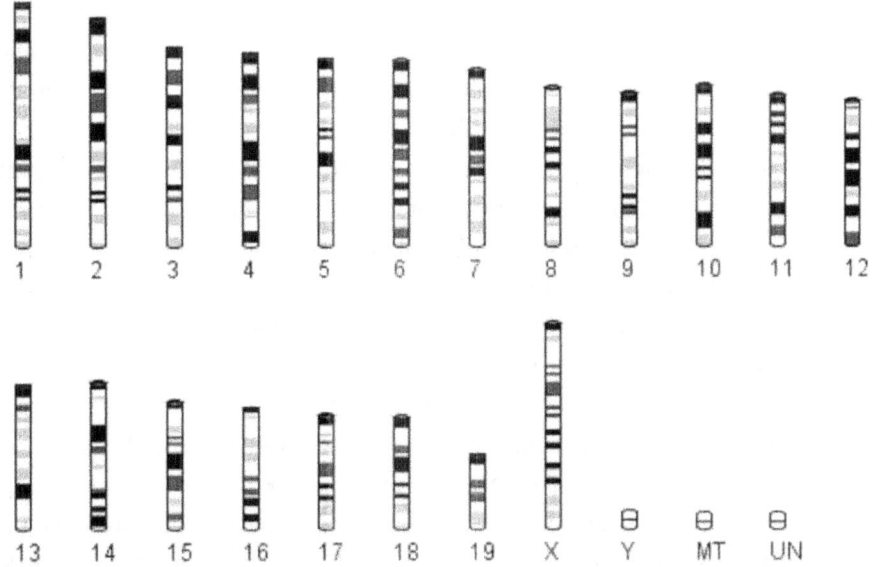

But when you compare the karyotypes of humans and apes, they look amazingly similar, as shown in figure 11. The tremendous degree of similarity between those four human and ape genomes, in stark contrast to the differences between the human and mouse genomes, is compelling.

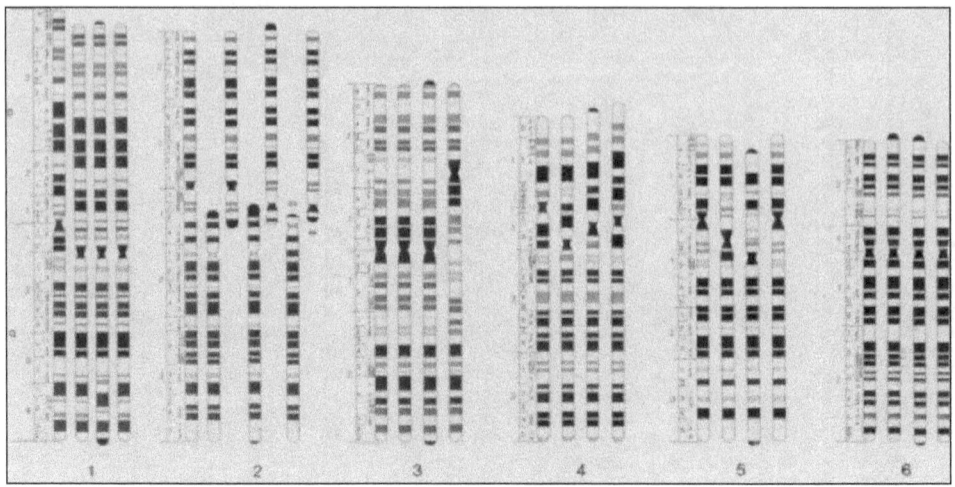

Figure 11. Six groups of chromosomal karyotypes, numbered according to the human numbering scheme (see bottom of panel). Each of the six groups portrays the human, chimpanzee, gorilla, and orangutan chromosomes, respectively. Note the similarity in banding pattern within each group. Also note that chromosomal group 2 includes two chromosomes from each primate together with the one human chromosome. With permission of Dr. J. Yunis and American Association for the Advancement of Science.

Even the differences between the human and ape genomes are equally exciting. Look at the karyotype in figure 11 for chromosome number five and compare the human and chimp images: they're almost identical, except for a segment approximately one fifth the length of the chimp chromosome. In your mind, divide the chimp chromosome into five equal segments, then flip the second of these segments upside-down, rotating around the pinched part of the chromosome (this is where the centromere is located), and the human and chimp chromosomes become essentially 100 percent identical. Geneticists refer to this as an inversion event: when a section of DNA gets flipped upside-down or backwards.

We now have lots of evidence of whole sections of DNA moving around throughout the genome. It sounds dangerous. Sometimes it is. But this shuffling is in part exactly what goes on when the DNA of the father and mother are brought together within the fertilized egg, and explains how any given parental couple can produce such different children. Also, scientists think this is one of the mechanisms by which evolution has shuffled the cards to produce entirely new species. This shuffling makes complete sense from an evolutionary point of view, but not from a YEC creationist point of view (why would God shuffle the cards like this?). In the next section, I present an even bigger DNA rearrangement event which also convincingly shows how humans originated from an ancestor shared in common with chimpanzees.

Human Chromosome 2 Inversion

Gorillas, chimpanzees and orangutans all have the same number of chromosomes—twenty-four pairs—as might be expected if they arose from a common ancestor. Humans, on the other hand, have only twenty-three pairs. Creationists were quick to jump on this difference as being consistent with us being unique from the animals: specially created. Both creationists and evolutionists would agree that it simply isn't conceivable that we originated from the common ancestor of the other primates but lost the genetic information contained in a whole pair of chromosomes in the process. That represents a colossal loss: hundreds or even thousands of genes would have disappeared forever. A loss of that magnitude could only result in cell death, which in turn would mean our species would never have developed. It seemed like the creationists had unassailable evidence that humans were created *de novo* and distinct from the other animals.

But a closer look at our chromosomes brought the story back to the evolution camp.[5] When the karyotypes of the human chromosomes are examined very carefully, not only does a simple counting show that we seem to be missing one pair compared to the apes, but also that one of our chromosome pairs has a very remarkable banding pattern. The pair in question is referred to as chromosome number two because chromosomes are numbered from largest to smallest, and it is the second largest of our chromosomes. Two separate chimp chromosomes can be made to line up exactly with the one human chromosome number two by virtue of having identical or matching banding patterns with the human one; see figure 11. For the rest of this section, I will refer to these three chromosomes as chimp A, chimp B and human #2, respectively.

As expected, human #2 has telomeres at either end, but surprisingly it also has parts of telomeric structures in its middle section, even though telomeres don't belong in the middle sections of genes. In fact, these partial telomeric structures in human #2 are located right where the two ends of chimp A and chimp B would overlap when they're laid side-by-side with human #2 while trying to match up the banding patterns, as shown in figure 11. Moreover, the arrangement of the sequences within the partial telomeric structures matches up with what one would expect if the two ends of the separate chimp chromosomes fused together imperfectly. That is, in this middle section, one finds first a unique sequence which normally attaches the telomere to the gene's coding sequences (the "pre-telomeric sequence"), next one finds the typical telomeric sequence, then a backwards telomeric sequence, and finally a backwards pre-telomeric sequence. This is exactly consistent with chimp A and chimp B simply fusing head-to-head.

And then there are the centromeres. The location of the centromere of human #2 matches up exactly with the centromere of chimp A. Further down human #2, at the

---

5. Yunis et al., "Striking Resemblance," 1145–48; Yunis and Prakash, "Origin of Man," 1525–30.

corresponding position where one finds the centromere of chimp B, there are the remnants of a centromere.

One interpretation of these provocative findings is that humans and chimps diverged from a common ancestor and that, in the process, two chromosomes in the ancestral genome fused to produce human #2.[6] Another interpretation is that God created humans distinct from all the other animals, but he made human #2 to look like this. I can't think of any good reason why he would do so? If he were indeed starting from scratch on a completely new and unique creation which reflects something entirely different from all the other animals, why start with a genetic sequence that is already full of flaws, and just tweak that a bit, in part by fusing together two chromosomes in such a way that broken pieces of telomeres and centromeres are left lying around? The same question could be asked about many other features that we share in common with the animals but which don't make sense for human function. For example, the remnants of tails, and hairs on our skin with erector pili muscles, and other design flaws which will be discussed in more detail below (section entitled "Intelligent Design," p. 269).

Creationists have tried to rebuff the common descent theory in various ways.[7] They will admit that chromosomes can fuse together, but insist that this is exceedingly rare. Agreed. But rare doesn't mean impossible.

They have also pointed out that this fusion would have to have occurred in a germ cell (sperm or egg) in order for it to have been passed down to the descendants. Once again: agreed, but this doesn't negate the possibility. It's still a legitimate explanation.

They go on to question the success of a fertilization event involving an unequal number of chromosomes (twenty-three from the one parent and twenty-four from the other). The fact is that fertilization under such conditions can still occur. In some cases, the offspring so produced may be infertile, as is the case when horses (which have thirty-two chromosomes) mate with donkeys (which have thirty-one chromosomes), producing a mule. But there are many other examples of individuals within a species having differing numbers of chromosomes and still being able to interbreed and produce fertile offspring. The Chinese muntjac deer has twenty-three pairs, while the congenic species Indian muntjac deer has only two pairs of autosomes and one of sex chromosomes.

Some day we may learn that this chromosomal fusion event contributed to the emergence of the *Homo* line. It alone would not have been sufficient to produce a new species. But this fusion event, together with other genetic changes, certainly could have helped to restrict us to a different evolutionary path unique from those of the other primates. Genetic analysis of Denisovan DNA has found the same chromosome

---

6. Ijdo et al., "Origin of Human Chromosome 2," 9051–55; Avarello et al., "Ancestral Alphoid Domain," 247–49.

7. Rana, "Chromosome 2."

2 fusion in those cousins of ours,[8] telling us that the fusion event would have occurred before we split from them. Perhaps it made cell replication a little easier by reducing the number of chromosomes which had to be separated and carefully sorted every time the daughter cells divided. Perhaps the recombination event led to the loss of a gene which otherwise tended to have a negative effect. For example, sickle cell anemia is a genetic disease which produces both positive effects (resistance to malaria) and negative effects (impairs oxygen transport). Environmental and ecological conditions might change such that the positive effects were no longer needed (in our example of sickle cell anemia, the climate becoming more desert-like or arctic in nature would eliminate the mosquitoes that carry the malaria), and the loss of such a gene would amount to a net gain (because of the loss of its negative effects).

Overall Sequence Similarity

As stated before, karyotypes don't tell us anything about the genes themselves. That requires a whole different technique: gene sequencing, which became easily and widely accessible in the last couple of decades. In this way, a substantial portion of the human genome was determined in 2003, that of chimpanzees in 2005,[9] of gorillas in 2012,[10] Neanderthals in 2014,[11] and Denisovans in 2014.[12] The much smaller mitochondrial DNA of *Heidelbergensis* was also determined in 2013.[13]

Actually, I should qualify all those claims. We've only sequenced the more important or interesting parts of those genomes, primarily those which code for proteins. There are still whole other sections of all these genomes which haven't yet been sequenced because of the limitations of gene sequencing. For example, there are whole sections which seem to contain only repeated sequences, such as hundreds of repetitions of ACACACACACAC. Gene sequencing involves cutting the long DNA molecules into short segments of a few hundred bases, sequencing those short segments and then using computers to stitch back together the derived sequence. But when many of those short segments contain such extensive repetitions, the computer becomes unable to decide where to place segments which contain shorter strings of ACACAC (together with other coding sequences), or how many repetitions to stitch together.

That having been said, we *do* have the sequences of the major parts of the genomes of many different animals including humans. As each of these genomes became available, geneticists immediately started making comparisons between them.

8. Meyer et al., "High-Coverage Genome," 225.
9. Chimpanzee Sequencing and Analysis Consortium, "Initial Sequence," 69–87.
10. Scally et al., "Insights into Hominid Evolution," 169–75.
11. Prüfer et al., "Complete Genome Sequence," 43–49.
12. Meyer et al., "High-Coverage Genome," 222–26.
13. Meyer et al., "Mitochondrial Genome," 403–6.

In doing so, they found considerable similarities between some, and marked differences between others. More importantly, they have found the three *Homo* genomes (humans, Neanderthals, and Denisovans) to be essentially identical to each other but slightly different from chimpanzees, even more different than gorillas, much more so than the other apes, and quite different from other mammals like dogs, cats and ancient cave bears.

Some will quibble over the actual percentage similarity between human and chimp DNA. Numbers vary all over the place, depending on which parts of the DNA one includes in the analysis (protein-encoding only versus all parts). The constant waffling back and forth may confuse some non-scientists.

Let me use an analogy to help see why this isn't so surprising and shouldn't be a problem: let's make some whole body comparisons between humans and apes. When you look at their height, they're very similar. But if you measure their weight, they're quite different. With respect to general facial features (placement of two eyes, nose and mouth), they're quite similar to each other relative to that of a squid, which also has eyes and a mouth. But according to brain size, they're quite different. Number of arms and legs: similar. Relative length of arms versus legs: different. Fingernails: similar. Overall body hairiness: different.

I could go on, but the reader should get the point: depending on exactly how you set the criteria for comparisons, the two can appear quite similar or quite different. The same holds for comparisons between DNA sequences. The two can be made to appear similar or different depending on whether you compare only regions that code for protein sequence (exons), or also include the intervening regions that regulate gene expression (introns); whether you include or exclude long repetitive sequences which don't seem to have any function in the genome (at least, we don't know at this time why they're there); and in several other ways.

But whenever the two debaters agree in advance on what criteria will be used to determine which portions of the DNA will be compared, and then proceed to compare the genomes of humans, chimps, apes, dogs, chickens, snakes and mushrooms, one will always find the humans, chimps and apes clustered very close together, and then the other species on ever increasingly distant branches of the tree (with dogs being much closer to us than the mushrooms).

For someone who is open to "a plain reading of the data" and who is not influenced by the possible theological ramifications, it's absolutely undeniable that we are highly related genetically to the chimps and apes: that our genomes are incredibly similar in many respects. In the next few sections, I'll focus on some key differences which also underline our common genetic origin.

Alleles

For any given gene—the equivalent of a single paragraph in my student-essay analogy above—you have only two copies, one from your father and one from your mother (who in turn each had two copies). These two copies may or may not be the same. Each type or version is otherwise referred to as an "allele."

For example, you are one of four blood types depending on which of two alleles you inherited from your parents: either you got one or two copies of one allele but not of the other (so you're either type A or type B), or you got one copy of both (type AB) or you got neither (type O).

So the existence of these two different alleles leads to four different combinations. If there happened to be four different alleles for a given protein—L, M, N, and O—the offspring would have sixteen different combinations (although many would be functionally identical: a baby having dad's M and mom's N would be indistinguishable from another baby having mom's M and dad's N). Other genes can have other numbers of different alleles. Many of the genes which code for the markers on our blood cells (the human leukocyte antigen complex) have dozens of alleles.[14]

I should clarify and emphasize that I'm not referring to characteristics which are controlled by multiple genes. Like hair color, for example, or intelligence, or the shape of one's nose. Traits such as those are defined by the sum total of inputs from alleles of a variety of different genes, in some cases hundreds of different genes, on different parts of a chromosome or even on different chromosomes entirely.

What I am saying, though, is that any individual person can have at most only two alleles for a given gene: one from their father, and one from their mother. For this reason, if one insists that we all began from a historical Adam and Eve, there should be no more than four different alleles for any given individual gene in the human genome: the most we could inherit is two different alleles from Adam, and two others from Eve.

And yet for some genes there are many more than only four alleles: again, the gene controlling the human leukocyte antigen complex has as many as fifty-nine different alleles.

You can't explain that happening in a population beginning with two individuals only six thousand years ago (the equivalent of three or four hundred generations), unless you choose to believe that the rate of gene mutation was massively higher in biblical times than they are today. That would be what we call an *ad hoc* explanation: one that doesn't have any underlying basis or rationale other than that it allows one to hang on to a certain preconception or hypothesis. Such a belief would conflict with many different lines of DNA evidence combined with radioisotope dating which tell us that gene mutation rates have been constant for thousands of years. Those lines of evidence come via samples extracted from: a wandering hiker in the Austrian alps

---

14. Mack et al., "Well-Documented HLA Alleles," 194–203.

5,300 years ago;[15] from ancient Egyptian mummies and Peruvian sacrificial victims; from bones of Denisovans that are many tens of thousands of years old,[16] and Neanderthals which are hundreds of thousands of years old.

Proviral Sequences

In chapter 4, I explained how certain viruses will integrate their DNA into the genome of a host organism and then commandeer the host's cellular mechanisms against it. Sometimes, integration occurs in a sperm or egg cell, and in that way all the progeny of that infected host can inherit that proviral sequence in the same way that they inherit any other gene from their mother or father. And this is where the story gets interesting from the point of view of this book.

Scientists have indeed found proviral sequences in the genomes of humans, apes, chimpanzees and orangutans.[17] In many cases, each of these species exhibit the proviral sequences in different places; this isn't surprising, since these different lines of species have had millions of years during which one or another could experience a unique viral infection. But some proviral sequences are found in the very same relative position within the respective genomes of humans and chimps, and some even in all four human/ape species. This is a provocative coincidence for which there's really only three explanations.

First, it's just simply that: a coincidence. Given that there are roughly three billion base pair locations, the chances that an insertion occurred at the same relative position in the humans and the chimps would be roughly one in three billion. Already pretty steep odds. For the sake of making the rest of the math easier to work with, let's just round this down to one in one billion, or 10^9 (this rounding down actually works in favor of those wanting to claim that the outcome could be simply a coincidence). The chance for that coincidence occurring independently in all four species would be one in 10^9 x 10^9 x 10^9, or one in 10^{27}. But then one has to multiply those probabilities by the number of different types of viruses, since there are many different types of viruses, each with unique proviral sequence signatures, and each of the species has to receive the same type. Then one has to take into consideration that we don't just find one proviral sequence insertion in common between the four species, but identical proviral sequence insertions at hundreds of locations in the genomes. The calculation of the odds for these kinds of coincidences happening are absolutely mind-blowing.

---

15. Handt et al., "Molecular Genetic Analyses," 1775–78; Owen, "5 Surprising Facts."
16. Reich et al., "Genetic History," 1053–60.
17. Steinhuber et al., "Human Endogenous Retrovirus," 188–92; Reus et al., "HERV-K(OLD)," 8917–26; Mayer et al., "Human Endogenous Retrovirus K," 1870–75; Jern, "Divergent Patterns," 1367–75; Romano et al., "Demographic Histories," e1026; Hughes and Coffin, "Genomic Rearrangements," 487–89; Marchi et al., "Neanderthal and Denisovan," R994–95.

Second, claims have been made that there might be hot spots in the genomes: places in which insertion is more likely than in others. Advocates for this don't have any direct evidence for such an *ad hoc* explanation. But even if this is true, this only increases the probability marginally above that of a random event. Let's presume, for example, that there is something about a particular site which makes it ten times more conducive for viral insertion than other sites. Those are fairly balmy hot spots. But even then, the probability for one occurrence drops from one in 10^9 down to one in 10^8, and the probability for the four species receiving any given virus at the same spot becomes one in $10^8 \times 10^8 \times 10^8$ (or 10^{24}): still an unfathomably huge number. If the hot spots make it one hundred times more likely for a virus to insert its package, then the probability for the four identical but independent insertions is still one in 10^{21}: still spectacularly steep odds. One could also ask: if such hot spots do occur, did God create them that way and, more importantly, why would he do so? Is this again an intentional deception on his part intended to test our faith?

Third, the retrovirus inserted itself into a germ cell (egg or sperm) of one of our common ancestors, and in that way became a part of the host genome which was then passed on to its progeny. If we find some of these sequences in human DNA and in chimp DNA, this tells us the insertion occurred at a point in time *before* our two different species diverged from our common ancestor. If we don't find that particular insertion in the ape or orangutan DNA, this suggests it occurred *after* those two species diverged from our common ancestor. There is also the possibility that the insertion occurred before all these divergences, leaving all four species with the same proviral insertion, and that the apes and orangutans simply lost them during the normal mixing and matching of maternal and paternal DNA that are characteristic of sexual reproduction. That wouldn't change the main inference that we're making here: that we all came from a common ancestor.

It's important to recognize that it is possible to lose a proviral sequence, because their absence has sometimes been the basis for anti-evolution rebuttals. For example, one prominent YEC website called attention to the discovery of a specific type of proviral sequence within a particular human gene: the one that codes for complement C4 gene, a protein which plays an important role in our immune defense system.[18] This proviral sequence was found in the exact same position in orangutans and green monkeys, but was absent in chimpanzees or gorillas. Given that chimps and gorillas are believed to be more closely related to us than the orangutans or green monkeys (that is, the former diverged from our common ancestor more recently than the latter), one should normally have expected the opposite finding. To the YEC authors, this clearly refuted the idea of common descent, or at least the idea that the human and chimp lines branched off together separately from those of the orangutans and monkeys. But an alternative explanation is that the sequence originated in a very ancient common ancestor, one from which the humans and all four primates arose (each

---

18. Dangel et al., "Complement Component," 41–52.

possessing that proviral sequence), but that the chimps and gorillas simply lost their copy of the proviral sequence in the course of the normal mixing and shuffling of maternal and paternal DNA during sexual reproduction. For example, not every child inherits a genetic disease from their parents, even if both parents exhibit the disease (two parents with cystic fibrosis, a genetic disease, can produce a perfectly healthy child). It's hugely more likely to lose a proviral sequence—because whole segments of chromosomes containing thousands of base pairs can be shuffled out of the mix—than it is for another species to randomly gain the exact same proviral sequence at the exact same position, for the very reasons given in the paragraphs above addressing the probability of this kind of coincidence happening spontaneously.

Pseudogenes

The discovery of several pseudogenes within the human genome also argues strongly in favor of a common descent with several other species.

One gene found in the genome of many animals is that which codes for the enzyme L-gulono-γ-lactone oxidase (GLO, or GULO), the enzyme which is necessary to make vitamin C. In humans, this gene is dysfunctional. For this reason, we need to continually eat citrus fruits like oranges in order to get our vitamin C and remain healthy. In ancient times, sailors would come down with scurvy if they went too long without citrus fruits like limes, which is why some people called them limeys. The GLO gene of other primates and in guinea pigs is also dysfunctional: what's particularly interesting, though, is that many of the mutations are the same in the human and primate pseudogenes, but quite different in the guinea pig genome. This is exactly what one would expect if the guinea pig line diverged a very long time ago from the other animals and then experienced one kind of mutation, while the common ancestor for humans/primates diverged from the other animals later in time and experienced a different kind of mutation.

Another example would be the genes which in egg-laying animals produces the major protein in egg yolk (vitellin), or the ones that make tails in the fetus of many mammals: those are still present in the human genome, but they're switched off (p. 164). One interpretation is that we inherited our genes from our predecessors, but some of them have since been mutated or otherwise turned off and yet we continue to pass them on (common descent). The alternative is that God just made it look that way (common design).

Many animals have a highly developed sense of smell. They need this heightened ability to smell in order to detect the presence of either predator or prey, or to select potential mates. In some cases, not only does their sense of smell greatly surpass our own, but they're even much more sensitive than any machine we can make: this is why dogs are often used to sniff out drugs, bombs or dead bodies. Their powerful sense of smell is not due to the shape or size of their noses, or some other such feature. It owes to one simple fact: they have a greater diversity and number of genes for proteins which actually detect a given unique smell. Smells are made up of different types of molecules. Most readers will know that methane (CH_4, a carbon atom surrounded by hydrogen atoms) smells very different from ammonia (NH_4, a nitrogen atom surrounded by hydrogen atoms). If you substitute just one of the hydrogen atoms in methane with an acid group, you get acetic acid (CH_3COOH), which smells entirely different yet again (it's the chemical which gives vinegar its distinctive smell). If we modified that acid group by essentially replacing the oxygen atom with a couple hydrogen atoms, you get ethanol (CH_3CH_2OH), which smells entirely different yet again. Many other smells with which we're familiar are comprised of various mixtures of molecules.

Animals have developed special detectors for each of these different molecules: we call these olfactory receptors. (For those familiar with the concept of receptors and ligands, the detector is simply a G-protein coupled receptor—called an odorant receptor—for which the ligand is a specific chemical configuration.) Some of the olfactory receptors are less discriminating than others, responding equally to several different but structurally similar chemicals. Others are highly specific for one particular chemical. Each olfactory receptor is encoded by its own unique gene. Humans have nearly a thousand genes for olfactory receptors, but more than 60 percent are inactivated and functionally useless.[19] Mice, in comparison, have almost 1400 such genes, and only 20 percent are pseudogenes. Apparently our ancestors had an exquisite sense of smell comparable to that of many animals today, but we lost that ability through mutations to many of our olfactory receptor genes, possibly because of adaptive changes which equipped us with other means to hunt prey and fend off predators.

The Quantum Leap: Paleontology and Genetics Synergize

I was absolutely floored by reading Svante Pääbo's book *Neanderthal Man: In Search of Lost Genomes*.[20] It gives a very readable account of his research career, with a particular focus on how it culminated with a sequencing of the Neanderthal and Denisovan genomes. It's an excellent and provocative read for any nonexpert who isn't immediately intimidated by scientific details. Here's my thumbnail sketch of the story.

After learning about the amazing powers of genetic sequencing done on living samples, Svante set about to applying this technique to dead samples. He knew that genetic material is rapidly broken down once the cell dies, and the cellular processes which maintain and repair its genetic material cease. The same thing happens to proteins, unless one takes care to fix the carcass with preservatives (many anatomy students will easily recognize the smell of formalin). But he also knew that it was possible to detect pieces of proteins in dead animals; scientists had already been doing

---

19. Niimura and Nei, "Olfactory Receptor Genes," 12235–40.
20. Pääbo, *Neanderthal Man*.

this with samples taken from museum specimens or mummified remains in pyramids and tombs. Therefore, he wondered whether it might be possible to still get small bits of DNA from those same dead specimens. Skip forward several chapters in his book and we find out that he takes the scientific community completely by surprise by developing the techniques to extract, clone and sequence long stretches of DNA from not only animals that have been dead for a century (the thylacine, or marsupial wolf), but for tens of thousands of years! In the process, he obtains samples of Neanderthal bones found in Croatia and elsewhere and sequences first their mitochondrial DNA, and later their genomic DNA. Later on, he makes contact with the discoverer of the tiny little Denisovan finger bone and sequences that DNA. Altogether, he showed that the human, Neanderthal and Denisovan genomes are incredibly similar. In fact, Neanderthal mitochondrial DNA differs from that of humans at only approximately 202 nucleotide positions, while that of Denisovans differ at approximately 385 positions[21] (to put this in perspective, the number of differences with chimp mitochondrial DNA numbers in the thousands[22]). We're also highly related in our genomic DNA (the DNA found in the nucleus, not the mitochondria). In Svante's own words:

> Together with the new data from the 1,000 Genomes Project, these two archaic genomes of high quality now allow us to create a near-complete catalog of sites in the genome where all people today are different from Neanderthals and Denisovans as well as from apes. This catalog contains 31,389 single nucleotide changes and 125 insertions and deletions of a few nucleotides. Of these, 96 change amino acids in proteins, and perhaps 3,000 affect sequences that regulate how genes are turned on and off. There are surely some nucleotide differences, particularly in the repetitive parts of the genome, that we have missed, but it is clear that the genetic "recipe" for making a modern human is not very long. The next big challenge is to find out what the consequences of these changes are.[23]

Comparisons of that mitochondrial DNA, as well as parallel comparisons made with chimpanzee mitochondrial DNA, tell us that humans and Neanderthals both emerged, in two separate migratory waves that came out of Africa, from relatively small populations of a few tens of thousands of individuals.

Why Are Humans So Special among Our Genetic Cousins?

One concept that all too often needs to be clarified for non-scientists is the idea that evolution is not exclusively linear. Too often, the image in their minds is of one type of animal producing another, which in turn produces a third, all the way up till we come

---

21. Ibid., 229.
22. Arnason, "Comparison," 145–52.
23. Ibid., 252.

to humanity, in a long *line* of evolutionary changes. It is this kind of thinking that leads some of them to point out in opposition to evolution theory that "we don't see monkeys turning into humans anymore, do we?" The global popularity of the widely recognized iconic image of human evolution—the one with an ape crawling on all fours on one side of a sequence of hominids and a fully walking human on the other side, with two or three intermediates in between—only exacerbates this misperception.

That iconic image misleads people about another important fact. We humans are not alone at the end of a continuum. We have many cousins: Neanderthals; Denisovans; other subspecies of *Homo* such as *H. antecessor, H. erectus, H. floresiensis, H. habilis, H.heidelbergensis, H. naledi*. We now know that the DNA of the first two of those was nearly identical to our own. They travelled in small groups, and made stone tools, and buried their dead, just like we did. We're genetic cousins in a long genealogy going back millions of years. Yet only one of us survived!

In case the reader doesn't understand exactly what I'm getting at, think of how many species of cats there are: leopards, lions, tigers, lynx, panthers, and of course, the housecat on our lap. Or how many different species of birds: not just obviously different ones like humming birds and ostriches, but even very closely related ones which are still completely different species. Like finches: zebra finch, goldfinch, Darwin's finch, Timor finch, hawfinch, rosefinch, and bullfinch. I could go on for a long time listing more. The same goes for whales, horses, worms, trees. Even among our closest genetic relatives—great apes—there are still six different species alive today: chimpanzees, bonobos, two species of gorilla and two of orangutan.

I don't think there's any other kind of living thing on earth for which there's only one single representative species.

Just to clarify what might be a point of confusion for some: we humans may have different colors of skin, but that just distinguishes different races within one human species, just like Sphynx, Siamese, Manx, and Persians are different breeds of the house cat species, while terrier, greyhound, shepherd and chihuahua are different breeds of the domestic dog.

We are the last remaining subspecies of *Homo* among several others that have since died out. That fact raises the interesting question: "What makes us so special?" Why did we survive, but not any of them? Was it simply a matter of superior evolutionarily fitness that allowed *Homo sapiens* to continue into the modern era but not any of our many genetic cousins? Or was it simply Divine protection?

It wasn't necessarily just our ability to make tools, because we have reason to think that some of the others also made tools. Perhaps it might have had something to do with our capacity for language: we really know nothing about that capacity in the other hominids, except that Neanderthals possessed a gene which in modern humans has been linked to speech.[24] Besides, we know that today many non-hominid species make and use tools and have some capacity for language, including many primates

24. Fisher and Scharff, "FOXP2," 166–77.

(chimpanzees, orangutans, Capuchin monkeys, macaques), elephants, crows, Bottlenose dolphins, sea otters and many others. I will cover this point in greater detail later on pages 215-219.

It wasn't just our brain size that saved us, because Neanderthals also had very large brains (in fact, larger than our own).

Maybe we were the most ruthlessly murderous of our clan, and managed to out-kill them before they snuffed us out. Sibling rivalry can be brutal. Neanderthals flourished throughout Europe and Asia for hundreds of thousands of years, but then disappeared forty thousand years ago, right around the time that we encountered them. Coincidence? A recent report[25] gave a bit of a coroner's report on a murder which apparently took place 430,000 years ago: grisly evidence that violence is an intrinsic part of the earliest human culture. Denisovans were around a little more recently but then disappeared for reasons that are not entirely clear.

The Family Photo Album, part 2

The general picture we now have is that hominids first appeared in Africa fifteen to twenty million years ago. An archaic human line and the chimpanzee line diverged approximately five to six million years ago (gorillas and orangutans having diverged well before that). The former gave rise to *Australopithecus* and *Ardipithecus* approximately three million years ago, and then from them the *Homo* line. *Homo* left Africa in three waves. Several hundred thousand years ago, one group of *Homo* expanded into Europe and gave rise to Neanderthals. Another group headed eastward into Asia and gave rise to Denisovans. Other *Homo* stayed in Africa and evolved into Cro-Magnons (our own ancestors), who then left Africa about forty thousand years ago and encountered Neanderthals and Denisovans. We obviously interbred with them (since we can still see traces of their DNA in our genomes) and possibly also lived with them in uber-cosmopolitan communities far more culturally diverse than our major cities today, beginning to dabble in the arts (musical instruments; wall paintings), and ponder philosophy and theology (buried our dead, sometimes with tools, personal items, food). Or maybe we just killed them off as we continued to expand and fill the entire globe. Finally, ten thousand years ago, we see an explosive growth of ancient technology in the anthropological and archaeological records.

This schema is not built up from just a handful of teeth, a few partial skulls and a couple femurs; it's even completely misleading to claim that the known fossil remains of man's ancestors are so few in number that they would fit on a billiard table or in a coffin, as some still do. In addition to a large pile of hard fossil evidence, we have the story written right into our genome, which we're now learning how to read like a book.

25. Sala et al., "Lethal Interpersonal Violence," e0126589.

Conclusion

Recall my analogy which opened up the introduction of this book: the one in which I was marking student essays on American history and finding unequivocal evidence that one student copied from another, or that the two collaborated to produce the one essay which they both submitted, either way hoping I wouldn't notice the similarity. But I did notice, because of the highly peculiar similarities between them. And there was only one obvious conclusion. To doggedly evade that conclusion would be cowardly and irrational.

In the exact same way, the accumulating evidence is unequivocal when it comes to the common descent of humans and other primates.

The overall distribution of our genomes and theirs is nearly identical, analogous to the overall lay-out of the two student essays.

The sequences of the individual genes are almost identical (just like the actual sentence structures within the student essays).

The human and ape genomes feature many of the exact same types and placement of mistakes (pseudogenes; proviral sequences), just like the typographical and grammatical errors within the student essays.

Many mistakes in the human genome—such as the inversion of chromosome 2—are best explained by a simple copy/paste error of our ancestral DNA.

How else can one explain these similarities between the human and primate genomes, and their stark differences with those of other animals, increasingly so as one goes down the Tree of Life? It's such an inconceivably long poker bet that it happened by itself. The only other explanation is that God made it look that way. But why would he be so deceptive?

7

Christian Objections to the Evolutionary Model

Two millennia ago, a cosmology that was built solidly upon "a plain reading of Scripture," and vigorously defended in the same way, was eventually discarded because its authors acknowledged that there was no point in resisting observations made from nature around it that did not cohere. They gave way to irrefutable evidence from the scientific world (chapter 2), despite the fact that the changes seemed to conflict with Scripture and raised theological consequences which were seen at that time to be quite dreadful.

Today, we're perfectly comfortable reading those Scriptures—the ones speaking of flat earths built on solid foundations, and of domes holding back the waters above, and of souls living in underground caverns—and attributing those ideas to the unsophisticated and misguided thinking of people at that time.

But we haven't stopped learning about the world around us. God has gifted us with curiosity and intellect and other abilities to probe the hard questions of life. And in the process, we've now collected mountains of hard data which directly challenge another origins question. This time: the origins of life, and species, and humanity.

Once again, the data we have in hand conflict with the scriptural account, or at least with certain interpretation(s) of that account.

"A Plain Reading of Scripture" versus a Plain Reading of Science

The general picture one might take from the Bible is that all humanity arose from one man, named in the first two chapters of Genesis as Adam. Genesis also names his partner as Eve. Furthermore, passages in Genesis have been used to conclude that these two lived roughly six thousand years ago somewhere around the Persian Gulf.

And yet far too many very diverse lines of scientific evidence are simply incompatible with that picture.

The relatively limited genetic diversity that we find in the human genome tells us that the ancestral populations of *Homo sapiens* never numbered less than a few thousand at any given point in history.

Examination of our mitochondrial and Y-chromosomal DNA indicates our earliest common ancestor(s) lived several hundred thousand years ago.

The fossil record, corroborated by genetic analysis of the DNA from those bones, as well as comparisons of our mitochondrial DNA, all indicate we originally arose out of Africa.

Finally, and most controversially, side-by-side comparison of the human genome with that of other animals shows that we are highly genetically related to other long extinct subspecies of *Homo* (Neanderthals and Denisovans) as well as to the great apes (chimpanzees, gorillas, orangutans).

Objections to This Evolutionary Model of Human Origins

Some theists will be able to accept that plants and animals arose by processes of evolution (though many of them will be quick to point out that it was not an unguided evolution: that God was involved in some way, actively guiding it to the final outcome we see today).

But a large number of those theists, and certainly all YEC believers, will draw a line when it comes to humanity. They will all insist that humans did not—indeed, could not—arise by evolution from an ancestor that we share in common with the great apes. They will raise several theological implications which they feel absolutely preclude it as an acceptable explanation.

Interestingly, many of these objections are the very same as those raised against the heliocentric theory, and I've reproduced them here verbatim from page 36:

Objection #1: "It contradicts Scripture." Once again, at the top of the list is the point that the scientific model of mankind's origin conflicts with the Genesis accounts. This cannot be stressed enough. Scripture paints a very clear picture of the origin of humanity. "And Scripture is God's word. There can be no debate on that. If science contradicts it, then science must be wrong."

Objection #2: "It goes against church tradition." We've always believed in the old, old story. If it were incorrect, why would God not have made that clear to us? How could our church fathers, his shepherds, have been so wrong for so long? The Spirit within us doesn't resonate with this model based on the theory of evolution.

Objection #3: "God would not work that way." Evolution theory absolutely requires death, predation, and a desperate struggle against dying or being eaten. It also requires a mechanism built upon making mistakes: mutations in the DNA leading to changes in phenotype, whether they be favorable or deleterious. This is not how a loving or perfect God would do things.

Objection #4: "It means we're not unique among all created things." We believe we are distinct from all other created beings: we couldn't possibly have derived from them through small incremental changes. Evolution theory teaches that all living organisms are simply slightly altered versions of some other organism which may be still extant or have since gone extinct. It says we are just one of the many different species on earth. Not even the first one, or last one, or most complex one, or best one. Just one tiny twig on a branch of a very highly branched tree.

Objection #5: "It challenges the idea that we are created in the very image of God." How can we reflect his image if we're simply glorified monkeys?

Objection #6: "When God finished creating on the sixth day, he declared it all 'very good.'" Parasitic zombie wasps laying their eggs inside caterpillars which then serve as living lunches for their baby wasps; weeds and thistles that infiltrate gardens; cancers in children. This just does not sound "very good."

Objection #7: "Paul and Jesus both clearly believed Adam was a historical figure." Paul specifically refers to Adam in Acts 17, Romans 5, 1 Corinthians 15 and 1 Timothy 2, as did Jesus when he confronted the Pharisees.[1] One can hardly find greater authorities than these two.

These seven objections are exactly the same as those levelled against the heliocentric theory. And yet we, the church, were able to eventually move past the perceived theological barriers to that theory and embrace the new science together with our faith. But there are many additional, distinct objections to the theory of evolution:

Objection #8: "Evolution isn't a fact, it's only a theory. 'Theories' don't have the certainty of a Scientific Law, or anything like that."

Objection #9: "If we evolved from monkeys, then we're no better than animals and should just throw all morals and laws out the window and do whatever we please."

Objection #10: "Why would God use a process that took millions or billions of years to fully play out?" Why not simply snap his fingers and bring it all into existence in six days?

Objection #11: "The Bible specifically says 'each according to their kind.'" It doesn't say anything about some kinds arising out of other kinds.

Objection #12: "Genetic change was brought on by the fall and the flood." Atheists point to the huge genetic changes which have happened as evidence for a long and gradual evolution. But they could also have been caused in the flood described in the seventh chapter of Genesis.

Objection #13: "There are too many gaps and 'missing links.'" The theory of evolution involves a long and gradual change, and so we should see all kinds of intermediate or transitionary forms. But we don't. All we have is either this animal, or that

---

1. Matt 19:4–6; Mark 10:6–9.

animal: either apes on the one hand, or humans on the other. But never something in between.

Objection #14: "I believe in micro-evolution but not macro-evolution."

Objection #15: "This takes away the urgency of evangelism."

These theologically and philosophically based rebuttals simply do not change the fact that we now have a profusion of physical evidence—simple facts obtained by direct examination of God's creation all around us—for an evolutionary origin of mankind. In the next many sections, I'm going to revisit each of these rebuttals in more detail.

Objection #1: It contradicts Scripture.

Some in this camp will even go so far as to claim that the Bible speaks against such an idea. Actually, it doesn't. It is true that Scripture does indeed provide a different explanation than the one proffered by evolution theory, but that isn't the same as "speaking against" other naturalistic explanations. The Bible also portrays God as having storerooms in the sky full of snow, and others full of lightning,[2] but it doesn't "speak against" any meteorological explanation which invokes ideas of solar energy, temperature gradients, atmospheric pressure and moisture content.

It can be possible for two or more concepts which appear to contradict to both be correct at the same time. Most theists accept the concept of the Trinity: that God is one, and yet exists in three persons. These two statements contradict, but the one doesn't speak against the other. Most Christians will confess that Christ has already accomplished all that is needed for our salvation, and yet we are being saved[3] and are also looking forward to our future salvation.[4] Physicists once battled over whether light constituted particles or waves. They now see that it is both. Again: apparently contradictory but not necessarily mutually exclusive.

And yet, despite all this hand-waving about how things may not be contradictory, or how contradiction doesn't matter anyway, it can't be denied that the traditional, biblical version and the current evolution model do clash in a way that is fundamentally unsettling for many Christians.

If the Genesis account is taken to be God's authoritatively historical version of the way events transpired, then we do indeed have a problem. Both versions can't be right. (Which is not the same as saying one is wrong and therefore the other is right: logically speaking, both could be wrong).

On the other hand, perhaps this is a category error. As Peter Enns put it: "This could invite a strong response: 'You are putting science over Scripture.' But that is not the right way to frame the issue, for it incorrectly assumes that Genesis speaks to

---

2. Job 38:22–25.
3. 2 Cor 2:15; Rom 13:11.
4. 2 Tim 4:18; 1 Pet 1:9.

scientific matters. It does not."[5] The Genesis account and other supporting passages of Scripture need to be taken less rigidly and more contextually.

Many now accept that the Hebrew account was written within the zeitgeist of the Akkadian, Sumerian, and Egyptian civilizations, and was written as a polemic against it.[6] Genesis taught that there is only one God, not many (a point which causes some to stumble over the plural form of the word *elohim* for the Divine Council: "Let *us* create mankind in *our* image"[7] [emphasis added]). That God created everything in a peaceful benevolent way (speaking it into existence), rather than through acts of violence (the Babylonian version had Marduk creating the world by splitting open the carcass of the goddess Tiamat, whom he had killed in a battle). That humans were created to be co-regents with God over a garden of paradise, rather than as servants or slaves of some fearsome dictatorial pantheon of gods. That God provided food for people, rather than creating the humans to bring food and sacrifices to the god(s). That the sun, moon, stars, water, plants, animals—in fact, everything—were all merely created things to be enjoyed, not powerful deities to be feared. Genesis doesn't even give those foreign deities the honor of having names: "God made two great lights—the greater light to govern the day and the lesser light to govern the night."[8] It doesn't use the obvious words "sun" and "moon" (in Hebrew, *Shemesh* and *Yereakh*, respectively), which would also be the names for those Egyptian and Babylonian gods, but opts instead for "greater light" and "lesser light." They were created to serve humans (as markers for festivals, or to indicate day versus night), rather than humans needing to serve the celestial bodies / gods.

This embodies the true meaning of the word myth. Too many people think that myths are false or untrue. Instead, the actual meaning of the word myth is a story intended to convey a deeper, greater truth than just a superficial reading.

Objection #2: It goes against church tradition.

Yes it does. Undeniably. But so did the heliocentric theory, and the idea that the universe is only six thousand years old, and that the souls of the dead reside in underground caverns. But scientific evidence convinced holders of those tenets that perhaps it was time to reconsider.

In the early days of the NT Christian church, there were rules based on a long Jewish tradition regarding the length of hair, or about the necessity of circumcision, or about eating unclean foods or meat offered to idols. Later, the Christian church relaxed and eventually discarded all of these rules.[9] (I know there are theological

---

5. Enns, *Evolution of Adam*, 57.
6. Pinnock, "Climbing Out," 143–55; Harlow, "Creation according to Genesis," 163–98.
7. Gen 1:26.
8. Gen 1:16.
9. E.g., compare Acts 15:20, 1 Cor 8:4, and 1 Cor 10:23–33.

reasons for this, but my point is that the church proved itself able to set aside tradition if there was a good reason to do so).

Other traditions have since emerged. In the relatively recent past, church leaders taught that it is possible to effectively pay for a loved one to get into heaven through the Purchase of Indulgences, a practice which finally led Martin Luther to nail his Ninety-Five Theses to the door of the Wittenberg Castle church and launched the Protestant Reformation. That too was discarded. Church traditions do come and go.

The YEC tradition has not always dominated. In fact, it is a relatively recent phenomenon, primarily a Western and even North American one. It was not held by all the great church fathers. Augustine and Origen both challenged the literal interpretation of the creation account that is promoted by many YEC believers today. It vacillated for many centuries, depending on a variety of influences, not the least of which were political motives, and then ballooned into prominence a few hundred years ago in response to the advances of science, particularly geological discoveries which challenged the story of Noah's flood.

Today, some value more highly a straightforward reading of the English words of the Bible, upon which they superimpose their modern North American viewpoints and certain theological assumptions, and shun the input of trained scholars. Those scholars have taken the time to learn the original language of the writers and readers of the texts, to understand the cultural conditions in which they were written, and to learn the history which preceded it and therefore would have influenced the writing and interpretations. For example, in the modern era, the Depression and the Second World War both separately had major impacts on the thoughts and actions of people for many decades after the fact. The scholars are in this sense "experts." And it is interesting how often trainees soften or even set aside their Fundamentalist beliefs once they go to seminary and learn those original languages, cultural conditions, and history. This is not an indictment against the "liberalizing" effect of scholarship. Instead, could it say something about the fragility of Fundamentalist thinking which cannot stand up against concrete facts and honest investigation?

Tradition is not a good enough reason to stay the course and reject change. When John's disciples asked Jesus why his disciples didn't follow a certain religious tradition, Jesus answered: "No one sews a patch of unshrunk cloth on an old garment, for the patch will pull away from the garment, making the tear worse. Neither do people pour new wine into old wineskins. If they do, the skins will burst; the wine will run out and the wineskins will be ruined. No, they pour new wine into new wineskins, and both are preserved."[10] Sometimes tradition has to give way to a greater, deeper understanding of things. Science hardly changed for thousands of years, and any advances that were made certainly didn't trickle down quickly to the various authors of Scripture. It moved slowly during the Dark Ages, and picked up pace considerably during the Renaissance. Today, massive changes in our knowledge and the applications to daily

10. Matt 9:16–17.

life are occurring at breath-taking speed. The old church fathers never had to deal with the kind of scientific evidence we now have all around us: evidence such as the geologic record, fossils, radiometric dating, and the genetic codes.

Perhaps it's time to change the wineskins.

Objection #3: God would not work that way.

The ancient Greeks and the early Christian church dismissed elliptical orbits and craters on the moon simply because they presupposed that the Divine would only make things perfect, and they elevated themselves to the position of presupposing the definition of what "perfect" was: to them, perfection was circular and spherical, not oblong. But that was merely their ignorance and biased thinking, and they were wrong. Most orbits are not precisely circular, and most celestial bodies are not spherical.

In the same way, some today dismiss death, predation, and struggle as being part of the process of bringing living things into being because these things make us feel uncomfortable, and we want to define what should be "the right way" for God to act. This too is just presupposition: constraining reality to fit our biases.

Many refuse to accept evolutionary theory on the basis that death is the penalty for Adam and Eve's sin in the garden of Eden, a major component of Paul's theology regarding the redeeming work of Jesus Christ.[11] As they rightly point out, natural selection is a fundamental component of evolutionary theory: random mutations occur in each generation of a given species, some of which eventually confer an advantage on holders of those mutations during some kind of environmental change or stress. Those advantaged individuals are slightly more likely to survive and pass on their new genetic information, while those without the mutations are more likely to die off. Survival of the fittest. So death becomes one of the sieves that separates out the fit from the unfit. And it takes thousands of generations—that is, thousands of rounds of dying—to see even a few changes in one animal species, let alone all the species that have ever existed.

However, those that reject evolution theory insist that death could not have existed until after the creation story, based on Genesis 2:17 . . . *"but from the tree of the knowledge of good and evil you shall not eat, for in the day that you eat of it you shall surely die"* (NASB). And if there was no death, there could not have been millions of generations of individuals gradually gaining selective advantages over other generations.

I disagree completely, and will give three arguments from Scripture, and two others from biology, supporting the idea that biological death had to have existed before the fall in the garden.

---

11. Rom 5:12–19.

First, God could not have been referring to physical death, because God's warning was that *"in the day that you eat of it you shall surely die,"* yet the Scriptures go on to tell of how Adam and Eve do in fact eat from that tree and live physically to see another day. In fact, a very great many number of days: according to the Scripture, they go on to live long enough to produce numerous generations of children, including Cain, Abel (murdered by Cain), and Seth (who was said to have been born when Adam was 130 years old),[12] before dying at the ripe old age of 930 years.[13] And it wasn't a mental, emotional or intellectual death, because the text says they went on to be: fearful of God himself,[14] or of other people;[15] and to be ashamed of their nakedness;[16] angry;[17] and murderous.[18]

Instead, another interpretation of Genesis 2:16 is that the day they ate of the fruit they died spiritually. I don't think anyone would contest the latter idea: many, many scriptural writings talk about it in one way or another. Confusing biological life/death with spiritual life/death is the same mistake that Nicodemus made when Jesus taught him about the need to be born again.[19] Nicodemus couldn't understand, saying that a man *"cannot enter his mother's womb and be born a second time, can he?"* Jesus corrected him by distinguishing between being born of the flesh versus being born of the Spirit. If Genesis 2:16 is referring solely to a spiritual death, then there's no problem with the idea that physical, biological death was indeed a fact of life before the fall.

The second point is a bit related to the first one. Some use scriptural passages like Romans 5:12, which says, *"Sin entered the world through one man, and death through sin, and in this way death came to all people, because all sinned,"* as evidence that there could not have been millions of generations of plants and animals dying over the billions of years of evolutionary history. But to be precise, that verse in Romans only refers to death in humans: it says nothing about death being absent in the plants and animals before that event in human history. Likewise, God's warning of fatal consequences in Genesis was directed at Adam and Eve alone, not at the whole living world in general. To put this point in a modern perspective: our own various levels of government impose all kinds of laws and taxes on people, but those are not enforced upon our pets or plants.

Third, Genesis describes two specific trees in the garden of Eden, one of them being "the Tree of Life,"[20] and also records God giving them permission to eat from that

12. Gen 5:3.
13. Gen 5:5.
14. Gen 3:10.
15. Gen 4:13–14.
16. Gen 3:10.
17. Gen 4:5.
18. Gen 4:8.
19. John 3:1–21.
20. Gen 2:9.

tree;[21] but then later, after the story of the fall, God is quick to say that Adam "*must not be allowed to reach out his hand and take also from the tree of life and eat, and live forever.*"[22] Unless you assume (without scriptural support) that the fall happened so soon after Adam and Eve were created that they hadn't yet gotten around to eating from that Tree of Life, even though it was "*in the middle of the garden,*"[23] it's entirely reasonable that they would have already eaten from that tree more than once. This leads some to conclude that people were subject to the processes of death even before the fall, but that eating from the Tree of Life reversed or prevented those processes.[24] This idea that the Tree of Life was necessary as a tonic against the ravages of death reappears in the book of Revelation when it describes a river of life flowing out of the new Jerusalem: "*On each side of the river stood the tree of life, bearing twelve crops of fruit, yielding its fruit every month. And the leaves of the tree are for the healing of the nations.*"[25]

Then there's also a perfectly logical reason from biology to say that God would have actually *wanted* to incorporate physical/biological death as a necessary process into his original creation, especially if there was the possibility that creation would carry on for more than a few seasons before someone took a bite from the proverbial apple. At the very least, there would have been a lot of dying of plants in order to feed all the animals and people that he had created. But more importantly, the Scriptures have God issuing commands to every living creature, including humans, to "*be fruitful and multiply*" and to fill all the earth. With all those animals being fruitful and multiplying but not dying, the earth would very quickly just be overrun with a seething mass of flesh that choked the skies, the land and the seas, and which the earth's plant life just could not possibly sustain. It would have been an irresponsible act on God's part: like building a car without brakes, or a bathtub without a drain.

Finally, we also know today that many normal biological events *require* death. The transformation of a caterpillar to a butterfly involves a whole cellular reorganization, and some of the caterpillar's cells must die in order to accomplish the feat. Likewise, in normal fetal development, some cells serve merely to act as signal lights and guides for other cells which need to shuttle from one part of the growing fetus to the other, or which have to grow in a very specific direction to some particular target (for example, a nerve fiber growing from the brain all the way down to the toe), after which those guide cells must die to make room for other cells and/or to stop consuming precious metabolic resources. Inflammation, immunity and tissue healing all involve certain cells "choosing" to die in order to protect and rebuild the body. Death is actually programmed into our genes! Scientists are actively studying all

21. Gen 2:16–17.
22. Gen 3:22.
23. Gen 2:9.
24. Walton, *Lost World of Genesis One*, 99–100.
25. Rev 22:2.

of these diverse forms of "programmed cell death," which they have given the Greek name "apoptosis." At the cell level, life without death is cancer.

Altogether, then, to say that physical/biological death could not have been part of the original creation absolutely demands a reconsideration of what God meant when he said, "*For in that day*," as well as a complete head-in-the-sand approach to the advances in our understanding of normal cellular biology.

As for predation, the very skeletal design of many animals, especially of the dinosaurs, with abundant canine teeth and claws designed for ripping and tearing, as well as defensive armor such as bony plates and horns on the head, neck and back, all clearly attest to a violent past long before humans could have lifted their hands to the Tree of the Knowledge of Good and Evil. The fact is, we have many fossilized remains which bear the marks of predation by animals—bite marks on bones—millions of years before humans came onto the scene.

Time and time again we've learned that viewing predation as something bad or evil and allowing that to guide our actions has led to catastrophic and unwanted outcomes.

A perfect example of this has been documented in Yellowstone National Park.[26]

During the late 1800s and early 1900s, wolves were progressively exterminated from that area because humans were afraid of them, because they killed our livestock, and because we didn't like what they did to nice, gentle elk. By 1970, there was no evidence of any wolf population at all. Which was great for farmers and hikers. But what the Park Rangers and farmers also noticed was that the elk population went out of control, and was getting increasingly unhealthy, with more and more elk exhibiting obvious signs of sickness and injuries. What's more, the groves of aspen, willow and poplar were being nibbled down to the ground because there were too many elk browsing on them, and those elk no longer had fear of forests and brush (where wolves could more easily catch them). The loss of those trees removed nesting places for songbirds as well as a food source for beavers, and the populations of both dwindled. And with fewer beaver dams being built and repaired, the whole aquatic ecosystem was changing: water was rushing through the area and eroding the river banks, and marsh-life was disappearing.

Then in the 1970s, the Forestry Service began reintroducing wolves, against much public opposition, and saw the reversal of *all* the changes mentioned above. The exact same sequence of events were noted in Zion National Park (Utah), Wind Cave National Park (South Dakota), Yosemite National Park (California), Olympic National Park (Washington) and Jasper National Park (Alberta) when the top predators (wolf, cougar, lynx) were decimated or eliminated, and then reintroduced.[27] Ecologists refer to these top predators as "keystone species" referring to the keystone of a stone

---

26. National Park Service, "Wolf Restoration Continued."
27. Stolzenburg, "Lords of Nature."

archway that holds all the other stones in place: remove that keystone, and the arch comes tumbling down.

Other detrimental population imbalances were caused by introducing organisms into an environment which didn't have natural predators to keep them in check.

When jack rabbits were brought to Australia in the 1700s, they quickly proliferated to staggering proportions because there were no natural predators, and soon caused crippling economic costs through crop damage. Open season was declared on them, but no amount of shooting could control them. Special fences were designed to exclude them: again, to no avail. Finally, in the 1950s, biologists introduced a virus (myxoma virus) into the population, which severely reduced their numbers. But they quickly developed resistance and recovered substantially. In the 1990s, another virus (calicivirus) was released: the jury is still out on this latest attempt at controlling the problem.

Many similar scenarios have often unfolded in the past—Asian carp in the Mississippi River watershed; purple loosestrife in North America wetlands in the 1800s, a sac fungus (Ascomycota) allowed into Europe and then North America in the early 1900s and causing Dutch Elm Disease; cane toads in Australia in the mid-1900s (in an attempt to control the beetle populations there); and zebra mussels in the Great Lakes in the early 1980s—to name just a few examples.

Another criticism that some have about the theory of evolution is its seeming wastefulness: the perception that 99 percent of all species which have ever existed have gone extinct. They wonder why God would "throw away" so much. But that figure of 99 percent only arises because we humans have categorized organisms into either this pile or that pile: a species with this arbitrary name, as opposed to another similar species with that arbitrary name. In actual fact, all the organisms which have ever existed have all been part of one big Tree of Life, and 100 percent of all the organisms that have ever existed have either died or will die soon.

To illustrate my point here, I'd like to use the Pothos plant. Almost anyone who has house plants in their home or office will have one of these because Pothos can put up with the most unbelievable kinds of neglect and abuse. It's a vine with heart-shaped leaves that just keeps growing and growing and growing. Almost everyone that you or I know will have had one of these at some point, but I'd be willing to wager that not a single one of them started theirs from seeds: instead, they took a cutting from somebody and encouraged it to send out new roots. And I would further venture to say that they took that cutting from another plant that was itself started from a cutting, which in turn was started from a cutting, and so on all the way back to some incalculable time in the past. Envisioning that sequence got me wondering if the plant that sits on my office desk now is actually part of one original timeless primordial plant that goes back to the very dawning of time! But to my point: I would not view those earlier cuttings and plants as having been wasted. They enjoyed their own existence, provided some number of cuttings for other plant owners to enjoy, and eventually died a normal death. That's how I view life in the past, both the individual organisms and their collective species. They had their day in the sun, propagated more of their kind, albeit with some minor changes, and then died. They weren't "wasted." Every organism that has ever died was indeed a member of a particular species, the latter having been given a specific name by humans to help them categorize all life into neat little compartments, but was also a member of a long continuum of life forms.

What about the idea of genetic mutations being a mechanism through which God would choose to work in order to accomplish his goals? Some find this unthinkable. As one who loves to build and tinker and modify, I completely disagree. I take great pleasure in building something and then progressively modifying it to make it even better. This is also the way the vast majority of the inventions we all enjoy came to be. As a scientist, I've heard many presentations of some gene or protein arising from subtle modifications of another such molecule. A little twist here, an extension there, or an addition of some extra piece now yields a molecule with quite different and useful properties. It never ceases to amaze me. I've even published a scientific paper which describes a teaching tool made from paper clips bent into a particular shape such that they capture random vibrations and convert them into forward-movement.[28] I used this to show how tiny little motors inside our cells use thermal energy to carefully move molecules and organelles in specific directions. In the body, some classes of motors run in one direction, but other classes of motors run in the opposite direction. More to the point, though, I showed how adding a very slight bend in one or two of the "legs"—which represents a single minor mutation in the gene for those proteins—reversed the direction of this machine. This illustrates how straightforward it is to produce an entirely different outcome with such a simple change. Geneticists can do such amazing things by slightly adjusting a gene sequence, splicing parts of one gene into another, or moving segments around. Why is it so unthinkable that God couldn't do the same on a much grander scale?

In this sense, just like the Greeks and early church who dismissed craters on the moon as imperfections in Galileo's telescopes simply because of their preconception that the divine would only create perfection, theists today are again trying to explain away the many changes in our DNA as post-creation bomb-craters brought on by catastrophic environmental upheavals: manifestations of the curse put upon us by God after the fall. Simply because of their own presupposition of how God should act.

Objection #4: It means we are not unique among all created things.

Generally speaking, humans are arrogant. Civilizations all through time have portrayed humans as the pinnacle species. The epitome of perfection. Our attitude now is the same as that when the Geocentric Model was being criticized and defended: we see ourselves as being at the very center of all of God's attention, and the universe

28. DoHarris et al., "Molecular Motors," 213–18.

literally revolving around us. The eighth psalm describes mankind as the crown of creation. We believe we're so distinct from all other created beings that we couldn't possibly have derived from them.

But the simple fact remains: we are not unique. We have the exact same physiology as the other animals, and use the same genetic coding. In fact, our own genetic sequence is peculiarly similar to those of the other primates (chimanzees, gorillas, orangutans) as well as to Neanderthals and Denisovans. That genetic record, which God also created within us, betrays a common history. In fact, the situation is exactly the same as the one I described in chapter 1 with my analogy of marking student essays on American history and finding two that are blatant copies of each other. Our DNA contains many of the same pseudogenes and proviral scars that are found in the DNA of other primates. Our chromosome number two is clearly an inversion product of two smaller primate chromosomes. It's absolutely undeniable that the human genome, with all its mistakes and dead ends, is directly related to that of chimps, gorillas and orangutans.

Those similarities could suggest that we were modified from previous—"less evolved"—species. How might one ascertain whether we were specially created using similar design principles (common design model) or descended from an ancestor we have in common with the other hominids and primates (common descent model)? Unless we actually find a cell with God's name scrawled on it, or "Made in Heaven" stamped on it, how could we actually distinguish something that arose by a gradual change versus a sudden snapping of the fingers?

A scientist would ask, "What predictions can we make based on these two fundamentally different models?" Let's think this through.

The slow gradual change involves building new features on top of old ones. Friends of ours have renovated a small old stone school building, turning it into a beautiful house. You can still clearly distinguish the old parts from the new, and the different building materials used. Buried in the walls there may still be pulleys and cables which were previously used to ring the old school bell, or possibly old lead pipes and aluminum wiring from a time before the building codes were updated to copper. In the same way, the evolutionary model (common descent) would predict that our bodies should still reveal certain old designs. Remnants of old mechanisms which are no longer needed, or newer and better ones built on top of them.

And if we are a unique, special creation, there should *not* be places where we see half of an old design and half of a new design. All the parts should be in their proper place and make sense. And if designed by an omnipotent omniscient intelligence, then everything should work perfectly, and make perfect sense (I only add this sentence to deflect comments from owners who bought a brand new house that had all kinds of problems).

How do those two predictions pan out when we look closely at the design of our bodies?

Although we are indeed "fearfully and wonderfully made," we still do find remnants of old mechanisms that make sense in the common descent model but not in the common design model.

As a species, we are relatively hairless, but not completely so. The sparse hairs we do have over parts of our body are equipped with a tiny muscle—the erector pili—which raises the hair up on end when we get cold or when the fight-or-flight response kicks in (when someone jumps out at you from the bushes late at night). Goosebumps. The interesting thing is that none of this seems to serve any useful purpose for humans. We don't have enough hair to keep us warm, and the muscularized hairs don't seem to enhance that function. They certainly don't make us look more threatening. I don't know how to explain them from a special creation point of view. On the other hand, I can easily accept Charles Darwin's explanation from a common descent paradigm: we evolved from ancestors which had thickly furred bodies and used those erector pili muscles to cause their fur to stand up on end either to stay warmer or to make them look more fearsome to a predator or a competitor, and those evolutionary adaptations waned as we hominids evolved.[29]

We also have tiny muscles which are loosely attached to our ears, and which a few people can still use to wiggle their ears or their scalp. These muscles are useful in cats and dogs and other animals that turn their ears toward sounds; primates and humans have learned to simply turn their heads so we can get our more perceptive eyes on the source of the sound. These small insignificant muscles serve no purpose and seem only to be remnants of our evolutionary heritage.[30]

Richard Dawkins has frequently pointed out another example of a vestigial hangover from evolution: the design of the recurrent laryngeal nerve.[31] This nerve passes down the neck and a little past the heart, but then does a U-turn, loops underneath the aorta and returns back up the neck to the voice-box. Almost as if the nerve got lost within the developing embryo but then corrected itself. In humans, this U-turn only adds a few centimeters to its overall journey, for no apparently good reason. But in giraffes, this U-turn adds almost a couple meters; again, for no good reason. To my knowledge, this design doesn't really create any significant problems for us or for the giraffe, but it certainly doesn't make sense from an Intelligent Design point of view. It makes more sense to me that the wandering pathway for that nerve was wired in somewhere in our distant evolutionary past (Dawkins relates this to an original design in the ancestors which gave rise to fish and then the rest of us). Nerves have

29. Darwin, *Descent of Man*, 18–20.
30. Ibid., 13–16.
31. Dawkins, *Greatest Show on Earth*.

genetic instructions to follow certain landmarks within the developing embryo to lead to their intended targets. "Go straight down the neck, take 10 steps past the lungs, hang a left at the aorta and go a few more steps till you come to the kidneys, then . . ." But as the body shape and design changed over millions of years, the target moved a couple millimeters, and it was simpler to just add a small diversion at the end of the journey—"take one step to the left"—than to give it a whole new set of landmarks and instructions. The problem is that as the target continued to move, the small diversion at the end became an increasingly bigger side-trip. Yet it was still easier to simply adjust that last line of code—"take ~~one~~ two steps to the left" . . . "take ~~one two~~ three steps to the left" . . . "take ~~one two three~~ four steps to the left" . . . and so on—than to wait for a mutation which gave a completely new more streamlined set of instructions.

We have genes for making tails at the bottom of our spine. These were useful to animals who swing through trees, but not to their descendants who started walking on the ground. These genes are normally turned off in us *Homo sapiens*, but in a few very rare individuals these genes are reactivated and the person is born with tails up to twelve centimeters long, which can then be removed surgically.[32]

We have many hundreds of genes for proteins which detect subtly different smells (odorant receptors). A very large number of these are never used in humans because they've been inactivated or rendered dysfunctional. Those genes are needed in animals that rely heavily on being able to smell their prey, predators and/or mates. Our dogs laugh at our pathetically poor sense of smell. It doesn't make sense that we'd be created with all those genes that are never used, but it does make sense that we inherited them from ancestors who needed them, and later ours accumulated mutations (which was of little or no consequence because we developed other traits which compensated for those losses).

There are also genes to make the major protein in chicken eggs, or to make vitamin C, or many other functions, all of which are present but dysfunctional in us humans. I find it hard to explain from an anti-evolution point of view why God would have created all humans with a defective gene that only leads to disease (scurvy), but is perfectly functional in the animals.

---

32. Dao et al., "Human Tails and Pseudotails," 449–53; Dubrow et al., "Detailing Human Tail," 340–44.

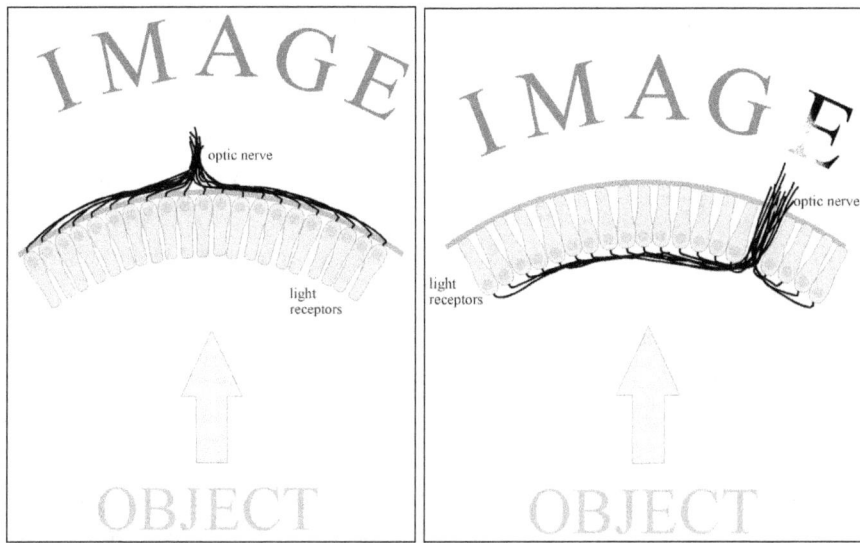

Figure 12. Structure of the retina of the octopus (A) and the human (B). Note how the blind-spot formed by the human optic nerve disrupts the image partially.

Another example which is frequently brought up is the design of the retina in our eye. There are sight-giving cells (rods and cones) which respond to light, turning that form of energy into an electrical signal which can be sent back to the brain. Those cells require metabolic and immunological support from yet other cell types, including the arteries and veins, and the light-detecting cells need to be in contact with the nerve cells which communicate their information to the brain (figure 12). One would think that the best design would have those receptor cells first in line to intercept the incoming light, and then the other support cells and nerve cells behind them (so that the latter don't distort the light signal in any way). Strangely, mammalian eyes have the order of these cells reversed: the in-coming light needs to first make its way through and around all the other cell types before it reaches the light-sensitive layer. This is completely counterintuitive: it otherwise decreases the resolution or sensitivity of our eyes.

A common apologetic rebuttal is that certain of those supporting cells (called Müller cells) are specially "designed" to compensate for this problem:[33] they act like optical fibers to direct the light around the other cells and down to the photosensitive cells. This is then taken to represent a wonderful example of Intelligent Design. However, this wonderful redesign was needed to compensate for what otherwise would have been a bad design.

Adding to the problem, this seemingly backwards cellular organization requires the nerves which transmit the visual signals to now penetrate through the retinal layer. To minimize problems associated with the entire retinal surface being punctured by all the individual nerves inserting through it, the nerves are all bundled together into

33. Labin and Ribak, "Retinal Glial Cells," 158102; Franze et al., "Müller Cells," 8287–92.

one tight cable before they insert through the retina. Again, a nice solution to a design problem, but it doesn't excuse the fact that there was a design problem which needed fixing.

Besides, in so doing, the retinal nerve cable now creates a blind spot on our visual field. A solution to that problem? Displace the blind spot off a little bit to one side where high visual acuity isn't quite as important, thus leaving the center of the visual field undisturbed. Yet another wonderful solution to a problem, but a problem requiring a solution nonetheless.

None of these problems are huge: we humans are able to see quite well despite those design flaws. *But it's still a flawed design*! It didn't need to be that way, and it would seem that a good designer developing a special unique creation could easily have come up with something better. For example, that optical nerve could equally have done its job of accessing those cells from behind. This is how the octopus solved the exact same problem after its ancestors and our own diverged down different evolutionary paths (figure 12). What makes more sense, though, is that the human retina is the product of an iterative process in which a sight-providing mechanism was modified, and then modified again, and modified yet again, over and over during the course of evolution as problems arose, culminating in the design we now have.

Most of us have had the embarrassing experience of getting food or liquid down our wind-pipe and then going through a major coughing fit trying to get that out. For some people, that experience ends up killing them: they choke on that piece of steak they tried to swallow, or drown in their vomit when they're unconscious or otherwise not in control of swallowing. There's no reason why our esophagus and trachea need to converge on the same opening on our face. A good designer could (and should?) keep those two functions and their related machinery entirely separate. However, if we descended from a common ancestor, it would be more economical to add a new design on top of an old one (economical in the sense that it didn't require whole new sets of coding for entirely new anatomical structures). Living organisms first developed various tubular systems to get food and liquids into the body and the digested products out, and otherwise breathed through their skin. Once they became sophisticated enough to require a tubular system to get gases in and out of the deeper parts of the body, it made sense to borrow certain structures which were already in place, and add a simple modification—essentially a valve—to make sure food/liquids went one way and air went another way. But there would be no need to combine these two different systems if one were making a special new creation from scratch.

The same problem and solution arises in considering how the anatomical structures used for getting waste products (urine) out of our body were co-opted to also serve as structures to allow sexual reproduction. Again, there's no intrinsic reason why a designer working on an entirely new creation would have to converge these two very different functions through the same anatomical structures. But evolution would find it to be a very economical solution to getting sperm from one organism into the body of another organism, using a tubular pathway already in place and designed for getting things out of the body of that organism.

Finally, the process of childbirth. In today's Western world, this generally proceeds without any problems or with only minor complications. But that hasn't always been the case in Westernized countries, and still isn't the case in much of the rest of the world. When it goes wrong, it can go very horribly wrong. Before the development of modern medicine, the chance of a mother dying during childbirth was as much as 1 percent. The death of the baby is mercifully a relatively quick one, usually one of asphyxiation. But it's a very different one for the mother. Sometimes, for a variety of reasons, it can become virtually impossible to deliver: the baby hasn't positioned properly or is unusually large, or the mother's spine and pelvis are too small or abnormally shaped. The outcome is painful but fruitless contractions until the mother is literally exhausted to death. Other frequent causes include bleeding to death or becoming infected. These are all horrible outcomes, and they were pathetically too frequent before we developed modern medical procedures to deal with them.

In theory, there are many other less life-threatening ways that a designer working on a special creation—the "crown of creation," in fact—could have solved the problem of delivering a baby. Perhaps one of the least life-threatening options would have been simply providing yet another opening in the belly that didn't require passage through the pelvic structure of the skeleton. Marsupial animals have solved this problem by having the highly immature baby exit when it was still incredibly tiny and finding its way to a pouch on the outside of the mother's body where it develops to maturity. Other animals have opted for laying their eggs in a nest.

On the other hand, this horrible situation of mothers and babies dying in childbirth makes more sense to me in the common descent paradigm. The main problem is that humans are unusually intelligent compared to the other animals, a product of having much larger brains. For animals, including our ancient ancestors, delivering babies wasn't much of a problem: their smaller heads containing their relatively puny brains could pass easily through the center of the pelvis of the mother, and it was easier from a developmental point of view to use an already existing exit point from the body (the urinary tract) than to create an entirely new one (for example, some new kind of opening in the belly), which would require huge genetic, anatomical and developmental changes to the programming of our cells and organs. But as humans evolved and became more intelligent, we reached a critical point at which the advantages of our increased intelligence (and bigger heads) were outweighed by the disadvantages of complicating childbirth. The final outcome is still lamentable in some cases, but it's easier to attribute the blame for this problem to an impersonal evolutionary process trying to find the easiest, most genetically and developmentally economical solution to the problem of giving birth than to think that God actually designed it that way, knowing full well that it would often go horribly wrong. When I share this with YEC proponents, many are quick to tie this back to the fall, when Eve was told that her pain in childbirth would be increased. Once again, the anatomical design preceded by millions of years a literal fall in the garden event roughly ten thousand years ago. As well, irrespective of whether one insists that this is related to a literal fall event ten thousand years ago or to a metaphorical fall event(s) millions of years ago, in both cases as a curse on all mankind (actually, on all mothers and babies), one needs to explain why God would impose this kind of lottery-of-death which strikes indiscriminately at some women but not others, and also squeezes the life out of the innocent baby.

I could go on about other examples. But for me, the conclusion is clear: gradual change and evolution (common descent) make much more sense of the observed data than special creation (common design). To insist on the latter should make one wonder why the designer made us with so many flaws, and who seemed to construct a very elaborate ruse intended to misguide us (making it look very much like we arrived by evolutionary processes). This long list of design flaws is not intended to challenge *who* made us, but rather *how* we were made.

We may not like that assessment of our place in the grand scheme of things. We may prefer the traditional view derived by the church fathers from the book of Genesis. But we can't ignore the facts in front of us, nor base our conclusions on feelings.

And we can't have it both ways.

Or can we? Perhaps the scientific evidence simply describes our biological status in creation, while the scriptural text tells us our spiritual status in creation.

Objection #5: It challenges the idea that we are created in the very image of God.

Theists have never agreed on what this term means. Many definitions of the *imago Dei* have been offered, but none has convinced everyone. In chapter 8, I discuss in much more detail a wide variety of ideas on what *imago Dei* might mean (pages 204–215), as well as what it means to be human (pages 215–219).

But suffice to say here that there's no need to be too literal about this.

I'm certain that very few readers, if any, believe our own bodily form reflects God's bodily form. That God actually has a literal face with eyes and mouth, nor arms with hands and fingers, nor legs with feet, despite the many biblical references to all of those very same body parts, as well as references to the smoke coming from his nostrils,[34] and another reference to his back[35] (which the King James translates as "back parts"). God is not physical, and our creation in his image doesn't reflect something physical or organic. *Imago Dei* is not referring to our physical bodies (which is

---

34. 2 Sam 22:9; Job 41:20; Ps 18:8.
35. Exod 33:21–23.

what is at question here when we're considering whether the human form and primate forms evolved from the same common starting point).

Nor is it referring to the material from which we are made. If certain Christians find it offensive, unacceptable and/or perhaps even illogical to say that God created us in his image from a hominid common ancestor, why do they not equally find it problematic that he created us in his image from a few handfuls of dirt? Whatever theological arguments that might be raised against having come from animate beings apply equally to us having come from inanimate dirt.

Finally, just because one finds it personally offensive to have a lineage in common with the great apes, as did Bishop of Oxford Samuel Wilberforce (see opening paragraph for chapter 5), our personal revulsion doesn't automatically mean it didn't/couldn't happen that way. As the saying goes: "You can choose your friends, but you can't choose your family."

Until the day arrives that we know exactly what *imago Dei* means, this can't be used as an argument against an evolutionary model of origins.

Objection #6: When God finished creating on the sixth day, he declared it all "very good."

Once again, in order to deal with this rebuttal, we need to agree on a definition of these descriptors "good" and "very good." Emotional pleas invoking cancers in children and parasitic zombie wasps and painful childbirth are not cogent arguments.

It doesn't seem that God calling everything "good" and "very good" means it was all as good as it possibly could have been. The text describes a garden which featured a prowling deceptive serpent (which some equate with the devil himself) determined to pit humans in rebellion against God. It also featured an Adam who grew to be lonely, and possibly jealous of the other animals (since they had a partner, but he didn't).

It also needs to be pointed out that we humans may describe some things as "not good" simply because we don't like them. The fact is that not everything is either good or bad: some things are simply ugly. As I explained above, predators perform a function that's critical to the maintenance of a healthy ecosystem. They've earned a bad reputation, though, because some of us don't like what they do to their prey. Why are carnivores considered to be not good or even evil when they take down their prey, but not herbivores when they nibble a young sapling down to its roots? Likewise with thorns and thistles (which the writers of Genesis attribute to God's curse on mankind for the fall in the garden). Thistles are simply another species of plant life and play a key role in many ecosystems, and thorns are merely deterrents to predation by herbivores; why would it be perfectly OK for an animal to eat the plant but not OK for the plant to defend itself? Many diseases are manifestations of yet other organisms simply going about the normal business of living and reproducing: fungal infections, malaria, intestinal worms, and flesh-eating disease, to name just a few, are just organisms trying to survive. We certainly don't like it when we're the ones contributing to their survival, no more than we like being the meal for a predator. But there's still a sense of beauty when you look into their design and the complexity of their integration into their environment. Other diseases are a result of normal cellular processes gone wrong, such as cancers, auto-immune diseases, and eating disorders. Yes, it's a profound tragedy for anyone afflicted with any of these, but those disorders aren't inherently evil.

For those who find this sympathetic characterization of predators and diseases offensive and unacceptable, just think how humans might be perceived by nature. There's no place on earth which hasn't been touched by our pollution. Think of the scars we've left on the face of the planet, the pristine vistas we've spoiled, and the Great Pacific Garbage Patch in the center of the northern Pacific Ocean. The animals we parasitize, and the thousands of species we've exterminated (mammoth; thylacine; Eastern elk; Caspian tiger; moa; dodo; quagga; just to name a few), some of them in the name of mere entertainment (passenger pigeons). From their perspectives, we humans are predators, thorns, thistles, parasites and disease all rolled into one.

What about death itself. Some would ask how that could be called good, let alone "very good"? But those who think God's original creation could not include death haven't thought through the logistics very well. Consider God's commands to the birds, fish, animals and even mankind to "be fruitful and multiply." If they were to follow through with this command but to never die, earth would soon be covered by a seething mass of flesh that no amount of vegetation could possibly sustain. Even if the death spoken of in the opening chapters of Genesis refers only to human death, God's command to physically eternal beings to be fruitful and multiply will ultimately create huge problems: just look at how we've overpopulated earth today simply by extending our life spans by a few decades through better agriculture and modern medicine. God would either have to be extremely shortsighted or careless, or he would have to have known that mankind would soon earn the death sentence (which leads some to question the concept of free will).

Objection #7: Jesus and the apostles clearly believed in the accuracy and historicity of Genesis.

It would seem that one could certainly make this kind of claim: Jesus and the apostles referred enough times to various passages in the OT, including Genesis. Luke goes through the trouble of recording the genealogy of Jesus all the way back to Adam. Paul made very clear references to Adam bringing sin and death into the world.[36] Jesus, Peter, and the writer of Hebrews referred to Noah's flood,[37] and Jesus also made a passing reference to the creation of mankind.[38] So it could be said that they all accepted Genesis as accurate and authoritative. What right do we have to call into question their understanding of things?

I can think of two reasons.

First: perhaps they didn't know anything different. That is to say: they believed in the accuracy and historicity of Genesis simply because they were people of their time.[39] All of them grew up as Jews, heard the Shema every day ("Hear O Israel, the Lord is our God, the Lord is One"), and attended synagogue regularly. Their families, relatives and neighbors all accepted Torah as central to their lives, given to them by Jehovah. They didn't know anything different. They didn't know about the rest of the world beyond a couple hundred miles from their home. Nobody from their time had ever seen earth from an airplane, or seen pictures of earth from space, let alone the rest of the planets and stars in the universe. They weren't aware of fossils of animals, or fossils of other humans, or even humanlike creatures; had no fossilized remains of Neanderthals or *Australopithecus*; didn't have scientific evidence which spoke of millions and billions of years; no clue about the genetic evidence which contradicted the creation stories they repeated among themselves. They didn't consider the possibility that one animal form could arise from another. They only ever saw each kind giving birth to more of the same kind, with each generation getting larger than the one before it. Working those observations backwards through time, they would eventually come to a primal pair for each animal, and then they could go no further. So in their minds, there must have been a scenario in which primal pairs came about through some completely different mechanism: creation by a deity. But we have seen accumulations of changes in successive generations of organisms, and do have the evidence of wholesale changes from one kind to another.

In some passages, Jesus and the apostles say things that we now know are not true.

When Jesus taught that the mustard seed "is the smallest of all seeds on earth,"[40] did he know that this isn't factually correct? There are many plants with much smaller seeds: begonia seeds, for example, are as small as specks of dust. Denis Lamoureux comments[41] on this scientifically inaccurate reference to the size of mustard seeds, as well as to two other sayings of Jesus which can't be taken to be scientifically or literally

---

36. Acts 17:26; Rom 5:12–19; 1 Cor 15:21–22; 1 Tim 2:13–15.
37. Matt 24:37–38; Luke 17:26–27; Heb 11:7; 1 Pet 3:20; 2 Pet 2:5.
38. Matt 19:4–6.
39. Peter Enns, *Evolution of Adam*, 93–117.
40. Mark 4:31.
41. Lamoureux, "No Historical Adam."

true: the fact that wheat seeds need to "die" before they can produce other seeds,[42] and the stars falling from the sky upon Christ's return to earth to set up his kingdom.[43] As Dr. Lamoureux points out, even one star falling to earth would completely obliterate the planet. Others, such as Dr. John Walton, attribute this to Jesus indeed knowing better but simply accommodating his audience.[44] That is possible, but it should be acknowledged that the biblical text does not explicitly state this, and so the need to attribute this to the lack of scientific sophistication of Jesus's audience rather than to a deficit in Jesus's own knowledge is simply to avoid bruising the divinity of Christ. However, the biblical text does tell us that Jesus did not know everything: not "the day nor the hour,"[45] nor the name of the demon(s) within the Gerasene man,[46] nor the number of loaves and fish which the disciples had,[47] just to name a few examples. Was Jesus remembering incorrectly when he referred to Zechariah being the son of Berekiah, rather than to being the son of Jehoiada as the book of Chronicles indicates,[48] or was he employing a rabbinic tool of equating fathers and grandfathers, or have modern scholars identified the wrong Zechariah?

When Paul talked about knowing a man who had been taken up into the third heaven,[49] did he know such a place doesn't actually exist. At least, the third heaven referred to within the ancient Hebrew, Sumerian and Babylonian cosmologies doesn't exist. And we can't simply equate Paul's third heaven with some other place in the universe, unless we can also accept that someday it would be theoretically possible for us humans to send a space probe to that heavenly place. I don't understand what or where heaven is, but I'm convinced it isn't a physical "place" somewhere out in the universe.

In other passages, they teach things that we no longer hold to be true for our own time. "Judge for yourselves: is it proper for a woman to pray to God with her head uncovered? Does not the very nature of things teach you that if a man has long hair, it is a disgrace to him, but that if a woman has long hair, it is her glory? For long hair is given to her as a covering. If anyone wants to be contentious about this, we have no other practice—nor do the churches of God."[50] Generally speaking, we no longer hold that kind of passion about the subject of hair length.

At that time, they were quite accepting of slavery, even teaching, "Slaves, obey your earthly masters with respect and fear,"[51] and of the inequality of women (too many

---

42. John 12:24.
43. Matt 24:27–29.
44. Walton, "Response from Archetypal View," 68–69.
45. Matt 24:36.
46. Mark 5:9.
47. Mark 6:38.
48. Matt 23:35 vs. 2 Chr 24:20.
49. 2 Cor 12:2.
50. 1 Cor 11:13–16.
51. Eph 6:5.

verses to list here). We certainly get much more animated over those issues today than was the case in the first century, and on the other hand, not nearly as animated about eating meat offered to idols[52] or about circumcision[53] as they did at one time.

In these examples and many more, I certainly think that they and other authors of Scripture believed completely that what they said and wrote was true. (I'll elaborate on that a bit more in a later section entitled "Why then are the Scriptures given to us in their present form?") They also believed in Leviathan, the chaos monster of the deep, and in a third heaven. The Greeks believed in cyclops and a heavenly place called Elysium. We don't believe in any of these. Just because Paul—a first-century Greco-Roman Judeo-Christian theologian—may have fully believed Adam was a specific, historical figure, does that make Adam's existence an irrefutable fact? Do his theological writings constitute sufficient evidence for us to have to reinterpret or discard all kinds of modern scientific evidence to the contrary? We have benefitted greatly from the thoughts and writings of Paul: but in some respects we now see much further than that giant did in his day.

Second, perhaps they were simply trying to best reach their audience who didn't know better.

Jesus himself said he was sent to the Jews, the lost sheep of Israel.[54] So to best get their attention, he used the scriptural writings which they valued above everything else except God himself; something they could relate to; something that was woven into their very lives. When Jesus referred to the story of Noah—"As it was in the days of Noah, so it will be . . . "—was he just using an image that his Jewish listeners would have visualized many times as children and as adults? An image of a depraved humanity focusing on the temporal, oblivious to the eternal, until it was too late? Today, when talking to a younger audience, one might refer to a movie clip to emphasize a point: "Just like when Skywalker was battling Darth Vader, and the imperial storm-troopers were closing ranks around them, and then Yoda . . ." Doing so doesn't imply that you believe George Lucas's *Star Wars* is actual history. It's just a useful communication tool to convey imagery and make a point.

In the same way, when Paul makes absolutely explicit references to Adam in Acts 17, Romans 5, 1 Corinthians 15 and 1 Timothy 2, and draws comparisons between Adam and Christ, was he doing that for the sake of his own audience which was very different from that of Jesus? In all four passages, Paul was speaking to non-Jews about a very Jewish message. A gospel involving a Jew sent to the Jews as their Jewish Messiah and prophesied about in the Jewish Scriptures was now being presented to the Greeks and Romans as their Savior too. And how better to bridge the gap between these different peoples, and to include them into this very tribal message, than to

---

52. Acts 15:20.

53. Acts 15:5.

54. Matt 10:6 and 15:24.

emphasize their common heritage through one common ancestor (Adam)? "From one man he made all the nations, that they should inhabit the whole earth."[55]

Or was Paul just using the concept of Adam and Eve and the fall in the garden as a teaching tool—an illustration—of sin in our lives and our human need for redemption?

It's often claimed that Jesus referred to Adam and Eve, and people use this to legitimize a YEC (or an anti-gay) viewpoint. But did he? The passages those people are referring to are the accounts in Matthew and Mark when Jesus is debating with the Pharisees: "Haven't you read, he replied, that at the beginning the Creator 'made them male and female,' and said, 'For this reason a man will leave his father and mother and be united to his wife, and the two will become one flesh.'"[56] In these passages, when Jesus refers to "them" he is indeed pointing directly at Genesis chapters 1 and 2, which use the Hebrew noun *adam* in two very different ways.[57]

On the one hand, it's used in the opening chapters of Genesis as the personal name of the first man created. *adam* translates literally as "man," and so the first man is given that name. Interestingly, he isn't introduced by that name, nor are we told whether he selected his own name or whether it was given to him until the fifth chapter of Genesis. The name "Adam" first appears in chapter 2 (and refers to humankind, not a person; see next paragraph) without any such introduction, after which we're told that "the adam" went through the exercise of naming all the animals and birds, later naming his own wife and children, and even having grandchildren who were named by their own parents. It's only in chapter 5 that we find out it was God who named Adam (and, again, the context of this passage indicates God was referring to humankind, not this one person). I often wish the original author(s) of Genesis could have been a little more imaginative in that naming, but I do also understand it's a bit of a play on words because *adam* also means ground or earth or dirt, from which he was made. So the *adam* named *Adam* is made from *adam*. The creature made from dirt was put in place to till the dirt and take care of the garden of dirt, and to later return to the dirt at the end of its life.

On the other hand, *adam* is sometimes preceded by the definite article—in Hebrew, *ha adam*—which means in English "the adam." In fact, there are thirty-four references to Adam in the first five chapters of Genesis, twenty-two of these employing the definite article.[58] The Hebrew language never puts the definite article in front of a personal name (like "the Joshua" or "the Mary"). Instead, these occurrences of *ha adam* are referring to a representative man—humankind or humanity—not "the guy named Man."

---

55. Acts 17:26.
56. Matt 19:4–6; also in Mark 10:6–9.
57. Brauch, *Abusing Scripture*, 127–29.
58. Walton, "Historical Adam: Archetypal Creation View," 91.

The ambiguity of the wording left translators of the Bible unsure as to when to translate the wording as "Adam" and when to render it as "the man." The first introduction of "Adam" as a personal name occurs at Genesis 2:16 in the Septuagint (Greek translation of the OT), while the American version takes this step at Genesis 2:19, the Orthodox Jewish Bible and the Complete Jewish Bible at Genesis 2:20, the Revised Version and Revised Standard Version at Genesis 3:17, the Today's English Version at Genesis 3:20 and the New English Bible at Genesis 3:21.[59]

The passage in Genesis to which Jesus refers when he says, "The Creator made them male and female," uses *ha adam*. So he's not explicitly referring to Adam and Eve as historical individuals. Likewise, when Jesus refers to the man leaving his father and mother and being united to his wife, he's not referring specifically to that primal couple, because they didn't have a traditional father or mother to leave behind, and Adam didn't leave God in a literal sense when he joined to Eve. Instead, he was referring to mankind in general, and to marriage figuratively.

Besides, it needs to also be recognized that Jesus is addressing a sociological matter—divorce and marriage—not teaching science or genealogy or ancient human history.

We do not need to think that Jesus's teaching here in his rebuke of the Pharisees demands a YEC interpretation of the origin of mankind.

Objection #8: Evolution isn't a fact, it's only a theory. Theories don't have the certainty of a Scientific Law, or anything like that.

This is a misconception I've heard far too many times. This sentiment was at the heart of a famous trial in the American justice system. I'm not referring here to the 1925 Scopes trial (more commonly referred to as the Scopes Monkey Trial) which focused specifically on a substitute high school teacher (Thomas Scopes) accused of violating Tennessee law by teaching the theory of evolution in a state-funded school. The 2005 Kitzmiller versus Dover Trial is a different landmark event in the conflict over teaching evolution in schools. In this case, the Dover Area School Board of Education had modified their curriculum, in part by requiring teachers to read out a statement at the beginning of their biology classes which essentially stated that evolution is "only a theory."

The fundamental problem is that the average layperson uses the word "theory" very differently than does the scientist. Exacerbating the confusion is the fact that both uses and meanings of the word are correct.

The Oxford Dictionary defines "theory" as, "a supposition or a system of ideas intended to explain something, especially one based on general principles independent of the thing to be explained."

---

59. Alexander, "How Does a BioLogos Model."

The average layperson usually means "theory" in the sense of a supposition, which in turn is defined by Oxford as, "an uncertain belief." A possibility. If they come across something puzzling that requires an explanation, they might say, "Hmmm. I've got a theory," and then they go and investigate whether they were right or not.

Scientists, on the other hand, use it to represent a system of ideas. And there are many examples of theories which are taken to be fully true. Quantum theory. Relativity theory. Gravitational theory. Acid-base theory. Cell theory. Germ theory. Plate tectonics theory. Big Bang theory. Evolution theory. These aren't things that scientists are unsure about. They represent collections of ideas and mechanisms which fit together, explain many things, and make useful predictions.

Laypeople need to recognize this different use of wording.

As an aside, a similar confusion between laypeople and scholars arises around the word "myth." The former use it to indicate something is not true factually or is merely a fairy-tale. "It's perfectly OK to swim after eating. You won't get cramps: that's just a myth." Or "It's OK dear, Dracula's just a myth." Scholars, especially literary ones, use the word to indicate a story which has explanatory power, and conveys a certain truth. So for them to refer to the biblical creation account or flood account as myth doesn't necessarily mean that they dismiss either outright, but rather that they feel there's a deeper meaning being expressed than simply the bald facts and details of the story. Myths were prevalent throughout the ANE,[60] and served a unique function but do not need to be taken as actual history.

Objection #9: If we evolved from monkeys, then we're no better than animals and should just throw all morals and laws out the window and do whatever we please.

This criticism is frequently raised, but is simply not a reason to dismiss evolution theory. The latter just isn't dependent upon any form of morality, nor does it enforce any kind of morality. Just because we may have originated from a state of lawlessness doesn't mean we must maintain that lawlessness or revert back to it once we realize that's where we came from. All of us have grown out of the undisciplined nature of childhood—complete with dirty diapers, temper tantrums, and selfish defiance—yet we don't feel an obligation to return to that kind of social immaturity once that fact is pointed out to us.

Besides, the statement being criticized above is really no different than saying "we're fallen creatures" or "we're broken image-bearers," statements which I also frequently hear coming from Christians. These two theological metaphors can be too easily used as an excuse when we slip up, and it was for this reason that the Apostle Paul, after writing how our past mistakes have been forgotten, had to write: "What

---

60. Hooke, *In the Beginning*, chapter 4.

shall we say, then? Shall we go on sinning so that grace may increase? By no means! We are those who have died to sin; how can we live in it any longer?"[61]

The key difference between claiming to have evolved and claiming to be a broken image bearer or fallen creature is the underlying assumptions made about what we once were at some point in the past, rather than what we are today. The phrases "fallen creatures" and "broken image-bearers" assume that we as a species were once perfect and are now less than that, while "evolved from animals" assumes that we were never perfect to begin with and are still less than perfect today. The idea that we were once perfect is only supported by certain *interpretations* of Scripture. The idea that we were never perfect, on the other hand, is consistent with actual observed facts.

The most important point, though, is that, irrespective of whether one believes or assumes we were originally perfect or not, we are called by God to now be perfect.[62] We're called to fight our base tendencies.[63] And we look forward to our ultimate divine natures to be revealed.[64]

Objection #10: Why would God use a process that took millions or billions of years to fully play out? Why not simply snap his fingers and bring it all into existence in six days?

First, I could ask the same question about the standard YEC view: why would God spread the creation event out over six or seven discrete days, with periods of rest in between? Both views are susceptible to the criticism of "why wouldn't he use an instantaneous mechanism?"

Second, this criticism implicitly assumes that what we see now is God's intended end-point. But what if the human world that we see today is no more complete or mature than the world of *Homo erectus* a couple million years ago? Who says we are the pinnacle species, any more than *Australopithecus* might have seen itself as the pinnacle species four million years ago (if it could think that abstractly)? Or *Tyrannosaurus rex* many millions of years before that? Maybe something much more evolved is still yet to be revealed in another couple million years?

Back in the days of *Australopithecus*, they lived in their own filth, and were subject to the elements. They lived in fear of being eaten by a predator or killed by a competing tribe. They were unable to protect themselves against many of the dangers which they faced, and eventually went extinct. We have evidence that they ate other hominid species; perhaps they ate each other, or perhaps they offered human sacrifices to the Great Being.

---

61. Rom 6:1–2.
62. Matt 5:48; Rom 12:2; Jas 1:4.
63. Rom 7:18–25; Rom 8:13; Col 3:5.
64. Rom 8:19–22.

Have we evolved any further? Most of our global population is living in squalor and filth (even within major Western cities), and subject to the elements. We're still living in fear of being obliterated by a meteor impact, or a devastating strain of avian flu, or killed by a foreign world-power with nuclear weapons. We hear constantly about wars, killing, and abortions. We engage in self-destructive and addictive behaviors. So how can we really claim any superiority over *Homo australopithecus*?

Perhaps even today the universe, the world, and life on earth are still not fully developed. Have we evolved as much as we possibly can? There's certainly room for further evolution. Who knows what kind of physical and/or cognitive changes could yet occur in our imperfect bodies and brains. We've already seen examples of humans developing phenotypic traits which were absent even just a few thousand years ago: the ability to digest milk into adulthood because of a genetic change for the enzyme that breaks down a milk sugar (lactose),[65] or the appearance of blue eyes.[66] Average heights and life spans have grown dramatically, mostly because of better nutrition and modern medicine. We've also lost certain phenotypic traits: it looks like thick bodily hairiness and wisdom teeth are on the way out, and the jury's undecided on organs such as the appendix and the smooth muscle around our airways (which seems to only contribute to airway diseases such as asthma).

But several things could potentially get in the way of us evolving any further. Our entire species might succumb to some new disease that sweeps the globe the way the Black Plague did, but this threat is now multiplied by our ability to fly any person to any part of the globe in hours and spread the disease there. Or we might engineer our own demise: trigger a global thermonuclear destruction, develop a supertoxin or super bug, or allow bacteria to develop resistance to all forms of antibiotics. In theory, our technology can also remove the primary driver behind evolution. Adaptive changes require an environmental stress, something that jeopardizes survival and/or reproductive success, and we're pretty good at removing all such stresses. In that sense, we may very well drive our own devolution. Also, we won't see humans diverge into distinct species as long as our various populations have the ability to intermingle and reproduce, thus continually stirring up the gene pool rather than cutting off certain parts of that pool and allowing that to evolve in a different direction.

All that having been said, some will still choose to believe that we humans are indeed the pinnacle species in God's creation, and in support of that they will refer to biblical passages like Psalm 8: "what is mankind that you are mindful of them, human beings that you care for them? You have made them a little lower than the angels, and crowned them with glory and honor."[67] They may be right. I won't flatly deny that. But I will point out to them that it was a human that wrote that passage about humans: dolphins might believe that *they* are the pinnacle species. Besides, what about the

---

65. Bersaglieri et al., "Genetic Signatures," 1111–20.
66. Eiberg et al., "Blue Eye Color," 177–87.
67. Ps 8:4–5.

many biblical passages in which the Israelites wrote about their own privileged status within world politics during their time: they too were eventually superseded by another group of people. Circumstances can change.

A third explanation I'll give as to why God would use a process spread out over millions of years is best illustrated by a metaphor I used in my first book: one of two gardeners. The first plants a garden every year by going out to the local nursery, buying all kinds of annuals produced by other growers, puts them into their discrete places in the garden over the course of a weekend, and enjoys that garden for the rest of the summer before having to rake it all up, throw it in the compost and start planning for next year all over again. The second gardener plants all kinds of perennials (that is, not annuals), taking into account the varying flowering times, colors, heights, and so on. Once that's in place, she then sits back and enjoys her garden unfurling itself every year all on its own. In my mind, the second gardener shows a greater understanding of her plants, as well as a greater foresight and creativity. In the same way, I see it to be more glorious that God creates a cosmic egg, lets it explode in the form of a Big Bang and unfurl itself into a tremendous universe teeming with all kinds of diverse forms of life, rather than that he set aside six periods of twenty-four hours a few thousand years ago and planted the final product all in the theological equivalent of one weekend.

Objection #11: The Bible specifically says "each according to their kind."

When one reads any legal document—a law-suit, a product guarantee, a merger between two companies, even a simple renter's lease—one of the first things such a document does is define terms: who the various participants are, what is entailed when the word "XYZ" is invoked, and so on. This is intended to avoid absolutely any confusion in the transaction at hand.

The same needs to be done in discussions between biblical literalists and scientists when it comes to words such as "kind," "species," "phenotypes," and "genotypes." Not only with respect to what those terms mean, but also how the objects referred to by those terms relate to each other. Confusion often crops up between creationists and evolutionists with regards to such terms, and this becomes very problematic when talking about the creation event and Noah's ark. The literalists often equate the words "kinds" and "species," and often don't know the precise meaning of the other two closely related words (genotype and phenotype).

I can quote many dictionary definitions of the term species, but the most important concept which they'll incorporate is that the organisms within a species can produce viable off-spring, while organisms between species cannot. One would think that this should provide a fairly easily demonstrable test to rigorously separate between species. Unfortunately, that isn't always the case. Horses and donkeys are two distinct species, but the two can mate and produce viable off-spring—mules—although the latter are infertile. Brown bears and polar bears can interbreed and produce viable off-spring,[68] as can many primates. Humans and Neanderthals are said to be two different species, and yet we have abundant evidence that they interbred and that we now carry a certain small percentage of Neanderthal DNA coding in our genes.[69]

Genotype refers to the specific gene or collection of genes that an organism has. For example, when it comes to blood type, you might find that you are defined as being A/negative, which means you do carry at least one A gene but not the Rh gene (nor any B gene). Biologists can easily test for the presence of this or that gene and therefore the organism's genotype.

Phenotype addresses how the genotype manifests itself to us. That might refer to the physical appearance of the organism, such as whether the bird has a long thin beak versus a short, bulky beak, or whether the person has red hair or blond hair. It might refer to something internal, like an enzymatic activity: one strain of bacteria showing a positive phenotype, meaning it is able to metabolize a certain drug, versus another strain which is phenotypically negative and cannot metabolize the drug. The range of possible kinds of phenotypes is boundless. It's important to point out that many phenotypes—such as height or intelligence—can be controlled by many different genes (in other words, many different genotypes can give the same phenotype).

This is where things can get confusing. Since genes are so important in defining who/what we are, two different species will of necessity have two different genotypes. But this does not mean that different phenotypes represent different species, even though the phenotype is determined directly by the genotype. That may be difficult for the nonexpert to digest: perhaps an example will help. Dogs can range tremendously in size and body proportions, color, hairiness, intelligence, and many other features: for example, compare the Poodle, Saint Bernard, Greyhound and Pit-bull. They are so incredibly diverse in their phenotype. They are as diverse as two distinct species could be: are a Dachsund and a Scottish terrier (two different phenotypes of the one species of dog) any less different than an otter and a raccoon (two different species entirely)? Some dogs are indeed effectively unable to interbreed (for example, the Chihuahua would need more than a ladder to attempt anything with a Great Dane), that you would otherwise think they must comprise different species. And yet they are all one species.

But now we must bring in the biblical term "kind" (the Hebrew word is *min*). That word is a generic term, and should not be confused with our word "species," The Hebrews had no concept of anything like species (that term was defined by Carl Linnaeus thousands of years later). The Hebrews would refer to cattle giving birth to other cattle according to their kind, while we refer to a particular species of cattle giving birth to another of that same species and not to some other species of cow which

---

68. National Geographic, "Grizzly-Polar Bear Hybrid."
69. Pääbo, *Neanderthal Man*.

might even look very similar. The same could be said for the many different species of sheep (the Hebrews would just refer to "sheep" in the collective sense), and goats and many other specific kinds of animals. Even if our concept of species had been possible in the mind of the author(s), there was no Hebrew equivalent for it. YECists are now frequently using the term "baramin"—a term coined by Frank Marsh in 1941[70]—as a surrogate for "species," but this term is not recognized in mainstream scientific circles.

The mismatch between the words "kind" and "species" is a problem not only in the creation account, but also the flood account. Proponents of the historicity of Noah's ark will often be asked: "How could the ark have held one pair of every species on earth?" And one answer frequently given in response is that Noah only needed one "representative pair" of any given species. Once the ark landed and was emptied, the representative pairs could then give rise to all the different species we have today. The example which is often given for this is Noah bringing on board only a pair of grey wolves, and they in turn would later produce the other canine species, including the domestic dog. Likewise, only one pair of finches, which then gave us the dozens (hundreds?) of other types of finches. One representative cat to give rise to all the cat-like animals. One representative cow for all the cow-like animals. And so on.

This explanation may satisfy nonexperts and their readers or listeners. AiG are currently strongly promoting this explanation, but the two authors of a recent article[71] outlining their position are simply not experts: one has degrees in theology, the other has two years of training as a veterinary technician and five years of experience as the zoo keeper at AiG's Creation Museum in Petersburg, Kentucky. Ken Ham—the founder, leader and spokesperson for AiG—also lacks scientific credentials to give this explanation any gravitas (he was previously a high school biology teacher). Such an explanation astonishes those with bone fide scientific experience in the area, and reveals a whole level of ignorance—nonexpertness, actually—on the part of the self-proclaimed experts and proponents of this idea. It betrays a complete misunderstanding of what is involved in speciation.

There are an incredible number of families, genera, and species of animals; and all of them have diverse genomes. According to all that we can measure and observe right now, such genetic diversity requires hundreds of thousands of generations to develop, which in turn requires many tens of millions of years.

So for a nonexpert YEC spokesperson to say, with a casual shrug of the shoulders, something like "Noah simply took representative species on board the ark," and to suggest or imply that those representative pairs then diverged into the diverse array of animals we have now, including mass migrations of certain species to very specific regions all over the globe, all in the course of only a few hundred years after the flood, involves an astounding amount of faith or ignorance. It requires a massive,

---

70. Gishlick, "Baraminology," 17–21.
71. Belknap and Chaffney, "Reimagining Ark Animals."

off-the-scale increase in the background mutation rate. There is absolutely no evidence for that kind of *ad hoc* explanation

In addition, this YEC explanation of "representative pairs" raises another big problem, pointed out as far back as 1777 by Eberhardt Zimmermann.[72] For many years following the flood, what would the various carnivorous species have eaten without wiping out whole groups of species, if in fact it were true that all these species groups arose from only the one pair of representative animals? If the hypothetical representative wolf/dog-kind happened to kill the hypothetical cat-kind, we would not have today's housecats, bobcats, lions, tigers, jaguars, cougars, pumas, ocelots, lynx, leopards, cheetahs, or many other species from the *Felidae* family. Likewise if they decided to enjoy some lamb or chicken the next week. The same threat to mass extinction of species would result if one of the representative pairs succumbed to disease or pestilence brought on by all the dead and rotting bodies lying around following the flood.

Creationists should just retreat from this position and drop any form of argumentation of "representative pairs" from their apologetics.

Objection #12: Genetic change was brought on by the fall and the flood.

The presence of mutations in our genome—the proviral insertions, deletions, pseudogenes and apparent inversion of chromosome number two, for example—raises problems for those with a YEC worldview, because it is uncomfortable embracing an idea that God created everything "very good" and yet full of mistakes. They also have difficulty explaining the large number of alleles we find for some genes: a YEC view can really only explain four of these at most (as outlined above).

To get around this problem, some will claim that everything was indeed perfect on the seventh day of creation, and for some indeterminate period of time after that until Eve plucked the apple from the Tree of the Knowledge of Good and Evil (the Bible doesn't tell us how much time transpired between the second and third chapters of Genesis), after which all the imperfections quickly manifested. Once God had finished pronouncing the various curses on the three central characters in this vignette, all the rest of creation descended into chaos. According to the text, animals turned on each other, predatory and carnivorous behaviors appeared for the first time, thorns and thistles grew up, weeds invaded everything, and disease and death took control. Some of these changes are ascribed to immediate and drastic changes in the environment which in turn caused a cascade of genetic mutations.

A similar claim will be made about the flood event. Some are convinced that there was some kind of canopy over the earth which held back "the waters above" and shielded us from harmful ultraviolet (UV) rays and other forms of radiation from

---

72. Browne, "Noah's Flood," 137.

the sun or from space. Their only scriptural basis for this concept will be the texts which refer to the "firmament" in the heavens. According to some YEC proponents, the flood event included a catastrophic collapse of that protective canopy and massive geologic events (earthquakes, volcanic eruptions, continental upheaval) the likes of which eclipse any major catastrophe we've experienced since. The sudden influx of mutagenic UV rays from the sky and toxic, radioactive chemicals from deep underground combined to cause major changes in the genomes of all life. This massively increased rate of mutation then settled back down to a quiet hum, to the same degree that we see today, and life has never been the same since.

All of this without a shred of scientific evidence or scriptural text to support it.

In one sense, this proposal is very similar to the Punctuated Equilibrium model proposed by Steven Jay Gould to explain the sudden appearance of many different life forms at the Cambrian Explosion. That too is a strained hypothesis because it's based on the absence of data. Scientific theories are normally based *upon* observations and can be repeated in the lab, but the Punctuated Equilibrium theory only explains an absence of observations: at one moment in geological time those species are absent, and the next moment in geological time they're suddenly there in abundance and ecological diversity. And, of course, we can't repeat that phenomenon in a controlled experiment, so the theory can't actually be tested. The same is true for this hyper-mutagenic environment theory being promoted by some YEC apologists. There's no evidence that the background mutation rate suddenly shot into the stratosphere and then quickly dropped down to baseline levels, but it seems to conveniently help to explain the genetic changes described above. At least, it may satisfy a nonexpert.

Genetic mutations just do not occur that quickly: six thousand years is just far too short of a time for such colossal changes in gene structure and sequence. In the case of humans, which can reach sexual maturity by age twelve, six thousand years allows for at most five hundred generations. That may sound like a lot of generations to a nonexpert, but it's totally insufficient as far as an expert geneticist is concerned. And the claim that such dramatic changes in ecology could occur within only a few handfuls of generations after the fall or the flood is just absolutely indefensible from a scientific point of view.

Objection #13: There are too many gaps and missing links.

Everyone knows that weapons-of-mass-destruction were hidden all over Iraq by Saddam Hussein early at the turn of this century. We heard about the evidence (although neither you nor I ever saw or handled the evidence ourselves, nor for that matter know personally anyone who did). TV talk shows were giving us information, albeit second- or thirdhand from unnamed sources or "White House insiders." There were reports on the TV and radio every day. People all around us were talking about it. Everybody *knew* of the existence of these weapons-of-mass-destruction.

And yet those didn't exist! It was all talk. People saying, "I *heard* that they found . . . ," and then other people quoting them for saying that, and in turn being quoted, and so on.

As Joseph Goebbels (Adolf Hitler's propaganda minister in Nazi Germany) said: "Repeat a lie often enough and people will start to believe it."

The same thing is playing out too frequently in religious circles, in the form of the many questions asked and statements made about "the missing link."

"Everyone knows that there are huge gaps in the fossil record."

And yet there aren't.

It's absolutely true that we don't have a complete record of bones documenting every discrete step in the transitions between all the different species that have ever existed. It's unreasonable to expect that amount of evidence before the transition can be acknowledged as the most likely explanation (it can never be proven, at least not until we invent time travel or can repeat the event in the lab). It's like having video footage of a person entering a building where a murder was committed, and having that person in custody with the knife in their pockets and blood on their hands, but still insisting that we need to have concrete moment-by-moment evidence of every single step they took through that building and during the time that they left the building and were taken into custody before we'll believe that they were suspects, let alone perpetrators.

"Why haven't scientists found the missing link between apes and humans?" I've already belabored this point above on page 125. But let me be clear once again: scientists do *not* believe there is a direct link between apes and humans: humans did *not* evolve from apes themselves. Instead, evolution theory claims that we evolved from an ancestor shared in common with the apes and our two species diverged from one another.

That having been said, we do have all kinds of transitional forms. *Homo naledi* may be one example, representing an early species which branched away from the human line soon after it branched away from the ape line. *Archaeopteryx* is another example, representing the transition between non-avian dinosaurs and modern birds. And we keep finding more.

To explain why the fossil record looks like it skips and jumps: imagine watching a sunflower seed planted in your backyard garden beside a pond containing some frog's eggs. You know for a fact that the seed slowly unfurls itself into a seven foot tall plant, and that the amphibian eggs produce tadpoles that gradually morph into frogs.

But now imagine that you only get snapshots of the seed and tadpoles whenever a lightning bolt strikes: a sudden flash and another frame is added to the movie. Sometimes you have to wait three weeks before the next set of pictures is taken, and at that time you get two dozen more frames. Then another two weeks before another dozen or three are taken. And so on till the end of summer.

What are you left with? A series of snapshots that accurately captures the growth changes in the sun flowers and frogs, but does so in fits and starts as you replay the movie.

Just like the striking of the lightning bolts, fossilization is a sporadic event requiring very specific conditions. Most dead bodies are eaten and/or rot away within a matter of weeks. But every now and then—very rarely—you get exactly the right conditions for dead bodies to become protected from scavenging animals, insects and bacteria and to become fossilized. This can't happen in the soils of rainforests, because these are too acidic (from all the decaying vegetative material) and quickly dissolve the bones before the organic matter can be replaced by minerals. It also requires rapid burial (for example, by a mudslide) in order to exclude the oxygen needed by bacteria to digest the remains. One also needs the right soil chemistry to promote replacement of all the animal material with minerals, a process which occurs very slowly. Then, once the bones (and rarely even some of the soft tissues) have become completely mineralized, the fossils have to be protected from erosion (by running water or even persistent wind-borne sand) or from shattering (by earthquake, frost upheaval, landslides and other such geological events) and to thereby last millions of years. It's all an exceedingly rare series of events. But over the course of many millions of years, all the right coincidences can sometimes happen and we're left with yet another snapshot of the movie we call *Evolution*.

Just the same, some people just want to believe that big gaps exist. They need gaps to be there, in order to hold on to an *interpretation* of Scripture. And so they'll listen to and quote from nonexperts who heard this or that evidence which supports their viewpoint. Despite the overwhelming testimony of many experts to the contrary. Firsthand evidence from experts being trumped by third- or fourthhand evidence from nonexperts. Sure, one could dismiss the experts as having an agenda (ignoring the fact that the creationist nonexperts equally have their own agenda). That kind of dismissive thinking helped float the flat-earth balloon and the sun-revolves-around-the-earth balloon for centuries. But those zeppelins eventually crashed and burned.

Objection #14: I believe in micro-evolution but not macro-evolution.

As if those were two completely different things. The terms have been repeated often enough that they're now in our lexicon. A standard figure of speech. But they have no practical value. They simply represent two extremes of a spectrum. Like trying to distinguish between a white lie and a real lie, or a minor sexual attack from one that really matters.

Microevolution is evolution. A single gene mutation can lead to a minor change or a major change: the latter is especially true if the gene in question kicks in early in the development of a fetus. Whole genes or groups of genes can become accidentally duplicated in a single step, allowing the one copy to carry out the normal cell function while the other one accumulates mutations and eventually leads to an entirely different function. Two groups of a single species can become isolated from each other (by getting stuck on two different islands, or either side of a mountain range, or otherwise in completely different ecosystems) and continue to evolve in different directions until they become so distinct that they can no longer inter-breed. At that point they become two different species, and will continue to diverge.

Evolution theory is much more nuanced than simply "micro is possible" versus "macro is impossible."

Objection #15: This takes away the urgency of evangelism. If humans don't have original sin, then Jesus didn't need to die on the cross, and we therefore don't need to spread the gospel.

Some would say the same about Christian theologies such as Calvinism. At least in the sense that adhering to the tenet of unconditional election calls into question the need to convince others of the gospel message. But this book is not about Calvinism.

While the scientific data might cause us to reconsider the tenet of original sin, they don't rule out the idea of sin in general. *Hamartia*, the Greek word often used in the NT texts for sin, means literally to "miss the mark," and borrows metaphorically from the action of aiming an arrow at a target. God can still have set a bar or goal for us to aim for (represented legalistically in the OT laws, and now exemplified in the life and teaching of Jesus Christ), and we may all have failed to do so in many ways. We don't require the concept of original sin, inherited from an original primal couple in the garden of Eden six thousand years ago, to need a Saviour.

A theological argument alone (in this case, the one pertaining to original sin) is not any reason to dismiss concrete data and observations. Belief should not trump fact. As I said with objection #9, the logic here doesn't have any merit. Even though we humans can no longer trace our lineage to a single primal couple who directly rebelled against God, this does not mean that the world does not still need a Saviour. Denying the historicity of "Adam" doesn't negate the universality of sin. The gospel message—the *euangelion*—is about more than just buying a life insurance policy. Jesus came to do more than offer himself as payment for sin. In the next chapter, I will show that we can still fall well short of God's ideal, irrespective of what any of our ancestors may have done, and will provide alternative perspectives on Christ's atonement (see p. 231).

Conclusion

"A plain reading" of the substantial body of scientific evidence that we've collected tells us that humanity is the product of a long and gradual evolution from an ancestor we share in common with Neanderthals, Denisovans and other primates. This idea is repulsive to many theists, and they've resisted it vigorously.

Many just simply resist it without evaluating it. They effectively put their hands to their ears, or the heads in the sand, and just repeat to themselves that they don't accept it. In all sincerity, I don't say this to mock them. I simply say that some choose to deny it but can't articulate an actual reason why, other than that it goes against their beliefs or wishes.

Others, however, have taken the intellectual effort to raise certain objections to this theory. I've identified the fifteen objections that I hear most often, and have given counterarguments to all of them.

In the end, though, it comes down to a matter of choice. Choice on the part of both theists and atheists, with or without scientific education. All of us are also free to choose correctly or incorrectly. Medical science has convincingly shown us that smoking cigarettes is linked to cancer: and yet there are still many physicians and scientists who choose to smoke. Most people understand that the odds tend to be stacked against them when they go into a casino or buy a lottery ticket, and yet many still do just that. Many people make educated choices that go against what makes the most sense.

For those who choose to accept that mankind has indeed evolved, there are going to be tremendous changes needing to be made in their theology. That is the subject of the next chapter.

8

Adjustments to Theology

STRUGGLING SLOWLY UP A sheer rocky cliff, a mountain climber searches for a small crack in the rock face into which he can hammer a piton. After firmly planting that anchor, confirming that it's solid, and attaching his safety line, he advances a short distance and looks for the next place to put another piton and begin the next advance. Behind him is a string of such pitons through which his safety line threads, giving him confidence against a deathly fall backwards.

In the same way, as we challenge the big question of where mankind came from, let's look at the anchors that are now planted firmly as a result of the scientific investigations into our genetics, as well as the bones and other artefacts that our *Homo* ancestors have left for us to study:

- Humans did *not* originate from one couple; in fact, we never numbered less than several thousand.

- Humans did *not* originate six thousand years ago: instead, we've been around for hundreds of thousands of years.

- Humans did *not* originate in Mesopotamia: we left Africa in several waves.

- Humans are *not* a completely different creation entirely separate from all the other created organisms: we are highly genetically, physiologically and anatomically related to other animals, especially so in the cases of chimps and apes, and are almost genetically identical to our extinct genetic cousins, the Neanderthals and Denisovans.

- Neanderthals, Denisovans and *Australopithecus* shared our ability to use stone tools, which indicates they too were problem-solvers and could think in abstract terms; they also seemed to share our value of close family social groups; we conducted extensive trade with them.

- Neanderthals and Denisovans also had an appreciation for the aesthetic (they made cave wall paintings, beads and jewelry), and a sense of the afterlife (they buried their dead, sometimes with valuable artefacts, and possibly even planted flowers at their graves).

- Musical instruments did not originate six thousand years ago in Mesopotamia, as described in Genesis 4:21. Although, to be precise, that text only refers to stringed instruments and pipes, we have nonetheless found three flutes in Germany dated to approximately forty-five thousand years ago,[1] and a bone whistle at Haua Fteah in Libya (Northern Africa) dated to the same period.

- The first eleven chapters of Genesis bear an uncanny resemblance to ANE myths and literature; they contain many details which are quite unlike present day reality (the description of the location of Eden; men living for hundreds of years; talking serpents); they include several paradoxes (two separate accounts of the creation of humanity; light appearing before the celestial bodies which produce it; Cain's fear of people when the only people around should have been a small number of his own siblings); it makes more sense that the texts were written as a polemic against the pagan beliefs endemic to the culture at that time than that they represent actual history.

If one can agree with the planting of these flags, then one has to make considerable adjustments to one's theology. These adjustments pertain to questions about: who then are Adam and Eve; the fall in the garden; the nature of original sin; the theological implications of death and disease; the flood; the various genealogies in the Bible, including the one that links Jesus directly back to Adam; inerrancy of Scripture; what does it mean for a text to be inspired; how does one read the Bible; what does it mean to be human (that is, "in Adam"). How can we conclude that the whole salvation story—the rescue from original sin—applies to all the descendants of Adam and Eve if there was no Adam and Eve? Or how can that salvation promise apply only to a certain genetic sequence that we call human when Neanderthals and Denisovans had essentially the same genetic sequence? If that very slight difference in gene sequence was enough to exclude them, then what do we do about humans with considerable genetic anomalies (for example, possessing an extra chromosome, or who have suffered horrible mutagenic injury from a massive blast of nuclear radiation), or who may in the future be treated for a variety of diseases using complex gene therapy rather than simple drugs? It won't be an easy journey.

On the other hand, one can cling to a belief that the theology we've inherited from our church fathers millennia ago, some of it from the Bronze Age, is absolutely rock-solid, certain and beyond questioning. If so, one will need to deny the findings of genetics and paleontology which I've summarized in chapter 6. To allow a worldview or "–ism" to determine how we interpret empirical testable facts. And to remain out of touch with the society in which we find ourselves, and with which we've been commanded to interact. And to remain unable to fulfill the command to make disciples,

---

1. Higham et al., "Testing Models," 664–76.

because we appear to be just as irrelevant as the Flat Earthers and the Domers (those that believe that the firmament was an actual physical structure).

The story has changed too much in the past couple decades (since the development of gene sequencing), let alone the past two hundred years (when we began finding hominid bones and artefacts at an ever-accelerating pace), let alone the past several thousand years (when we still believed in the Geocentric model of the universe). The changes once again require critically thinking theists to reevaluate their theology. Simply changing cosmetic details while hanging on to the original underlying theology won't work. As Cardinal Bellarmine cautioned believers a couple millennia ago when they were confronted with a different set of scientific facts: "One would then have to proceed with great care in explaining the Scriptures that appear contrary."

This chapter will focus on several points on which careful reconsideration of one's theology may now be necessary. As pointed out in the opening chapter of this book, these recommendations are aimed at those who have shifted from a YEC point of view to an OEC one, as well as to those who are already quite comfortably OECs but who consciously or subconsciously hold a YEC-influenced theology.

Adam, Eve and the Other Hominids

I have already commented above (p. 61) on the remarkable similarities between the flood account recorded in Genesis chapters 6 to 8 and the Babylonian account found in the *Epic of Gilgamesh*, and asked whether it might be possible that the writers of Genesis borrowed from the latter, or were influenced by the zeitgeist from which *Gilgamesh* came. In the same way, the two Genesis accounts describing the creation of humanity contain particularly unusual imagery which bears striking resemblance to other ANE texts describing the creation of humanity, including those of the Sumerians (Song of the Hoe; Hymn to E'engura; Enki and Ninmah; KAR4), the Akkadians (Atrahasis; Enuma Elish) and the Egyptians (Pyramid texts; Coffin texts; CT spell; Instruction of Merikare).[2]

The image of YHWH creating Adam from the dust of the ground isn't new. Several of these ANE texts have comparable imagery, including those which refer to clay (Enki and Ninmah), clay on a potter's wheel (some Egyptian Pyramid texts), and seeds planted in the field breaking out to produce humans (Song of the Hoe; the Hymn to E'engura).

The same can be said about the image of YHWH breathing the breath of life into the lump of clay. Several other ANE texts also describe the gods incorporating elements of their own bodies when creating humans: divine breath in the Coffin Texts and Merikare; tears in the Coffin Texts; blood and flesh in Atrahasis; blood in Enuma

---

2. Lamoureux, *Four Views*, 57–59 and 98–100; Middleton, *Liberating Image*.

Elish and KAR4. These are all images of the Divine imparting something of itself into the created humans.

The authors of the Genesis accounts—whether one holds them to be Moses and his contemporaries, or priests in the Babylonian exile—would have seen and read those other non-Hebrew texts, and the culture in which they lived was saturated with these ideas.

Irrespective of whether or not one holds that the creation accounts of human origins in Genesis represent a reworking of ideas borrowed from those ANE texts, it is clear that the church needs to now rework once again its understanding of human origins. There is a growing recognition of the need to reconcile that understanding with the new data coming in from scientific investigations, especially in the fields of genetics and anthropology. Many individuals and groups have attempted to do so in a wide variety of ways.

Dr. C. John Collins has recently collated and evaluated numerous proposals of who Adam and Eve might have been, in light of this new evidence.[3] Unfortunately (from my point of view), Dr. Collins prefaces this by listing a number of presuppositions: provisos which must be met in order for the proposals to be considered valid. These arise from theological considerations alone, which is exactly the kind of manipulation and tunnel-vision that has prompted me to write this book. These preconditions include:

1. There had to have been a supernatural element to the appearance of any Adam and Eve, because "this follows from how hard it is to get a human being" (by this, Collins, a theologian, means humans are so complex and distinct from animals that they could not possibly have evolved from a common ancestor of the other primates). This reveals an overly simplistic definition of exactly what it is that separates humans from other species: I'll challenge this assertion below on pages 215–219 in the section entitled "What does it mean to be human." This version of the special creation argument is also essentially the same as "God would not do it that way" (Objection #3 in the previous chapter).

2. There could have been no other humans around prior to this original Adam and Eve. This precondition is given in order to account for all humans having received God's image and inherited original sin through Adam. This raises a considerable problem for theists, especially those of the YEC persuasion, given that the oldest *Homo* is two or three million years old.

3. Any new proposal for an Adam and Eve must include them directly disobeying God in some way (that is, the fall must be historical), in order to account for the "universal sense of loss" all mankind has experienced since the fall.

---

3. Collins, *Did Adam and Eve Really Exist*; Collins, "Adam and Eve as Historical People," 147–65.

4. If other humans existed at the same time as Adam and Eve (note, this condition doesn't contradict point #2), then they all must have existed as a single tribe, with Adam as chieftain (again to account for the solidarity we all share in their original sin).

With these preconditions in mind, Collins then evaluates a variety of proposals from various authors, philosophers, theologians and organizations.[4]

For Collins, the best models describe Adam and Eve as a unique couple created *de novo* by God. These obviously include the YEC view (the proverbial "plain reading" of Genesis) as well as some OEC ones which describe God choosing and modifying a pair of already existing hominids into Adam and Eve.

He then considers other models which attempt to accommodate more of the anthropological and genetic data.

This includes the model embraced by RTB, one of the most well-recognized and influential Christian apologetics organizations. One of its spokesmen, Dr. Fazale Rana, writes:[5] "The RTB model views Adam and Eve as historical individuals—the first human beings—originating by God's miraculous intervention approximately 70,000 to 50,000 years ago. Adam and Eve's descendants formed a small initial population that eventually gave rise to all human population groups around the world."

A similar model from Gavin Basil McGrath[6] has God "refurbishing" two hominids into the first human beings approximately forty-five thousand years ago. After so creating Adam, God took new genetic material from one of Adam's ribs to make Eve fully human.

John Stott and Denis Alexander take cues from the fourth chapter of Genesis and propose Adam (and Eve) as a Neolithic farmer approximately ten thousand years ago.[7]

The final model considered by Collins is that formulated by C. S. Lewis in *The Fall of Man*.[8] Lewis makes reference to the process of evolution ("God perfected the animal form which was to become the vehicle of humanity and the image of Himself"), but then has God giving one of those creatures "a new kind of consciousness" including those cerebral qualities that separate us from the other animals (such as self-awareness and a sense of abstract qualities like "truth, beauty and goodness"). But then "Someone or something whispered that they could become as gods" and they fell: "We have no idea in what particular act, or series of acts, the self-contradictory, impossible wish found expression. For all I can see, it might have concerned the literal eating of a fruit, but the question is of no consequence."

4. Collins, *Did Adam and Eve Really Exist*, 121–31.
5. Rana and Ross, *Who Was Adam*, 248.
6. McGrath, "Soteriology: Adam and the Fall," 252–63.
7. Alexander, *Creation or Evolution*, 237.
8. Lewis, *Problem of Pain*, ch. 5.

I didn't find any of these fully satisfying. I feel that one of their biggest problems is that they try to have their cake and eat it too. They want to incorporate the new scientific data into the original theological concepts with little or no changes to either, and yet, I feel, the two mix as well as oil and water. It isn't merely that the two have quite different timelines, the one measured in thousands of years and the other measured in hundreds of thousands or even millions of years. More importantly, my reservation is with the precondition that all humans must derive from one primal pair, in order that all of mankind inherits sin and guilt "in Adam," as Paul puts it in Romans and 1 Corinthians. But the genetic data tell us that *Homo sapiens* never numbered less than a few thousand.

Collins addresses this conundrum, but admits that he is not qualified to do so. He says, "I am not sure how to assess this DNA evidence; I do not know whether the evidence is only compatible with these conclusions, or if it strongly favors them."[9] From my own perspective as a scientist who sometimes uses genetic techniques in the normal course of my research, it is not this kind of either/or dichotomy. The DNA evidence is only roughly compatible with Collins's conclusions, and certainly does not strongly favor them. The current evidence paints a very different evolutionary historical picture for *Homo* than it does for the other primates. The data say that primates must always have existed in exceedingly large numbers, but that *Homo* could not have: instead, they tell us that *Homo* was nearly wiped out, perhaps by the Ice Age, but made a spectacular recovery after our numbers dipped as low as ten thousand.[10]

After grappling a bit with this genetic modelling, Collins admits he is out of his league, being a professor of Old Testament theology with a PhD in Hebrew linguistics. He writes, "I do not know whether these positions make use of assumptions that we really ought to be questioning. In view of my limitations, this is not the place, and I am not the person, to say whether these inferences are good or bad—though I wish that there were more critical discussion in the popular literature, laying out strengths and weaknesses."[11] My book now in your hands attempts to answer this call for critical discussion.

Another recent and popular compendium of disparate views on the question of "Who was Adam?" is that edited by Barrett and Caneday.[12] This provides an interesting and easily read discussion between four highly academically accredited scholars:

- Dr. Denis Lamoureux: holds that Adam and Eve were not historical individuals;
- Dr. John Walton: affirms that Adam and Eve were historical individuals, but more importantly serve as archetypes of all humanity;

---

9. Collins, *Did Adam and Eve Really Exist*, 118.
10. Li and Durbin, "Inference of Human Population History," 493–97.
11. Collins, *Did Adam and Eve Really Exist*, 119.
12. Barrett and Caneday, "Adam, to Be or Not to Be."

- Dr. C. John Collins: largely reiterates the points I've already summarized on the previous pages;
- Dr. William D. Barrick: affirms the standard YEC view.

This compendium provided by Barrett and Caneday is quite useful in providing a wide breadth of the range of views, but I feel it suffers from the same two major criticisms that I raised for the volume authored by Dr. Collins.

First, the panel of debaters collectively have very little scientific training. All four have exceptionally high theological credentials—three are professors of OT—but only Dr. Lamoureux has scientific credentials (he is an associate professor of science and religion). Collins also has a Master's degree in computer science and systems engineering.

Second, the two editors who compiled this work outline the parameters and ground rules for the discussion, and in so doing reveal, once again, that a presuppositional framework was imposed upon the entire debate. In particular, they explain in detail on pages 27 to 28 how they asked each author to answer three specific questions while defending their understanding of Adam and Eve:

1. What is the biblical case for your viewpoint, and how do you reconcile it with passages and potential interpretations that seem to counter it?
2. In what ways is your view more theologically consistent and coherent than other views?
3. What are the implications your view has for the spiritual life and public witness of the church and individual believers, and how is your view a healthier alternative for both?

Notice how all three preconditions force a theological framework on the views expressed by these four participants.

While each participant might be able to provide well-researched theological arguments for one view or another, three of them are simply not equipped to deal with the extensive genetic and anthropological evidence that tell us that humanity did not in fact originate from an individual couple, but instead is the product of an evolutionary process spread out over millions of years and involving thousands of breeding couples. Vague allusions were made to genetic evidence, but the overwhelming body of scientific data which I've summarized in chapter 6 were kept out of the discussion.

Many treatises on this question (of the historicity and theological significance of Adam, Eve and the fall in the garden) such as the two which I've summarized above begin with theological untouchables and *a priori* conditions to which any scientific explanation must conform, and then sometimes decorate that framework with a few scientific data which were carefully selected by unqualified advocates. Others have examined the question from the opposite side of the table.

Dr. Davis A. Young (professor emeritus of geology) began with a number of scientific findings—some of the paleoanthropological and genetic data which were available a couple decades ago—and then tried to adapt various theological models to those.[13] Young admitted defeat and concluded that "the biblical and scientific data pertaining to the antiquity and unity of the human race force us toward positions that are fraught with serious problems. The weaknesses of all these positions are sending a signal that careful reexamination of some basic exegetical and confessional premises is in order."

Drs. Hugh Ross and Fazale Rana have very recently provided an exceptionally large and up-to-date volume which examines in detail the genetic and anthropological evidence from a creationist perspective.[14] Both authors have high scientific credentials—doctorates in astronomy and chemistry, respectively—and in their book address perhaps every aspect of the science which bears on this question, doing so in a way that is easily understood by a motivated but nonexpert reader. Their conclusion is that Adam and Eve were historical individuals who were uniquely created by God, and all other hominids arose through the standard evolutionary process. However, they make very little attempt at addressing the theological aspects of their model.

It isn't only our understanding of Adam and Eve which may need to be reconsidered: there is also the question of what to do with Neanderthals, Denisovans, *Homo naledi* and other species of *Homo* which we now know about, as well as possibly other genetic cousins which may yet come to light. These were completely unknown to Moses, Paul, Augustine, and all the other biblical authors and influential church leaders up until the nineteenth century.

Many YEC theists believe that Neanderthals represent a distinct fully human clan—descendants of Adam—which was scattered after the fall and then was killed off in the flood. Some would even equate Neanderthals and other hominids with the Nephilim in the sixth chapter of Genesis, simply because the Nephilim are described as exceptionally undesirable (wicked to an extent that the *elohim* needed to come down to earth to investigate and intervene) and Neanderthals are often incorrectly perceived in very negative terms (as brutish, unintelligent cavemen). This seems to me to be very naïve, uninformed and contrived. But in either case, these views can't explain the degree of genetic diversity which is now known to exist between humans, Neanderthals and Denisovans. Until now I have been emphasizing the substantial similarity between these three genomes: they can indeed be said to be incredibly similar in the context of comparisons with chimpanzees, apes and primates, let alone with dogs, cats and mushrooms. But the differences which do exist (summarized on p. 146 of chapter 6) between the three of our *Homo* species simply could not have arisen within a mere ten generations (the number from Adam to Noah, according to Genesis five), or even merely one or two thousand years as per the timeline in the fifth chapter

---

13. Young, "The Antiquity and the Unity," 380–96.
14. Rana and Ross, *Who Was Adam?*

of Genesis. Any readers who think that a couple thousand years is enough time to produce such dramatic changes in the general appearance of humans should take a look at ancient Greek statues, or even older Chinese engravings, or yet older Egyptian wall paintings of humans: these do not show any dramatic change in our phenotype for many thousands of years.

It's also hard to contemplate the widespread distribution of Neanderthals throughout Europe and Asia, and Denisovans thousands of miles away from the Middle East (the presumed location of the garden of Eden), within the time-frame of a YEC model. There are other more detailed aspects of the genetic differences between the three species which are beyond the intended level of this book. Suffice to say, these YEC explanations of Neanderthals and Denisovans as descendants of a fully modern *Homo sapiens* named Adam are severely lacking in credibility.

Some who hold a more OEC view simply count Neanderthals and Denisovans among the animals: lacking a soul and other qualities pertaining to the image of God, which we jealously keep to ourselves.[15] There aren't any scientific data or biblical passages which would justify this kind of inter-species bigotry. Nor are there any to refute it. And for that reason one could choose to hold that view, but I would hope that those that do so recognize it is based solely on a preconceived bias. What is this prejudice based on? Is it theologically necessary? Is it a need to be superior to or more God-like than all other species? What is lost by giving up this exclusiveness?

Why not consider instead the possibility that the biblical references to humans in the garden of Eden simply reflect the limited understanding of Jewish authors—from Moses to the Apostle Paul—who did not have access to the kind of scientific data that we now have. In the same way that those early authors and theologians embraced a faulty Hebrew cosmological model of the Solar System based on their limited uninformed perspective, they also pondered the origin of mankind and came up with a story that made sense to them based on the bits of data available to them at that time.

Not surprisingly, reaction to this proposal—that humans are merely products of a naturalistic evolutionary process—can be harsh and visceral. For some, letting go of a historical Adam and Eve and the fall in the garden means letting go of the entire salvation story. To them, it would seem that I've betrayed my colleague who asked that I do whatever I wanted to do with the anthropological data as long as I left Jesus dying on the cross for our sins. (I'll explore other interpretations of the crucifixion story on pages 233–234).

But it isn't at all necessary to be so indiscriminate: to "throw out the baby with the bathwater." Neither should one skirt scientifically proven facts just because this appears to be a slippery slope. This is not a question of *whether* God used evolution. An overabundance of facts tells us that evolution did happen. Instead, it's a question of *how* he used that mechanism and how we then adjust the theology we inherited from our forefathers who didn't have the benefit of this new knowledge. Any aspect of our

---

15. Rana, "Neanderthal Brains."

theology which is founded in one way or another on an original primal couple living in a garden of Eden and a choice they made which led to a fall of mankind and original sin needs to be reconsidered.

This happens all the time in science. Sometimes a theory just doesn't work. The scientists involved with the question at hand just keep finding that the model they've been working with doesn't explain all the data. At that point, they drop the model, start with what they know as fact and which they can demonstrate with certainty, and then rebuild something new.

In the opening paragraph of this chapter, I used a metaphor of a mountain climber driving in pitons into the cliff wall as he struggled to his goal. At that point in the mental image, the climber needs to look up and plan the best path to the top. In the same way, with our own pitons now in place, we need to start considering where the next pitons might need to be placed. Later in this chapter, I'll propose a reinterpretation of sin and the human need for redemption which takes into account an evolutionary perspective.

If Adam and Eve Aren't Historical, What about Jesus?

The gospels of Matthew and Luke both recount the genealogy of Jesus Christ. Both trace him back to Abraham, while the writer of the Gospel of Luke continues on from there to connect Jesus with Adam (maybe it was the Greek side of him that wanted to find a connection further back than just at Abraham, father of the Jews). If one holds that Adam isn't historical, is one forced to conclude that Jesus must also be fictional?

Not at all.

There's certainly an abundance of historical evidence for the personhood of Jesus Christ of Nazareth. Not only biblical evidence, but also many other writings from inside the church and by non-Christian historians such as Josephus and Tacitus. Also, it's ludicrous to think that the NT church could grow out of a self-imposed delusion in the existence of a phantom of their own era. Why would they subject themselves to deprivation, hardship, torture, and even death in the Roman arenas solely because of their allegiance to a fictional character who they claimed outwardly to be intimately familiar with, but inwardly knew never existed? And why would their listeners and followers, who heard the stories of events in their own villages and cities which they could confirm or refute, also commit their very lives to this story?

A common rebuttal to questions like these often makes reference to other examples of people martyring themselves for something they too believe in. Today, the suicide bomber is the classic example, but one could easily come up with dozens of other examples of the same kind of thing. It's crucially important, though, to consider that key phrase: dying "for something they believe in." In this case, why would the disciples, the apostles, and their converts die for something they knew deep down was a ruse? A dogged faith built squarely upon someone—an individual person—that they knew never existed, or at least didn't do the things they claimed he'd done. Clearly there is something compelling about the establishment and rapid growth of the NT church. But I've digressed.

What then of the genealogy given in the Gospel of Luke? How can it at the same time be part mythical (with respect to Adam) and part historical (with respect to Jesus)?

That genealogy undoubtedly relied, at least in part, upon several genealogies given in Genesis, especially for its more ancient parts. Depending on one's view regarding the authorship of Genesis, the latter was either written by priests around the time of the Babylonian captivity a few centuries prior to the gospel writers, or was written a couple millennia earlier by Moses himself shortly after the Exodus from Egypt. Even if Moses wrote them, that is still half a millennium after Abraham, and approximately two millennium after Adam. That's a long time for a writer to try to reconstruct a genealogy (think how difficult it is even today to go back just a few hundred years, despite our computer technology and record-keeping).

Let's look more closely at that problem: how did the writer(s) of Genesis obtain these genealogies, and how much faith can we put on their reliability and historicity?

If one holds to the view that God essentially dictated the Scriptures to the author(s) (as Muslims believe regarding their Quran), or even handed them the texts (as the Mormons believe for their own Scriptures), then there could be reason to hold ultimate confidence in its accuracy and historicity. But that is not at all intended when most Christians refer to the inspiration of Scripture. Those that do hold to a view like that must also attribute to God all the inaccuracies and contradictions that we know exist in the OT books, as well as the morality which is sometimes hard to accept such as the slaughter of Canaanite women and children, and the killing of boys for hurling names at a prophet,[16] or the rules about how to treat an attractive woman captured in war.[17] I'm not going to divert the flow of argument by listing and dwelling upon them all here, but there are many detailed sources which list historical, moral, ethical, scientific, and logical problems such as these.[18]

On the other hand, if one acknowledges the human element in that data transmission over a couple thousand years, then one must accept that there will be inaccuracies and inconsistencies. It's hard to know how genealogies were recorded during the time between Adam and Abraham.

If it relied upon some physical form of data storage, what would that look like and how reliable was that? Obviously they didn't have anything like the computers, cameras and videos which we use to record information. It might surprise the reader that they also wouldn't have employed any kind of writing on paper. The earliest form of information storage that we know of, the one which was in place at the time that

16. 2 Kgs 2:23–5.
17. Deut 21:10–14.
18. Seibert, *Disturbing Divine Behavior*; Enns, *Inspiration and Incarnation*.

Abraham left Babylon (actually, he was then named Abram[19]), was scratching lines into clay tablets. We have detailed knowledge of their language and the way that these clay tablets were made and stored: in fact, we have mountains of their clay tablets documenting financial transactions and legal disputes, some dated to four or five millennia ago. It was the Egyptians who invented paper a couple millennia later. Don't forget that when Moses came down from Mount Sinai carrying the Ten Commandments, he carried them in the form of two stone tablets. There might be theological arguments which can be made to explain why those laws were introduced in that physical form, or it could simply be something that Moses and the Israelites were comfortable with: that's how people in that ANE milieu read things and transacted business. Or it could have been a logistical outcome: something to do with the fact that reeds—from which the Egyptians made paper—were hard to find in the desert or on top of a mountain (admittedly, they could have used animal skins, as was sometimes done).

Some will insist that we can have incredible confidence in the perfect accuracy of Scripture, and will base this on references to various forms of priestly/rabbinic systems in which every freshly copied manuscript was carefully scrutinized—some proponents even refer to one or more people peering over the shoulder of the copyist even as they were writing—to check for errors and if any were found, the manuscript would be destroyed. Irrespective of the veracity of such priestly/rabbinic systems, they would not have been in place until several millennia after the time of Abraham (again, Moses himself, the very founder of the Jewish religion and the rabbinic systems it accrued, appears two thousand years after Adam, and hundreds of years after Abraham). Up until approximately the second century BCE, "the texts were not yet completely fixed; their transmission was still fluid. Copyists made mistakes, wanted to improve or expand a text, or adapted the spelling of certain words. Sometimes the copies could be quite different versions of the same text—for example, the Book of Jeremiah. Thus, the biblical text in the Dead Sea Scrolls is not quite the same as the version that later became official in Judaism and Christianity."[20]

Clay tablets may have been a workable solution for a stationary civilization with massive stone libraries like the Mesopotamians had, but would have been impossibly impractical for a desert-wandering group of Bedouins such as the Semites were. To claim that Abram left Ur with sacks of clay tablets containing these genealogies, let alone all the other stories recorded in Genesis, as they travelled hundreds of miles from Ur of the Chaldees (on the Euphrates River, close to the Persian Gulf) to the Land of Goshen in Egypt (the eastern delta of the Nile River) really strains belief. Such sacks would be heavy, and the clay tablets prone to chipping and fracturing.

And that's only talking about any clay tablets that Abram might have possessed (we don't know that he did in the first place). We don't know that Adam ever created

---

19. Gen 15:7.
20. Sanders, "Missing Link in Hebrew Bible Formation," 46–52.

any clay tablets, or recorded information in any other way, nor that Cain or any other of his children then updated those records, nor the next generation, nor the next.

So there must have been some amount of oral transmission involved: stories passed down from generation to generation at gatherings in the temple, or in the marketplace, or around the campfire. If so, then the reliability plummets. Orally delivered (and remembered) stories would have to remain intact for at least two thousand years—the time from Adam to Abram—or even tens of thousands of years if one accepts newer proposals made by organizations such as RTB, who place a historical Adam at fifty to a hundred thousand years ago.

I know that some people today claim that oral transmission in those days and/or in that culture was impeccably reliable, but such claims are made without any foundational evidence. It's simply a claim that one might want to make, in fact one that is needed in order to have unquestioned acceptance of the content of those writings. But the fact remains: there is no evidence that it was any more error free than oral transmission is today. Many studies have been done of oral transmission in contemporary illiterate societies. While these studies have indeed documented an amazing ability of oral tradition to maintain a high degree of accuracy, they have also demonstrated that this accuracy is not perfect: and the errors which they've noted accumulate within a few decades. This level of accuracy is not enough to keep a story intact for centuries, let alone tens of thousands of years. Small errors accumulating every few decades will expand into huge errors over the course of thousands of years.

It's also true that people are able to memorize large volumes of text. Even children can memorize chapters from the Bible or Quran after a few hours or days of dedicated effort. You may have heard of someone reciting the whole Bible or Quran from memory, but only after they've spent significant portions of their lifetimes focussing solely on that goal through endless repetition. But there are a couple reasons why that is more possible today than it was in ancient times.

First, that may be possible in societies where those individuals can separate themselves from the struggle to survive and devote themselves to the handling of texts: societies that have monasteries, convents, yeshivas and ashrams. But those institutions are not found in the small groups of Semitic families herding sheep and wandering across hills and deserts.

Second, in the modern era, when we check the accuracy of the one doing the reciting, we compare their words against a written or electronic document and inevitably will find an error here or there which we can then correct them on. But in the premodern setting, that just wasn't the case. The only quality control on the memory of the one doing the reciting was the memories of others in the group doing the listening; human memory relying on human memory. And human memory is known to be faulty. Errors will occur, and if not caught early will reinforce themselves in the memories of the listeners and eventually become a part of the story. Even a small amount of error every decade repeated over two thousand or fifty thousand or one hundred thousand years results in a substantial loss of fidelity.

All of the above pertains to the reliability of the texts when efforts are made to keep them as accurately as possible. But a few words should also be said about how the stories and the genealogies might have been modified intentionally for theological purposes. The ancients saw nothing wrong in pruning an individual or two from a family tree for the sake of numerology and symbology (a family tree with very symmetrical branching looked more well-kept, perhaps more ordained by YHWH) or to emphasize the importance of certain ancestors. Matthew's genealogy of Jesus is specifically broken up into three groups of fourteen: "Thus there were fourteen generations in all from Abraham to David, fourteen from David to the exile to Babylon, and fourteen from the exile to the Messiah."[21] It seems that Matthew has exercised editorial license, because comparison of his genealogy with the stories given in 2 Chronicles[22] shows that he skips several generations to maintain this symmetry. Like many ancient societies, the Jews often attached numerical meanings to letters, and would add up the numbers represented by the letters in a name: fourteen was the number for David.[23] So for them, three groups of fourteen was a clear reference to the house of David and spoke volumes about divine providence in that family line. Borrowing an analogy from my previous book: it's like an author of a novel introducing a character using no more information than that he "lives in apartment #007 in the Double-Oh-Seven Condominium on Zero-zero-seventh Avenue." Without being told anything more about this person, you can make a very good guess about who he is, what he does and how important he'll be in the developing storyline. The same idea applies in a thoroughly Jewish mind when one says that Jesus's genealogy involves three groups of fourteen.

I don't doubt at all that Matthew and Luke fully believed those genealogies were accurate and historical. Both authors were steeped in a Jewish tradition which was handed down to them from their religious leaders for centuries. (Some claim that Luke was a Gentile, based on early church tradition; even if he was not actually a Jew, he was certainly motivated by Jewish thinking). All their relatives and neighbors believed it. They would have heard the characters in those genealogies referred to frequently in the synagogue. They knew nothing different. That doesn't mean that those genealogies have to be historically accurate.

For all these reasons, the genealogies in Matthew and Luke don't need to create any problems for believing in a mythical Adam and a historical Jesus.

---

21. Matt 1:17.

22. Compare 2 Chr 21:4—26:23 with Matt 1:8.

23. The Hebrew word for "David" is *DVD*, with *D* and *V* being the fourth and sixth letters of the Hebrew alphabet: therefore, "David" = 4 + 6 + 4 = 14.

Adam as the Forefather of All Humanity or Just of the Semitic Races

The Adam of Genesis is central to all three Abrahamic faiths (Judaism, Islam, Christianity). These faiths comprise a substantial fraction of the world's religious people today. A large number of those, and indeed of their predecessors over the entire course of history of those three major religions, hold Adam as the father of all humans. This tenet is based on passages in various Scriptures, including the book of Genesis in the Christian Bible, the Sefer Bereshit in the Hebrew Bible, and the Quran and Hadith for Muslims.

I believe the authors of Genesis wrote what they did out of a sincere belief in its historicity. I have no doubt that they saw the writings to be 100 percent accurate and historical, and that many of the parts that we might label as metaphor or allegory they saw as simply fact.

Otherwise, I don't know why the authors would bother with such precise details as the geographical descriptions of where to find the garden of Eden:

> A river watering the garden flowed from Eden; from there it was separated into four headwaters. The name of the first is the Pishon; it winds through the entire land of Havilah, where there is gold. (The gold of that land is good; aromatic resin and onyx are also there.) The name of the second river is the Gihon; it winds through the entire land of Cush. The name of the third river is the Tigris; it runs along the east side of Ashur. And the fourth river is the Euphrates.[24]

The writer mentions not only the names of some places which could be mythical (there's no place on earth today where these rivers and lands converge, although it is true that global geological events could have markedly altered the landscape), but also tells the reader what valuable substances can be found there! Those kinds of details aren't necessary for a metaphor. In fact, some might say that such level of minute detail is not consistent with the telling of a fable or myth (although the *Epic of Gilgamesh* and the stories of King Arthur also have a great deal of minute detail).

The details contained within the genealogies in Genesis[25] also hint how strongly they believed in the veracity of these accounts. These not only name specific male individuals, generation by generation, but also record the precise age at which each male fathered his first-borne and later died.

But those very details themselves also raise considerable doubt in my mind about their veracity. It isn't just the fact that many of these individuals are said to have lived for hundreds of years: that alone I find very hard to accept.

---

24. Gen 2:10–14.
25. Gen 5, 10, and 11.

What is far more unbelievable to me than their unusually long life spans is how old each one is said to be when they have their first son. For the first ten generations,[26] many of the fathers are well over one hundred years old before siring the next generation; the youngest ages given are for Mahalel and Enoch, both of whom are said to be sixty-five years old before this milestone event in their lives. During the next ten generations,[27] the patriarchs are said to all be in their third decade before fathering their first sons. Today, any sexually active couple not using some form of birth control finds themselves pregnant within a year or two of enjoying intimacy. During the time referred to in the first eleven chapters of Genesis, on the other hand, God had given them the *command* to "be fruitful and multiply,"[28] it is said that "the daughters of men were beautiful,"[29] there were no laws against any kinds of sexual activity (those laws would not have been written until two thousand years later), and there was no form of contraception available other than the rhythm-method (assuming they would even try to control their birth rate, given God's direct *command* to them to be fruitful and multiply). We have no other historical example of human populations living under conditions like that and being able to avoid pregnancy for even a few years, let alone till they reach a hundred years of age. We're just too prone to that primordial urge to reproduce and too capable of doing so successfully.

Setting aside whether or not those passages are in fact accurate, it is details such as those (the description of the location of the garden of Eden, and the genealogies) which lead me to conclude that the authors and the readers of that time clearly believed it to be true that Adam was the father of all humanity. But that doesn't mean it has to be true. And many findings from various fields of science (particularly archaeology, anthropology, linguistics and genetics) now tell us that *Homo sapiens* long preceded the Adam of Genesis.

This revelation forces some to reject the entire Genesis account as entirely fictional, and some even go so far as to then reject the entire Bible.

Others try to reread the Genesis accounts in more of a metaphorical manner.

One other possibility is that Genesis recounts only the origin of the Semitic races among all the other races of humans.[30] Could these Semitic writings reflect only a Semitic story: *their* history, and *their* origins, rather than those of all humans? Could Adam be merely one of the many *Homo sapiens* who appeared through the normal process of evolution, and was the father of many races including the Jews? Isaac La Peyrère, the one who rocked the theological world with his heavily researched pre-Adamic theory (see p. 66), was one of many who over the past several centuries believed this to be the case. Many other peoples and races have recorded their own

26. Gen 5.
27. Gen 11.
28. Gen 1:28.
29. Gen 6:2.
30. Hooke, *In the Beginning*, 15–38; Enns, *Evolution of Adam*, 65–70.

cherished origins stories, some of them as unique and intriguing as the ones recorded in Genesis. One can still see Genesis as divinely inspired, and dig deeply to find that inspiration within its pages, without requiring it to be absolutely historical and factually accurate.

Imago Dei

The first few chapters of Genesis make it clear that mankind is distinct from the other animals. After the various acts of creation described for days one through five, and after beginning day six with the creation of "the creatures that move along the ground,"[31] we then read about God making a unique and separate decision to create a new being—*ha adam*, or mankind—"in our image, in our likeness, so that they may rule over" all the other creatures.[32] There are several references in Genesis to creation in the image and/or likeness of God,[33] always applying this to humans and never to the other things that God created. (Logically speaking, though, this does not mean that other species didn't also have this image and likeness: just because it isn't stated explicitly doesn't mean it wasn't the case. The garden of Eden is said to have contained trees and fruit and green plants, but just because it never mentions vegetables doesn't mean they were necessarily absent. "Absence of evidence is not evidence of absence.[34]")

What might the text mean by this concept of humans being created in the image and likeness of God—the *imago Dei*—and what impact does that concept have for contemporary Christian discipleship, teaching, and ministry?

First, let's look closely at the words that are translated into "image" in Scripture.

The passages in Genesis use the synonymous Hebrew words *tselem* ("image") and *demuth* ("likeness"). *Tselem* is often used elsewhere in Scripture in the context of idols (for example, the numerous references to graven images, or the many references in Daniel to the golden image in King Nebuchadnezzar's dream). *Demuth* is used many times in Ezekiel and one other time in Isaiah, referring to various beings or elements in their visions having a resemblance to other things on earth.[35]

The NT, on the other hand, most often uses the Greek word *eikon* ("image": that which resembles an object and represents it, as a copy represents the original). In addition to Paul's use of *eikon* to refer to the image of God (or of Christ),[36] it is also used to refer to Caesar's image on a coin,[37] or to the image of a god or idol.[38]

---

31. Gen 1:24.
32. Gen 1:26.
33. Gen 1:26–7, 5:1, 5:3, 9:6.
34. Generally attributed to Carl Sagan.
35. Ezek 1:5, 10, 13, 16, 22, 26, 28; 8:2; 10:1, 10, 21, 22; Isa 40:18.
36. Rom 8:29; 1 Cor 11:7; 1 Cor 15:49; 2 Cor 3:18; Col 1:15; Col 3:10
37. Matt 22:20; Mark 12:16; Luke 20:24.
38. Acts 19:35; Rom 11:4; Rev 12:14, 15; 14:9, 11; 15:2; 16:2; 19:20; 20:4

Interestingly, the writer of Hebrews alone uses the Greek word *charakter*, which is derived from another word used to refer to the tool of an engraver, implying the concept of a stamped image, and is translated as "exact copy," "exact representation" or "exact image."

What the Imago Dei Is Not

Theologians have never agreed universally on what the term *imago Dei* means, although there is a fair degree of unanimity that it does *not* refer to any physical likeness. Although we might say that a child "is the image of his father" (or even use the colloquial expression "spitting image") to refer to a seemingly exact physical likeness (usually in the face) between a father and son, there is no reason to suggest that the biblical reference to "God's image" means that any aspect of our own bodily form reflects that of God's bodily form. Theologians do not hold that God actually has a face with eyes and mouth, nor arms with hands and fingers, nor legs with feet, despite the many biblical references to all of those very same body parts, as well as references to his nostrils spewing smoke,[39] and another reference to his back[40] (which the King James translates as "back parts"). God is not physical, and our being created in his image does not reflect something physical or organic.

The *imago Dei* is also not gender specific; it seems to be something that both males and females can exhibit, since both are said to have been so created.[41] In fact, Brauch[42] and Middleton[43] argue that the *imago Dei* is not something that an individual possesses, but is something that emerges from the "male-female complementarity and partnership," and is only realized "in the context of human relationships, in human communities of love."[44] Numerous biblical passages describe God using imagery that is either masculine or feminine in nature. Exodus 15:3 refers to him as a warrior, and many other passages describe him as going into battle, which in that era was an activity for males only. Interestingly, God is also frequently referred to as a king,[45] but never as a queen (a modern Western mind can view a queen who is completely sovereign, but an ancient Hebrew mind could not). Yet God is also given a very feminine image

---

39. 2 Sam 22:9; Job 41:20; Ps 18:8.
40. Exod 33:21–23.
41. Gen 1:27.
42. Brauch, *Abusing Scripture*.
43. Middleton, *Liberating Image*.
44. Brauch, *Abusing Scripture*, 128.
45. There are too many references to list comprehensively here, but one such occurrence would be Psalm 95:3.

as a hen gathering her chicks under her wing,[46] and there are references to God giving birth[47] and mothering his nation.[48]

The Ancient Hebrew Understanding of "Imago Dei"

Middleton provides a comprehensive and enlightening study on the *imago Dei*.[49] One particularly valuable contribution is his study of the ANE context for this concept of "image," including:

> (1) a reference in the *Gilgamesh Epic* that describes Enkidu as an image (or double) of Gilgamesh the king; (2) two references in Egyptian wisdom literature to the creation of humans as the images of a god; (3) the widespread practice of Egyptian and Mesopotamian kings setting up statues or images of themselves in lands where they are physically absent; and (4) Egyptian and Mesopotamian references to kings (and sometimes priests) as the image of various deities.[50]

These concepts of humans bearing the image of their god(s), and serving as the representative of the divine toward humans, were prevalent in those Egyptian and Mesopotamian societies which preceded the birth of the Hebrew nation, and which subjugated the Hebrew people. For example, Middleton recounts an Egyptian statue of King Adad-iti found in modern-day Syria which bears an inscription which "contains the Aramaic equivalents of both *ṣelem* and *děmû* and the Akkadian equivalent of *ṣelem* as synonymous terms designating the statue."[51] That worldview certainly influenced the writing of the Hebrew sacred texts. "It is now generally agreed that the image of God reflected in human persons is after the manner of a king who establishes statues of himself to assert his sovereign rule where the king himself cannot be present."[52] Likewise, John Walton[53] interprets the creation story as God building a temple and installing his image in it. In particular, YHWH took a universe which was "formless and void," then created forms or spaces on the first three days (the heavens, the sky, the seas, the land), before filling those empty spaces on the corresponding days four to six (birds, fish, animals). And then, following the common practice of temple-building in that ancient era, YHWH is said to consecrate his own temple with his image: mankind. Whereas the other ANE civilizations would use a lifeless repre-

46. Ps 91:4; Matt 23:37.
47. Job 38:29.
48. Isa 66:13.
49. Middleton, *Liberating Image*.
50. Ibid., 94.
51. Ibid, 107.
52. Brueggemann, *Genesis*, 32.
53. Walton, *Lost World of Genesis One*.

sentation of their god made of stone, metal, and jewels, God here lovingly places an "image" made out of living flesh, and having a mind and spirit, to rule over it all. In this sense, the creation story can be taken as a polemic against the religious thought of that time.

Imago Dei during the Apocryphal, Apostolic and Patristic Periods

There is very little mention of the concept of *imago Dei* in apocryphal or deuterocanonical literature. Sirach maintains the ancient Hebrew thinking that it is related to dominion and authority over creation.[54] In the Wisdom of Solomon, it is identified with immortality and incorruptibility.[55] Finally, in 2 Esdras, the *imago Dei* is put forward as the reason that God should spare Israel.[56]

The Apostle Paul seems to bring an additional perspective to the dimensions of the *imago Dei* when he writes about head coverings, saying that "a man . . . is the image and glory of God; but woman is the glory of man."[57] Elsewhere, he writes that neither gender yet possess the fullness of the image which Christ bears, saying we: are "predestined to be conformed to the image of his Son";[58] "are being transformed into the same image with ever-increasing glory";[59] must "put on the new nature, created after the likeness of God in true righteousness and holiness";[60] and will yet "bear the image of the heavenly man."[61] Interestingly, the writer of Hebrews uses an entirely different word—*charakter*—to refer to Christ as "the exact representation" of God.[62] Altogether, these passages suggest a spectrum of degrees of possessing or displaying the *imago Dei*, or suggest a progressive spiritual metamorphosis. Subsequent church fathers developed this spectrum further, finding evidence for lesser aspects of God's image in other nonhuman creations: Dionysius found aspects of God's image in "irrational creatures," in a solar ray, and even in the universe as a whole, while Boethius found this in the whole world.[63] Aquinas later challenged this thinking in his lengthy excursis of the subject.[64]

Valentinian thinking tried to separate image and likeness, claiming that God created a likeness which was spiritual and masculo-feminine in nature, and then encased

---

54. Sir 17:3; Middleton, *Liberating Image*, 16.
55. Wis 2:23; Geyer, *Wisdom*, 66–67; Clarke, *Wisdom*, 27; Middleton, *Liberating Image*, 16.
56. 2 Esdras 8:44; Middleton, *Liberating Image*, 16.
57. 1 Cor 11:7.
58. Rom 8:29.
59. 2 Cor 3:18.
60. Eph 4:24 RSV.
61. 1 Cor 15:49.
62. Heb 1:3.
63. Aquinas, *Summa Theologica*, article 2.
64. Ibid., articles 2 and 4.

that within a male image fashioned out of clay.[65] Origen saw that initial creation as occurring prior to the creation events described in Genesis, and the encasing occurring as described in the second chapter of Genesis.[66] Irenaeus challenged the Valentinians with his own view that the *imago Dei* referred to Adam being shaped, physically from mud as well as spiritually according to the body and spirit of Christ.[67] "Thus we are in the image of God because our bodies have been shaped after the pattern of the body of the incarnate God."[68] Adam lost the fullness of that image when he fell in the garden, but humanity will regain it in the fullness of time, when the story of salvation has reached its fulfillment.[69] Origen likewise saw the image of God within us as having been corrupted, that when we indulge in earthly desires we put on an earthly image which is perishable, and it is our mission to strive to put on the image of God (which is incorporeal and eternal) and put off the earthly image.[70] Augustine saw the image and likeness as distinct characteristics,[71] and Thomas Aquinas echoed that view nearly a millennium later: "it is clear that likeness is essential to an image; and that an image adds something to likeness—namely that it is copied from something else . . . for instance, an egg, however much like and equal to another egg, is not called an image of the other egg, because it is not copied from it."[72]

As time passed, other aspects of *imago Dei* were focused upon or new ones introduced, and its localization also began to be shifted: away from an interconnected mankind in general comprising both genders, and instead becoming individualized, "as though each individual bears the image of God within him- or herself."[73] In fact, Middleton describes how theologians and interpreters during and since the Patristic era have altered the meaning of *imago Dei* to suit their developing theologies, quoting one OT scholar as saying they "made the word mean just what they choose it to mean."[74] Middleton writes:

> Although various candidates were suggested for the content of the image, David Cairns can comment that, as a bare minimum, "in all the Christian writers up to Aquinas we find the image of God conceived as man's power of reason."[75] This notion of the rational, substantial soul mirroring its divine

---

65. Holsinger-Friesen, *Irenaeus and Genesis*, 116–17.

66. McGuckin, *Origen*, 83–85, 131; Tripolitis, *Origen*, 19–20.

67. Minns, *Irenaeus*, 59; Osborn, *Irenaeus of Lyon*, 196, 213; Surin, "Atonement and Christology," 144.

68. Minns, *Irenaeus*, 60.

69. Ibid., 61; Osborn, *Irenaeus of Lyon*, 93 and 213.

70. McGuckin, *Origen*, 131–34; Tripoliitis, *Origen*, 24–31; Smith, *Ancient Wisdom*, 36–45.

71. Sullivan, *Image of God*, 125.

72. Aquinas, *Summa Theologica*, article 1; see also article 9.

73. Barr, *Life*, 475.

74. Middleton, *Liberating Image*, 18.

75. Cairns, *Image of God in Man*, 110.

archetype—which is part of the pervasive influence of Platonism on Christian theology—is nuanced or supplemented in the Latin West by notions such as conscience, spirituality, immortality, freedom, and personhood and by Augustine's famous proposal of various intrapsychic trinitarian structures (particularly memory, intellect, and will), which correspond to the triune nature of God. In the Greek East the substantialistic image was often understood dynamically, as the progressive conformity of the soul to God or a salvific partaking of the divine nature, a process typically called "divinization."[76]

The seeming drifting of the Patristic interpretation of the *imago Dei* (that it pertains to various intellectual aspects of an individual) away from the ancient Hebrew thinking (that it pertains to the Divine ambassadorial responsibility and authority of mankind generally) may be related to the fact that the former did not have access to the archaeological artefacts from the Akkadian, Sumerian, Egyptian and Babylonian empires. Contributing to this loss of ancient cultural context was the change in their own contemporary context: the ancient Hebrew texts were now being exegeted by Greek and Roman theologians living outside of the geographical and cultural borders of Israel, many (most?) of whom could not read or speak Hebrew.

Imago Dei during the Present Era

We today have been able to rediscover both the ancient ANE context and the Hebrew language. Scholars such as Middleton[77] and Walton,[78] among many others, have brought that ancient Hebrew perspective back to life for us using texts which have only been discovered within the past couple centuries. Also, we have become aware of how our own current worldview affects our interpretation of Scripture. Nonetheless, in the past few decades, there has been a large and growing number of advocates engaged in Christian apologetics continuing in that Patristic tradition, making definitive statements about the nature of the *imago Dei* as some kind of intellectual ability, in order to identify the point in human evolutionary history at which we acquired it, or to state definitively whether a given hominid species possessed it or not.

One of the earliest of these, Henry Morris (founder of the Creation Institute in 1970), wrote: "In any case, there can be little doubt that the 'image of God' in which man was created must entail those aspects of human nature which are not shared by animals—attributes such as moral consciousness, the ability to think abstractly, an understanding of beauty and emotion, and above all the capacity for worshiping and loving God."[79]

---

76. Middleton, *Liberating Image*, 19.
77. Ibid.
78. Walton, *Lost World*.
79. Morris, *Genesis Record*, 74.

The suggestion that the *imago Dei* refers to our sense of morality, especially an awareness of God's moral laws, would conflict with the Genesis accounts, in which we find humans imprinted with the *imago Dei* from the very moment they were created, but much later in the storyline seeking to bite into the Tree of the Knowledge of Good and Evil because they desired that ability (either because they felt they didn't have any of it, or didn't have enough of it).

Many others have added to Morris's list of the various intellectual qualities which are claimed to be absent in animals: a capacity for conscience, empathy, love, volition, reason, freedom, language, creativity, and/or symbolism. Here, though, I feel it becomes too easy to conflate two very different questions: what it means to be created in the image of God, and what it means to be a human. For some of these proposals, it becomes quite difficult to extract a theological element (as should be possible, one might think, when addressing a theological concept such as the *imago Dei*): how might it be that the ability to speak or to think in abstract terms has some kind of theological implication? For this reason, I've chosen to deal with those proposals in the next section which explores "human-ness" specifically. Besides, we have now found all of these characteristics to varying degrees in many higher animals such as primates, dolphins, whales, dogs, elephants, birds, and many other species. Aquinas had already recognized this nearly a millennium ago when discussing whether angels were also created in God's image: "Unless we presuppose the first likeness, which is in the intellectual nature; otherwise, even brute animals would be to God's image."[80] Even the claim that only humans can worship God could be countered with the nineteenth psalm ("the heavens declare . . .") and the common reference to "the birds singing God's praises." As for the claim that only humans can "love" God, we have no evidence or understanding whatsoever of what goes on inside the minds of animals, so how can we make such a claim?

Could the *imago Dei* be behind the universal need in humans to reach out to some "Great Being"? Throughout the scope and period of human history, in every part of the globe, every society, and every demographic subset of the human population, some form of religion can be found. Jared Diamond, a renowned scientist and Pulitzer Prize-winning novelist, wrote, "Virtually all known human societies have had 'religion,' or something like it. That suggests that religion fulfills some universal human need, or at least springs from some part of human nature common to all of us. If so, what is that need, or that part of human nature?"[81] Even Neanderthals and Denisovans seemed to share that propensity, since they buried their dead with jewelry, food or tools, indicating an awareness of an afterlife.

By definition, every image of an object requires an original source. The former is inextricably linked to the latter; there is a conceptual vector which always points from the image to its source. We are not self-defining or self-referential; we might choose

---

80. Aquinas, *Summa Theologica*, article 3.
81. Diamond, *World until Yesterday*, 324.

to believe that we are self-actualizing, but theologians would beg to differ (and the honest among us will admit that we're not really the masters of our own destiny, but are instead all through life merely responding to the curve balls that life throws at us). Even those who consider themselves to be atheists can sometimes turn to God or lash out against him when disaster strikes. But nobody strikes up a deal with Superman when they're about to die, or blames Zeus when their loved ones are killed in a car accident, or asks Darth Vader, "Why me?" when they're diagnosed with a deadly disease. So despite lifelong claims that God is only mythical, they clearly put God in some category separate from Superman, Zeus, and Lord Vader. It seems that this expectation or searching for a "Higher Being" is a fundamental core of our being. Something that we yearn for; at least as long as we're not distracted by the constant sensory input from TVs, radios, cell phones, portable music players and other forms of digital distraction which deprive us of introspective moments.

As one last possible consideration on a point like this that so easily resists being restricted to a certain definition or to boundary setting, perhaps it is worth asking whether the *imago Dei* is simply a product of the ancient Hebrew poetic mind. The authors of Genesis saw humans as being so completely distinct from the rest of creation that we must have some kind of extra ingredient that the animals don't have. Just like the Sumerian and Egyptian stories of the gods incorporating their own flesh or bodily fluids into the human beings that they had created (see p. 190). But we can now see that in so many respects we're very little different from the animals. In the next couple of sections, we'll see that it's also extremely hard to define exactly what distinguishes humans from animals.

Implications of the Imago Dei

In the first creation account, God commanded mankind to subdue the earth and rule over "everything that has the breath of life in it,"[82] and this is echoed in the Psalms.[83] In the second creation account, he created a human and "put him in the garden of Eden to work it and take care of it."[84] In all three passages, humans are given responsibility over creation. The definition of *imago Dei*, as originally interpreted by the ancient Hebrews, was to partake in God's authority and care over his creation, having been given "the commission to extend God's royal administration of the world as authorized representatives on earth."[85]

Our intellect and technology—two aspects which some have recently attributed to the *imago Dei*[86]—have now powerfully enabled us to care for creation. Unfortunate-

---

82. Gen 1:28
83. Ps 8:6–8.
84. Gen 2:15.
85. Middleton, *Liberating Image*, 289.
86. Rana and Ross, *Who Was Adam*, 99.

ly, those have also enabled us to damage or even destroy it. In many respects, we have failed repeatedly in our husbandry of God's creation. Middleton writes, "It has now become popular wisdom to make the historical claim that the modern environmental crisis, which is the direct result of the exploitative stance toward nature characteristic of modern, Western science, can be traced back to the culture of Western Christianity in which modern science arose. This culture, notes White, was informed paradigmatically by the creation story in Genesis 1."[87] Fortunately, there is a growing awareness of this failing, as well as a growing effort to reclaim our responsibility and to reverse the damage we have done.

For many, the concept of the *imago Dei* is the primary reason many people stand against the practices of abortion, euthanasia, and eugenics. Theophilus of Antioch wrote: "When God said, 'Let us make man after our image and likeness,' he first reveals the dignity of man. For after making everything else by a word, God considered all this as incidental; he regarded the making of man as the only work worthy of his own hands."[88] The Epistle of James reprimands even the mere cursing of our fellow humans, because they were created in God's likeness.[89] God himself imposed a steep penalty on the crime of murder: "Whoever sheds human blood, by humans shall their blood be shed; for in the image of God has God made mankind."[90] Altogether, the Bible commands deep respect for the sanctity of human life.

Another tremendous implication of the *imago Dei* is something that I'll refer to as theological bigotry. Questioning whether it might be possible for one to possess less or more of the *imago Dei*—the progressive spiritual metamorphosis or spectrum to which I referred above—may sound like merely an esoteric academic question, like arguing how many angels can fit on the head of a pin (a question which apparently had the scholarly world in Medieval times tied up in knots). But history has shown that the tendency of the church to claim total and exclusive ownership of the *imago Dei* has had great impact upon its interactions with others. More precisely, it has spawned an ugly and destructive form of theological bigotry directed not only toward our fellow humans, but also to extinct hominids as well as to extraterrestrial aliens (please bear with me).

In his book *Adam's Ancestors*, David Livingstone opens with a couple interesting and yet embarrassing and appalling chapters of church history describing the theological response to first encounters with other groups of humans in distant lands for which there was no possible explanation within the first eleven pages of Genesis. He extends this history as far back as the time of the first-century Roman encyclopedist Pliny the Elder who described a fictitious race of people known as the Plinians and

---

87. Middleton, *Liberating Image*, 273; also see Barr, *Life*, 474.
88. Holsinger-Friesen, *Irenaeus and Genesis*, 113.
89. Jas 3:9.
90. Gen 9:6.

who evoked "theological nightmares for Christian writers."[91] Augustine felt compelled to weigh in on this matter to stem the theological bigotry he witnessed: "Whoever is born anywhere as a human being, that is, as a rational moral creature, however strange he may appear to our sense in bodily form or color or motion or utterance, or in any other faculty, part or quality of his nature whatsoever, let no true believer have any doubt that such an individual is descended from the one man who was first created."[92]

But these "theological nightmares" became reality, and Augustine's declaration in turn was questioned, when the church began to hear that "the Plinians" were not all so fictitious after all. Europeans encountered Eskimos in northern Greenland, and indigenous peoples when exploring ships landed on the coasts of North America, South America and Australia, and pygmy tribes in Africa. All of these triggered a storm of debate and more theological nightmares among the church fathers. Of course, there were academic questions addressing the logistics of how those people survived the flood and/or how they came to be in those distant lands. But more to the point of this chapter, many church leaders insisted that those foreign peoples could not possibly have derived from Adam: if they had, they would have been Caucasian (just look at any painting from the seventeenth century)[93] and would have been destroyed in the flood anyhow. Since they were not of Adam, they were deemed to be not human. Not worthy of evangelization, nor of extending basic rights such as personal liberty or religious freedom. Livingstone writes of the storm that erupted within the church over the discovery of the American aboriginals: "Pope Julius II decreed in 1512 that the Indians were descended from Adam," but this did not quell intense debate on the matter, and half a century later "the papal legate had been dispatched from Rome to Spain to determine once and for all whether the Indians shared the *imago Dei* or were a distinctly other species, whether they were fundamentally bestial and fitted only for slavery or sufficiently advanced that they should not be considered barbarians."[94]

History bears sad testimony to how this question was so often answered for subjugated indigenous peoples.

It isn't just contemporary aboriginals who are victims of our theological bigotry. I've found that the concept of the *imago Dei* arises frequently when theologians (both with and without academic credentials) discuss anthropological findings of Neanderthals, Denisovans, *Australopithecus, Homo erectus* and many others, some of whom we now know had essentially identical genetic coding and exhibited advanced forms of tool use, a sense of the aesthetic and abstract (they made jewelry, and painted cave walls), and seemed to have a belief in the afterlife (they buried their dead, sometimes with tools, weapons, and food, and planted flowers around graves).[95] Wilcox has pro-

91. Livingstone, *Adam's Ancestors*, 11.
92. Ibid., 14; Alexander, *Creation or Evolution*, 237.
93. Livingstone, *Adam's Ancestors*, 135.
94. Ibid., 19.
95. Pääbo, *Neanderthal Man*; Livingstone, *Adam's Ancestors*; Leroi-Gourhan, "Shanidar," 79–88.

vided a very interesting and well-researched review of the evolution of the *imago Dei* in human ancestors.[96] Despite these similarities with our own species, the categorical statement is often made that those species did *not* have the *imago Dei*. I don't know on what basis that claim can be made, and I wonder why we guard that quality so jealously to our own species.

Some rationalize this by saying that something remarkable happened in human evolutionary history approximately fifty thousand years ago. A cultural Big Bang. Prior to that point in time, we had some crude manufacture and use of stone tools and fire, possibly some form of culture (living in small, collaborative communities) and some sense of abstract concepts like the afterlife (burial gifts). Then all of a sudden, the archaeological record shows a sudden acceleration in our technologies. Our tool kit was upgraded to include ropes, fishhooks and harpoons. We shifted from being hunter-gatherers to being agriculturally based; plants and animals became domesticated and farmed. Inter-tribal trade became established, and raw materials (different types of rocks, shells and foods) were exchanged: from these, we see the development of metallurgy and architecture. Cave art, beads, jewelry, musical instruments and sculptures fashioned out of wood, bone, stone, ivory, antler, horn and baked clay made their appearances. This sudden change in human abilities is taken by RTB[97] and others as a marker or indicator of the conferring of the *imago Dei*. Not the image of God itself, but a by-product of that image.

It is undeniably true that in terms of technological sophistication (manufacture and use of tools; medicine; art; culture), those early humans were far superior to the animals. There seemed to be an exponential rise in the level of technological sophistication, well above even the smartest animals like chimps and dolphins today, and a whole category above the level of most other species. But does that mean that technological sophistication is a proxy for the *imago Dei*?

An inherent property of technology is that it expands exponentially: each and every new discovery catalyzes several others. The culture and technological sophistication of the Egyptian empire would be exponentially higher than that of early humans fifty thousand years ago, and that of the Roman empire exponentially higher above the Egyptian, the European nations of the Renaissance period exponentially higher yet, and the technology of the mid-twentieth century much more so, which in turn pales against the technology of today. Why or how then do we draw a line at some arbitrary point in that steeply exponential curve and declare anything above that to have the *imago Dei* and anything below that to not have the same? How can technological sophistication be used as a proxy or benchmark or indicator of the *imago Dei*?

This theological bigotry has even been directed against extraterrestrials. I had to laugh, and then hang my head shaking, when I came across some discussion on the blogosphere claiming that any extraterrestrial intelligent life would, along with all of

---

96. Wilcox, "Proposed Model," 22–43.
97. Rana and Ross, *Who Was Adam*, 99.

creation, also be subject to the sin of Adam but would have no recourse to salvation, being on a different planet and therefore unaware of the good news. AiG have pontificated on this matter: "The Bible makes it clear that Adam's sin affected the whole universe. This means that any aliens would also be affected by Adam's sin, but because they are not Adam's descendants, they can't have salvation."[98] In this, they are in stark disagreement with theists going as far back as the Dominican philosopher Tommaso Campanella who, while also contemplating life on other planets, wrote in his 1622 book *Defense of Galileo*: "If the inhabitants which may be in other stars are men, they did not originate from Adam and are not infected by his sin."[99]

Humans generally have an arrogant perspective by which we see ourselves as the pinnacle species, the crown of creation. This is reinforced in part by biblical passages such as the eighth psalm (written by a human, albeit divinely inspired). We need to keep in mind that Scriptures say nothing at all about the possibility of "Plinians" on other planets (a possibility that scientists say is highly likely), just as much as it gave no indication to the early church fathers of aboriginal non-Adamites living on continents beyond the Middle East.[100]

This section of my book may seem facetious, trivial, or irrelevant. It is not at all my intent to rehabilitate the maligned reputation of Neanderthals, nor to preach the gospel message to the inhabitants of Alpha Centauri. Instead, I feel it is necessary to point out how an unsubstantiated, inflated or unfocussed view of the *imago Dei* simply reinforces our theological bigotry, magnifies our own reputation of being intolerant and anti-science, compromises our apologetics, and otherwise embarrasses the church and maligns the name of Christ.

What Does It Mean to Be Human?

As already mentioned in chapter 4, Carl Linnaeus was the first to standardize the way species and genera are named and defined. In the first edition of his 1735 book *Systema Naturae*, he named each animal genus and carefully noted various defining physical features which distinguished them from the other genera. Under *Homo* he simply wrote "*nosce te ipsum*," which is Latin for "know thyself."

The writer of Ecclesiastes wrestled with this question thousands of years ago. In chapter 3 of that book, the writer surveys the full breadth of what is the human condition. From the pitiable on the one hand ("What do workers gain from their toil? I have seen the burden God has laid on the human race"[101]) to the admirable on the other

---

98. Ham, "Find a New Earth."
99. Livingstone, *Adam's Ancestors*, 24.
100. Barr, *Life*, 478–9.
101. Eccl 3:9–10.

("He has also set eternity in the human heart"[102]). He sifts through his observations on the matter and eventually concludes,

> As for humans, God tests them so that they may see that they are like the animals. Surely the fate of human beings is like that of the animals; the same fate awaits them both: as one dies, so dies the other. All have the same breath; humans have no advantage over animals. Everything is meaningless. All go to the same place; all come from dust, and to dust all return. Who knows if the human spirit rises upward and if the spirit of the animal goes down into the earth?[103]

There is certainly a lot of depth in that well which can be plumbed by theologians and philosophers! But from a scientific point of view, what exactly does it mean to be human as opposed to animal?

Is it a capacity for conscience, morality or empathy? The fall in the garden centers around eating from "the Tree of the Knowledge of Good and Evil."[104] Many owners of dogs will testify with absolute conviction and certainty that their pets "know" when a rule has been broken: there's a distinct change in the animal's behavior. Studies have shown that dogs[105] and Capuchin monkeys[106] have a sense of fair and unfair (in other words, when one has been wronged). Similar claims are made for dolphins and whales. Rowlands has written a great perspective on the morality of animals.[107] Dogs[108] and bonobos[109] have also been shown to exhibit forms of empathy. So we're not completely distinct from the animals in these respects.

Is it a sense of free will? Again, many pet owners are convinced that their pets sometimes choose to do certain things, whether those be right or wrong, selfish or altruistic. I recently heard a radio interview of a dolphin trainer addressing the question of how animals (in this case, dolphins) communicate, and he testified that he would give a hand signal to the dolphins to "do something, anything, that I've never trained you to do before," and they would go to the bottom of the pool, appear to confer for a short while and then pop out of the water doing some carefully synchronized maneuver. Although the radio show was exploring how those animals communicated beforehand, I also marvelled at how this showed the dolphins had the free will to plan and decide: those dolphins chose to do that particular stunt. There are many examples of animals choosing to do something that goes against the better judgement of their

---

102. Eccl 3:11.
103. Eccl 3:18–19.
104. Gen 2:9, and 17.
105. Range et al., "Absence of Reward," 340–45.
106. Markey, "Monkeys Show Sense of Fairness."
107. Rowlands, *Can Animals Be Moral?*
108. Palagi et al., "Rapid Mimicry and Emotional Contagion," 150505.
109. Palagi et al., "Yawn contagion," e519.

instincts: going into a burning building to save a human or even another animal, or inserting themselves between a human and some dangerous situation, for example.

Are humans uniquely self-aware? Are we uniquely able to distinguish ourselves as a separate entity? Do all animals lack any sense of the demarcation between self and the rest of the world? Many lines of evidence suggest the answer to these questions is no.[110] Marine biologists have also demonstrated that dolphins are keenly interested in looking at themselves in mirrors hung in the pool. Some might interpret this as the dolphins simply perceiving the mirror reflections as other dolphins—trespassers in their territory—and responding with a certain level of curiosity and wariness. However, when the marine biologists strapped objects to the backs of the heads of those dolphins, the latter immediately swam over to the mirrors and maneuvered themselves to be able to see what the objects were: so they clearly saw those reflections as representing themselves, not strangers. Likewise, many primates will also use mirrors to investigate parts of their bodies that they otherwise can't see directly. Who knows what else is going on inside their heads regarding their personal existence? Experiments done to explore communication between humans and various great apes (for example: Kanzi, a bonobo chimp; Koko, a lowlands gorilla; and Washoe, a common chimpanzee) reveal many signs that these animals have a very distinct sense of self.

Can the difference between humans and animals be found on a more practical level? It was once widely held that only humans use tools. In fact, this was the very point on which Louis Leakey concluded that *Australopithecus* must be human (page 111). However, most people today have seen videos of chimpanzees using sharp sticks to harpoon termites from their burrows, or elephants painting pictures with paint brushes, or crows using certain objects to open a device to get a food prize. Bottlenose dolphins will hold sponges in their mouths to protect their snouts from rocks and spiny urchins as they stir up the sea floor looking for food. Sea otters will use rocks to open oysters. Many online videos are available of birds using pieces of bread to catch fish: the birds are very intentional and methodical in the way they place the bread and protect it from other swooping birds:[111] if a picture paints a thousand words, these videos write books on the question of whether even birds use tools. We now know that a large number of other animal species use tools, including orangutans, Capuchin monkeys, and macaques.[112]

Some will emphasize that the important difference is that humans are uniquely able to *make* tools while the other animals only *use* preexisting objects as tools. However, chimpanzees, bonobos[113] and crows can modify the sticks they use according to the circumstances (make them shorter, or sharpen them, or add a kink). Dolphins

---

110. Bekoff, "Awareness: Animal Reflections," 255.

111. One of many examples is "Clever Bird Goes Fishing," YouTube video, posted August 24, 2012, www.youtube.com/watch?v=uBuPiC3ArL8.

112. Wilcox, "Proposed Model," 22–43.

113. Roffman et al., "Varied Natural Tools," 78–91.

will collaborate to blow curtains of bubbles to corral a school of fish into a tight ball to make it easier to prey upon (bubbles are just as much a tool as a solid object: they're created to serve a specific purpose which is quite distinct from simply exhaling).

Could it have to do with our proclivity for language? I doubt that any reader doesn't already know that a very great many different types of animals exhibit varying degrees of language skills. Many of those readers may have even seen or heard of primates carrying on extensive conversations with humans through the use of images and symbols on picture boards or through sign language (such as Kanzi, Koko, and Washoe). There is also a fascinating thirty year study done by Professor Con Slobodchikoff of Northern Arizona University, who found that gophers employ an incredibly complicated language to warn each other about very different kinds of threats and social encounters, even to the point of distinguishing the different colors of shirt worn by the researcher.[114]

Some point to creative expressions, such as beads, jewelry or hand-paintings on cave walls, as distinctively human. However, crows and certain other birds will collect brightly colored objects for their nests. Bowerbirds build elaborately decorated nests which often include shells, flowers, and other colored objects (stones, feathers, plastic, coins, rifle shells, glass). Again, there is the example of elephants painting pictures: although this is admittedly a trained behavior which doesn't seem to appear in the wild, the exceptional quality of their paintings does indicate some kind of sense of abstract thought, appreciation of form and an artistic ability to reproduce the images they see.

"Ross," a character on the very popular TV series *Friends*, in trying to convince someone about evolution and the origin of humans, cried out in exasperation: "Opposable thumbs!" I suppose the writers of that sitcom didn't know that many primates also have them, as do koalas, opossums, giant pandas, birds, chameleons, and certain frogs.

Perhaps the one thing that distinguishes us from the animals is certain ways in which we interact with each other and work together. Thousands of years ago, Aristotle said, "Man is nothing but a political animal."[115] Sir Isaac Newton is often quoted as saying: "If I have seen further [than my peers] it is by standing on the shoulders of giants" (although it seems he borrowed that metaphor from Bernard of Chartres). What really sets us apart from animals is our ability to use language to pass on new information from previous generations and in that way to build knowledge on top of existing knowledge. This ability allowed us to build up a huge encyclopedia of engineering, medicine, biology and other sciences—as well as religious knowledge and insight—that continues expanding and has helped us to dominate the planet.

Otherwise, the more we look, the more the lines which once seemed to separate humans from animals become blurred and even fade from view. Increasingly, it seems that humans and animals exhibit the same characteristics and abilities, it's just that we

---

114. Crew, "Catch the Wave."
115. Aristotle, *Politics*, 9–12.

often do so to a greater degree. We simply occupy the extreme end of the spectrum when it comes to certain measures of ability. But we're not always superior to them: our ability to do certain other things pale in comparison to that of the animals. We marvel at how Monarch butterflies can navigate thousands of miles using the stars, earth's magnetic field and the far off sounds of oceans as navigational clues, and how the "maps" appear to be pre-wired into their tiny brains. We're amazed at the abilities of some animals to hear frequencies of sounds, see wavelengths of light, and pick up trace scents, which completely elude us. The ability of a bat to navigate through the forest and chase insects solely using echolocation. Some of the animals might find us quite inferior because we so easily get lost in the woods, or are incredibly weak or fearful, or have such poorly developed sensory skills.

In many ways we are not all so different from the animals, nor necessarily any better than them. But even this is not a new revelation: the writer of Ecclesiastes said as much thousands of years ago.

Might claims of human exceptionalism, and even of how that pertains in some way to the *imago Dei*, be only emphasizing characteristics that we hold in high esteem? Ones on which we put greater value because we're particularly good at them? Perhaps lions would describe God in more lion-like terms. A dolphin might see the *imago Dei* in being able to dive to incredible depths, to distinguish the different types of fish based on their sonar signature, to catch fast moving prey in the dark, and to have tremendous acrobatic abilities both above and below the water's surface.

Special Creation?

As mentioned above on page 161 (objection #4), many Christians can accept that almost all life on earth arose by evolutionary processes, but they draw the line at humans. In their minds, humans *must* represent a special creative act of God. This may in part be due to the arrogance we humans have: all human societies, irrespective of their theistic or atheistic beliefs, have seen humans as the pinnacle species: the one species which the gods would favor. It certainly is also done in order to rescue various theological concepts that would otherwise need to be discarded in an evolutionary paradigm.

Again, there's no scriptural reason to insist that only *Homo sapiens* were made in the image of God. One can choose to believe this, but there's no biblical passage which can be invoked to justify that decision.

In the same way, the passage in which God is said to breathe life into the human he had just formed out of the dust of the ground[116] doesn't imply that the man received

---

116. Gen 2:7.

anything that the animals didn't also have,[117] and in the flood story God refers to taking away from humans and animals alike the "breath of life" that he'd given them.[118]

Finally, I've already elaborated at great length in my response to Objection #4 how the scientific data are simply not consistent with humans representing an entirely novel design distinct from the animals. Instead the data are completely consistent with a Common Descent model full of many design flaws and defects. These data include a long list of genes which we can't or don't use (for synthesizing vitamin C; olfactory receptors; tails on our spines), the muscles attached to our skin hairs or our ears, the retinal layer in our eyes, the recurrent laryngeal nerve, the convergence of our trachea and esophagus, the convergence of our digestive and reproductive tracts, and the problems raised by human childbirth.

For these reasons, I would seriously call into question the idea of "special creation," particularly at the hands of someone with omniscience and omnipotence.

Those who choose to still hold to that idea must come up with explanations for those design anomalies or flaws, and also ask themselves why they feel the need to insist on special creation.

Morality, and the Way We Think

Morality has a deep theological aspect to it. Some theists would say it is exclusively a theological issue: if there is no such thing as a God, then there is no such thing as an absolute right or wrong. This is referred to as the "Moral Argument" for the existence of God. Many atheists would challenge that claim. But this section is not about that debate.

This section *is* about how Christian theology might need to be adjusted in light of certain findings from science. And there are several areas of morality which are informed by our theology but which may also need to take biology into account.

One of the greatest dangers to us today, something that is tearing our world apart, is tribalism and racism. We see it in the Middle East, in riots against police forces in America, between various religions and even between denominations within a given religion. Within our neighborhoods, grade schools, work forces and prisons, we quickly define the "in-group" and the "out-group," and establish pecking orders.

Where does this come from? In part, it's a product of a long history of competition for survival. Evolution has given all species deep and pervasive behavioral wiring which makes us hostile toward other organisms which challenge our hold on a given ecological niche. This often emerges as fierce conflict when two animals or groups occupying the same niche encounter each other. That competitiveness helped us survive and thrive through the millennia of evolution. We've learned to drive off or eliminate any threats to our hold on our niche or territory. In this modern era, however, we no

117. Gen 1:30.
118. Gen 6:17; 7:15, 21, and 22.

longer need to be so inimical of other occupants in our niche. Yet we still have those ancient feelings in us, and we so easily channel them against other groups of people for no good reason. We view certain people as "the other" or even "the enemy" simply because they have different skin color, wear different styles of clothing, have different political persuasions, or even allegiances to a different sports team than our own. In society today, this can affect who we allow our children to marry, or who we hire into a certain job. Or it drives us to herd people into containment camps and slaughter them in the name of ethnic cleansing or religious intolerance. We need to recognize that tendency for what it is: a primordial biological reflex which no longer has a place in today's world. We need to work against that: we can do better.

It wouldn't be profound to say that selfishness is a natural outgrowth of the instinct for self-preservation. All through earth's history, all organisms have had to struggle tooth and nail for their existence. How many times haven't we seen sea gulls fighting over a scrap of food, stealing it from each other. Or had the family dog warily creep toward the dinner table or a plate of food carelessly left unattended, and seen them quickly snatch the food and run. It's been bred into us to think first about self and to steal from another if it helps us survive. Today, we may not actually resort to thievery, but most of us will deprive others of what we could easily give away because it goes against our inclination to give it up. How many times don't we turn our faces away from an outstretched hand in the street as we walk quickly to the electronics store to get the newest generation of cell phone. We hear of a major catastrophe in some other part of the globe which has destroyed the lives and livelihood of whole populations, and our best response might be to donate twenty dollars. How many of us will gather at Thanksgiving or Christmas around dinner tables groaning with the amount of food on them, while suppressing the knowledge that others elsewhere are starving. Why do we do this? Because there's an incredibly strong competitive motivation and survival instinct within us to keep things to ourselves, and to horde as much as we can.

In the same way, anxiety, wariness, superstition and the fight-or-flight response are all also hardwired into us to help us avoid dangerous situations: we needed those inclinations in order to survive through the millennia. Every night looking into the darkness, or every day while crossing an open savannah, we were constantly looking all around us, looking for predators, other hominids who might kill us, or other sources of danger. Today, we still live in a constant state of fear: of our neighbour, or an asteroid strike, or a new strain of bacteria for which we have no protection, or a global economic collapse or global thermonuclear mutual destruction. This fear can take on bizarre proportions: fear of invasion from aliens or the rise of flesh-eating zombies, and even a variety of social phobias.

Humans have a tendency to find patterns in apparently unrelated things, and seek explanations for the unexplainable. Every now and then one reads in the newspaper headlines of someone finding the face of Jesus or the Virgin Mary burned into a piece of toast or on a rocky cliff face, or point to the classic "man in the moon." These phenomena even have a name: pareidolia. Gamblers will talk about "winning streaks," and sports fans will refer to a team having "momentum" going into a tournament. A circle and a few squiggly lines are immediately interpreted as a face or as a whole person complete with emotion. These examples of pattern recognition can even lead to rituals: most baseball players going up to bat will go through a litany of motions because they've perceived that doing so tends to lead to a favorable outcome, or not doing so will definitely lead to a negative outcome. This ritualism too is a product of our evolution. Some butterflies and moths will have large spots on their wings which look like eyes: predators suddenly confronted with these "eyes" immediately recognize them as a potential threat and will react instinctively, which sometimes buys the winged insects enough time to escape. We hominids didn't evolve to eat butterflies, but we too learned to recognize a pair of eyes peering at us through the grass and in the darkness as a potential threat. We learned to associate a sickness which nearly kills us with whatever it was that we ate the previous day, and learn to avoid that in the future. Those and many other examples of pattern recognition account for many of our behaviors, superstitions, and ritualism today. But sometimes those shapes of faces that we see in the clouds are just random shapes in the clouds.

Related to pattern recognition is a sense of causation and agency behind every event. When a herd noticed a movement of the grass, some individuals might immediately infer that the grass was being moved by some agent, and might further assume that the agent was dangerous, and would therefore instinctively flee. These individuals who fled every time the grasses moved might look silly to the others who waited to see if there was in fact some agent there: until the day that the grass moving was indeed caused by an agent—a predator—and it was too late to escape. As a result, it would be the more cautious and flighty ones who were consistently more likely to survive and pass on their genes than those which didn't. That instinct for agency detection still exists in us. When we humans started developing religions and theologies, the assumption was always that the Great Agent(s)—the god(s)—were dangerous and needed to be appeased with sacrifices. Even the early depictions of God in Genesis portray him as always angry and destroying things, or bringing disease, or closing women's wombs. Today, when something bad happens to us, we tend to look for an explanation—who or what it was that assaulted us, and what was their motive—rather than conclude that it was just a random occurrence.

A similar argument could be made about sexual urges. It's undeniable that many of us struggle to contain those urges. Without getting into sordid descriptions or personal confessions, suffice to say that most men and some women, if they were honest, will admit to having thoughts they'd prefer not to share with their parents, spouses or children. This too is wired right into us (in the form of testosterone and estrogen, together with some elaborate neural networking). Evolution has long ago instilled into all living beings a need to reproduce and perpetuate the species far and wide.

This doesn't legitimize the behavior today. It just takes it out of the realm of "the devil made me do it," and puts the responsibility squarely on our own shoulders. We need to resist those urges within us. We've also been wired by evolution to be selfish, to eat as much as we can, and to kill if threatened, but we do our best to curb those impulses too. Similar things could be said about gambling and other addictive behaviors.

Perhaps the greatest conflict within the church today is over the issue of homosexuality. Some Christians are quick to point to many different passages which speak directly against homosexual practices, as well as to other passages which can be interpreted that way. I'm certainly not in a position to declare them as being right or wrong in holding this view. But I certainly think it's fair to point out two things that might cause them to reconsider, or at least make sure that they've taken into consideration. I ask the reader not to draw lines in the sand and divide with me over this issue: my intent here is simply to present facts for the reader to keep in mind.

First, there is the underlying biology: there is evidence for a genetic linkage for homosexuality.[119] No one claims the genetic link *fully* explains that preference: no expert in this area concludes there is a gene for homosexuality, any more than that there's a gene for being left-handed or for collecting stamps, or that it might be possible to use gene therapy to eliminate any of those tendencies (if it were ever deemed necessary to do so, that is). Nonetheless, it seems that there is an influence that comes straight out of our genetic makeup. Many homosexuals will declare sincerely that they were "born that way." We need to be careful to listen to that. Some critics will point out that homosexual tendencies can't be genetic because they can't be passed down genetically: "homosexuals can't/don't reproduce." This simplistic inference fails to take into account complex genetic and environmental interactions (some phenotypes arise from multiple gene inputs; pleiotropic genes; and so on). If such an inference were valid, then there would be no such thing as diseases which are 100 percent lethal before a child afflicted with one of them reaches sexual maturity (without modern medical interventions, that is), such as certain forms of leukemia. Those children would otherwise never reach an age at which they could pass on their genes, and we should therefore never see those diseases.

The second thing to keep in mind is the sociological context of the passages in Scripture which speak against this practice: they were all written by ancient Hebrew authors. Most of the passages and their authors come from an ANE milieu which had a certain view of those practices. Those same authors had views on other matters which very few of us find any need to pay attention to in this day and age: laws pertaining to diet (not eating pork, rabbit, shrimp or lobster), cleanliness (touching dead bodies; menstruation; skin diseases), sacrifices and offerings (for newborn children), holidays (Sabbath; planting and harvest festivals; spring cleaning rituals), farming (planting mixed crops; shmita, or the practice of letting our gardens and fields lie fallow), and slavery (who one could own, and for how long), just to name a few. Some would justify

---

119. Sanders et al., "Genome-wide Scan," 1379–88.

ignoring those other laws by saying Christ has done away with the law. Others would acknowledge that the laws were written to/for a different people and a different time. Irrespective of the exact reasons why one finds it acceptable to set aside the laws, that same body of legislation which is being put on the shelf also contains the references to homosexuality.

Not all the behavioral reflexes and urges which we might be able to attribute to our evolutionary heritage are negative.

How can one not recognize the profound tender feelings most people have toward a newborn baby, especially their own, as something deeply rooted in the core of our being. There's no need to explain from an evolutionary paradigm why that urge would have been inherited.

Hominids also developed a sense of empathy. We began to learn how to read faces and body language. "Is that other hominid facing me angry or happy?" "Are they a threat, or an ally?" "Is that tiger hungry or well-fed?" We learned how to read certain clues which allowed us to essentially crawl into the mind of the other and discern their thoughts. Birds and cats and cows can't do that; some debate that dogs and horses can. Today, we can use that to good as well as to bad. We can comfort someone who is suffering, or anticipate a hardship that a distant group might be experiencing and send help in advance of their request. Unfortunately, we too often suppress those sympathetic urges.

These behavioral and instinctive adaptations protected us as we moved forward through our evolutionary journey. But we don't need many of those hardwired instincts to survive today, and it's too easy to let these things get highjacked and distorted in today's social settings. Again, examples include racially motivated crimes, interreligious wars, political attack ads, riots, thefts, famines that go relatively unnoticed, apartheid, and the Ku Klux Klan; the list goes on and on.

Fundamentally, none of what I've written here is different from what Jesus himself taught when he said: "For it is from within, out of a person's heart, that evil thoughts come—sexual immorality, theft, murder, adultery, greed, malice, deceit, lewdness, envy, slander, arrogance and folly. All these evils come from inside and defile a person."[120] The only thing that is new here is that we can now see *how* these tendencies came to be in the heart of mankind.

Knowing the source of these tendencies isn't enough. We need to do something about them: we need to actively work against them. We've also been wired by evolution to eat as much as we can and store the excess on board as fat to help us through the lean months on the open savannah plains of Africa millennia ago: but we no longer live on those plains, and so we accept the need to curb that urge. In the same way, we need to work against the other negative or destructive urges which are deep inside us, from the relatively minor ones like selfishness to the truly dangerous ones like tribalism.

120. Mark 7:20-23.

Although we can't work our way into heaven, we have been called by God to be better than what our innate impulses direct us to be. Jesus himself taught this, but framed it in very Jewish vocabulary. In the passage quoted above from Mark 7:20–23, he was talking about being clean versus unclean. And his prescription was simply to clean up our act: "Woe to you, teachers of the law and Pharisees, you hypocrites! You clean the outside of the cup and dish, but inside they are full of greed and self-indulgence. Blind Pharisee! First clean the inside of the cup and dish, and then the outside also will be clean."[121] In the same way, Paul teaches in many places about how we need to put to death an old way of thinking, an old nature, and strive for something better.[122] But that's a personal battle, not a public one, and I think Christians should reconsider the shame and judgement that they attach to or impose on this struggle (on themselves, as well as on others).

The Fall and Original Sin

The third chapter of Genesis clearly describes two individuals in a secluded garden being tempted by a talking serpent to bite into a piece of fruit from the Tree of the Knowledge of Good and Evil, something which God had expressly forbidden them to do.[123] The Apocrypha picks up a little bit on this event, Paul develops it further in the NT.[124] A few hundred years later, Augustine of Hippo develops this further into the concept of original sin, which every human is said to inherit. It's worth pointing out that Judaism does not subscribe to the concept of original sin, and has a completely different view on the problem for which the Mosaic laws and sacrifices were a solution. *They* (the Hebrews) are the original authors and original target audience of those Scriptures, but we (modern Westerns) have taken those Scriptures and reinterpreted them into something completely different!

A lot of Christian theology hangs on the history packed into the first three sentences of that opening paragraph. But what does one do upon coming to the conclusion that Adam and Eve aren't historical figures? How does one revise thousands of years' worth of theological tradition, held by authorities no less than Jesus, Paul and Augustine? And yet, the physical evidence in our hands simply discounts the idea of any primordial couple in a garden six thousand years ago.

It's worth pointing out that "Adam" is only mentioned three times outside of the first five chapters of Genesis, and in all three cases it's just a passing reference.

One of these occurrences is in 1 Chronicles 1:1, which begins a "postexilic genealogy that strives to connect the returnees from exile to Israel's primordial

---

121. Matt 23:25–26.
122. Rom 6:6; 1 Cor 5:7; 2 Cor 5:17; Col 3:9.
123. Gen 2:17.
124. Especially Rom 5:12–19 and 1 Cor 15.

beginnings."[125] There is no mention of a history-making fall from grace that would forever change the trajectory of mankind, and even of nature and the cosmos itself. No mention of the introduction of sin or death. Adam is simply the first in a very long series of names of forefathers.

The other two occurrences use this name to refer to a place rather than a person or people. One is found in Joshua 3:16, and its reference to a town is straightforward. The other occurrence, in Hosea 6:7, is a bit more difficult to parse out.[126] God accuses Judah and Ephraim of breaking a covenant. There is great uncertainty here as to what the word *adam* in this passage refers. Many English Bible translations, including the NIV, indicate this event occurred "at Adam," although no breaking of a covenant in that town is described. Many other English translations, including the King James and the Complete Jewish Bible, render Hosea's wording in the collective sense to refer to humanity as a whole, opting to use phrases such as "but they, like all men, have broken the covenant," or some variant of that. Other translations do indeed refer to "Adam" as an individual, but there is no scriptural support for a covenant existing between God and Adam prior to the biting-into-the-fruit passage. The same sort of uncertainty occurs at Job 31:33, where some translations render the phrase as "as people do" (Complete Jewish Bible, NIV and others) or "as Adam did" (KJV, American Standard Version and others).

The only reference in the OT to the fall in the garden incident is Genesis chapter 3, and Eve is never mentioned again in the OT beyond Genesis chapter 4.

Why do I find it important to point out that Adam, Eve and the fall in the garden are never revisited in the OT outside of the first few chapters of Genesis? It says to me that maybe they weren't all that theologically central to the writers of the OT; at least not important with respect to any explanation of sin or death or redemption.

As an analogy, consider an encyclopedic series comprising thirty-nine volumes, all addressing World War II. Hundreds of pages written about the details of the timeline, the countries involved, geographical details, the names of the generals and other leaders, famous battles, technological developments, the code-breaking, et cetera. But in all those thirty-nine volumes, Hitler—the very instigator of that war, the supreme leader of one side of the conflict, and arguably the key figure in this major episode in modern human history—is only ever mentioned in a few specific incidents in the first ten pages of the first volume of that series of thirty-nine volumes, and then never again.

Wouldn't that seem like a bit of an oversight?

Much of Protestant theology is built upon the concept of original sin (neither Jewish nor Greek Orthodox theologies share this view, and the Roman Catholic view on this is also quite different). But is this concept really found within the Jewish Scriptures? Extensive sections of the OT describe an incredible variety of sin(s)—conscious

---

125. Enns, *Evolution of Adam*, 83.
126. Ibid., 83–84.

and subconscious; intentional and accidental; trivial and egregious—and the various sacrifices and offerings needed to be given to atone for them. "Sin" or "sacrifice" are mentioned in essentially all of the thirty-nine books of the OT, and some of them mention both terms over and over again. Covering four thousand years of history, and yet not a single mention of these two key individuals within that major historical event beyond Genesis chapter 5. How can that be if they constitute the very root of the human problem?

Some have relegated the creation passages, as well as the rest of the first ten chapters of Genesis, to being merely reinterpretations of pagan origin myths circulating at that time. The discovery in the nineteenth century of Sumerian/Babylonian clay tablets containing the now famous *Epic of Gilgamesh*—complete with their version of a global flood, an ark built by a righteous character carrying animals, and the sending out of three different birds before it landed on a mountainside and the ark-builder offered sacrifices in thanksgiving—knocked the theological world on its side. Who copied from whom? (The archaeological evidence is heavily on the side of the Babylonian version preexisting the Hebrew one.) Further investigation of these clay tablets revealed stories having incredible parallels with the biblical accounts of the creation, the Nephilim, and the Tower of Babel.[127] Likewise, the lists of Semitic patriarchs living many hundreds of years and fathering various sons parallels the lists of Babylonian kings who lived and ruled for unbelievably long lives (measured in thousands of years).

Interestingly, there are similarities between vignettes in the *Gilgamesh Epic* and elements of the Genesis account describing the garden and the fall. Tablets VII-XII contain a lengthy story in which Gilgamesh's deeply beloved friend and constant companion Enkidu dies. This is Gilgamesh's first close-up encounter with death, and it impacts him deeply. He goes on a quest to find the secret to immortality from the one survivor of the global flood—Ut-napishtim—who was granted immortality by the gods. "He learns from his ancestor that immortality is no longer within the reach of mortals, but he is told where he may find a magic herb which makes the old young again. He obtains the herb, but it is stolen from him by the guile of a serpent."[128] In tablet XII, "Bilgames and the Netherworld," the goddess Inanna plants a tree in her "pure garden," but the tree is infested with malevolent beings, including a "Snake-that-knows-no-charm": Bilgames (a Sumerian variation of "Gilgamesh," and the hero of the story) helps the goddess by overcoming and destroying the snake and the other beings, and destroying the tree. Do these vignettes not reverberate with the story of the Tree of Life, mankind being deceived by a serpent and losing immortality in the process, and the promise given that the seed of the human would ultimately triumph over the serpent?

---

127. Heidel, "Gilgamesh Epic"; Damrosch, "Buried Book"; George, "Epic of Gilgamesh."
128. Hooke, *In the Beginning*, 29.

Many Christians have been able to turn the Genesis accounts of creation into an allegory. Most of us will admit we've made bad choices. Against our best intentions and the whisperings from our conscience, we've metaphorically reached out and grabbed that apple for ourselves and bit in, knowing it wasn't right. So we've all "fallen short" in that and many respects. What exactly is this thing that some Christians call original sin? (Bearing in mind that Eastern Orthodox Christians, Roman Catholics and Protestants have very different ideas on this topic, including whether we inherit the guilt of Adam and Eve's sin.) And can evolution theory shed any light on such a thoroughly theological concept?

I think yes.

Contrary to what some believe, the Genesis account of creation doesn't explicitly describe God making things on each creation day *ex nihilo*, snapping his fingers as each item suddenly appears out of thin air.

There are three Hebrew verbs which we translate into English as "create" or "make": these are *bara*, *yatsar* and *asah*. These three overlap significantly in their meanings, and are often used interchangeably: in other words, the use of one of these in a particular verse does not have to necessarily exclude the possibility that the other two could equally have been used. *Bara*, the verb used in Genesis 1:1 which describes the origin of all things, can take on the meaning of creating something entirely new or pristine, as if out of nothing, although many linguistic scholars point out that it doesn't have to take on that added meaning of *ex nihilo*. The other two verbs are used more in the sense of reshaping one thing into something else—as in a potter forming a pot out of a lump of clay—and are used far more frequently in the first three chapters of Genesis than *bara*. In some cases, these verbs are sometimes used in combination within the same verse, including the one describing the creation of mankind: "When God created (*bara*) mankind, he made (*asah*) them in the likeness of God."[129]

So the creation account is not solely about the making of stuff out of nothing, but also about the rearranging of stuff into something more complex: the creation of order out of chaos. In this sense it is no different than the current cosmological model which has all the energy and matter in the universe morphing and reorganizing into the various objects we see today. The very opening verses of Genesis have the Spirit of God hovering over the waters, and darkness covering "the deep" (which, by the way, is also how other ANE creation accounts begin, including the Babylonian *Enuma Elish*, the *Chaldean Cosmology*, and the Egyptian Memphite cosmogonic tradition): in other words, there was some "stuff" in existence prior to the reorganizing actions described in the creation accounts.

The verses which follow this opening describe God shifting matter around and reshaping things out of what was already there, which is exactly what evolution is all about.

Separating light from darkness.

129. Gen 5:1.

Separating waters above from waters below.

Letting the waters *gather into one place*, and land into another.

Letting the land produce plants (notice the active role played by the land here, while God's role isn't active in the sense that he isn't actually "making" anything).

And when the text refers to the creation of celestial bodies and of animals, it doesn't actually say those were created *ex nihilo*: nothing in the wording precludes God also creating those from preexisting materials.

Then he creates the man: out of the dust of the earth.

And then the woman: out of the man's side.

All of these creation events are metaphors for a process that we theists have been fighting over for too long: making something new out of preexisting material. A process otherwise known as evolution. And perhaps the process isn't finished yet. Perhaps the final product God intended is still being formed. Perhaps, after the various acts of separating order out of chaos described in Genesis chapter one, there's been yet one more reordering from chaos going on for the past many millennia. Here I'm referring to something inside humanity, but more on that in a moment.

What is this thing that we now call "original sin" and that God wants to separate out of us, and what new characteristic might he want produced in us? I think the answer to both questions has to do with the driving force that is within us.

There's been one major driving force behind all of evolutionary history: selfishness. Eat everything you can in order to be able to reproduce. Run away if you can, kill if you have to, but do anything and everything in order to save yourself and perpetuate the species. This has been the central theology of evolution: "it's all about me." That's what motivated *Homo sapiens* and all of our genetic cousins for millions of years, and their ancestors for billions of years before them, right down to the first amoeba. It's basically written into our DNA. Hence Richard Dawkins's book entitled *The Selfish Gene*.

So the Evolutionary Tree of Life gradually grew and branched, producing all the different species of the past and present as God supervised and possibly also watered and pruned it. Eventually we humans reached a point at which God could nudge us toward something better, something that reflected his image. You might even say his primary intent for creation all along was to reshape selfishness into selflessness. "Do unto others as you would have them do onto you."[130] In that sense, it's another example of God creating order out of chaos: out of the savage and ruthless violence of nature "red in tooth and claw" comes his kingdom built on a love ethic.

We came to realize the consequences of our actions. We perceived what would happen if we stole from someone else (they wouldn't be able to eat), or hurt them or even killed them with our own hands. Or, on the other hand, what would happen if we helped them, shared our food with them, protected them from a danger, and helped them when they were sick. At that point, we were able to make a choice.[131]

---

130. Matt 7:12.
131. Wilcox, "Proposed Model," 22–43.

From this perspective, Romans 8:19–22 takes on a whole new meaning for me: "For the creation waits in eager expectation for the children of God to be revealed. For the creation was subjected to frustration, not by its own choice, but by the will of the one who subjected it, in hope that the creation itself will be liberated from its bondage to decay and brought into the freedom and glory of the children of God. We know that the whole creation has been groaning as in the pains of childbirth right up to the present time."

Today, millions of years later, each of us is able to make that choice: either to think only about self, or to be other-minded. And far too many times, each one of us has made the wrong choice: we've fallen from God's ideal for us. There wasn't one fall in the garden six thousand years ago: instead, there've been hundreds of billions of falls over the past several million years. We've all failed. It's not an original sin, but an original flaw.

Rather than maintaining that we inherit sin and guilt "in" Adam or "from" Adam, saying that we inherit sin and guilt "with" Adam avoids these problems. Perhaps there's been too much emphasis on sin as something you do: killing, stealing, lying, lusting, hating. Eating forbidden fruit from the Tree of the Knowledge of Good and Evil. The deeds we do are the outward evidence of something within us: a sinful nature. Or better yet, they indicate the absence of something which isn't in us (yet). Most Christians are probably comfortable with that idea. It fits well with passages in which Paul refers to the old self or the old nature in contrast to the new ones.[132] But they would still see this quality as something we get *from* Adam. Passed on to us like a disease, or like a genetic mutation. But what if it refers instead to something that had always been there inside us, and inside Adam and Eve, and inside much of the rest of creation for that matter, for millions of years, and that God gave *Homo sapiens* the ability to overcome it but we chose to do it our own way?

What do I mean by that?

Sin means to fall short of God's ideal. The Greek word used by the NT writers is *hamartia*, a term borrowed from archery which simply means "to miss the mark." The imagery brought out by using that particular term is one of God setting a goal or target at which we should aim. That target is being his hands and feet on this earth, and reflecting his image (both of these refer to the *imago Dei*). He's given us the ability to rise above the animal nature within us, above the urge to think only about self, and to instead be selfless. And the fact is, both in archery and in our daily life's choices, we can miss by a little and we can miss by a lot. We repeatedly choose to reflect something less than his ideal. All of us can see that streak within ourselves. Too often, our dealings with our fellow human are less than honourable. We don't love enough, respect enough, or care enough. And in some of us, that leads to other things: cheating, lying, hatred, theft, rape, torture, murder, war, et cetera. I'm not a murderer simply because my great-great-grandfather was. True justice would not hold me accountable for some

---

132. Rom 6:6; Eph 4:22; Col 3:9; Titus 2:2.

sin or crime that one of my ancestors committed. But I do have a selfish streak simply because I'm human (evolution beat that impulse into all living species). And that selfish streak has blossomed into a unique set of sins to which I will have to admit and be held accountable. So it isn't that we're responsible for something that we didn't do (the sin of Adam), but responsible for not trying to be what-we-are-not-yet (once we're made aware of that).

Atonement

Christ's work on the cross is absolutely central to the Christian faith. But trying to fully understand it hasn't been easy: it seems to have been the subject of discussion in the church ever since the first meeting of the apostolic church. In particular, why exactly was it necessary and what exactly did it accomplish? Some theists, without a moment's hesitation, will blurt out: "to pay for our sins." But perhaps it's much more nuanced than that. After all, these questions have mystified scholars and theologians for two millennia.

One particular aspect of Christ's atonement which is difficult to understand is why God can forgive sin, but first requires blood to be spilt and someone to die. A news report of the beheading of Coptic Christians in disgustingly brutal fashion by ISIS brought this into stark relief for me one day. It was the incident that the whole world watched in horror as ISIS militants calmly marched twenty-one Coptic Christian captives up along a beach, had them kneel in the sand, spoke some words in Arabic and then beheaded them. A few days later, the media reported certain grieving family members simply forgiving those ISIS killers. The family members didn't put any conditions on that forgiveness, such as "we forgive you, but we need to see you hurting in some way." They just simply forgave the killers. I know that other people faced with equally horrifying circumstances have also chosen to simply forgive. It made me wonder: if humans are able to simply forgive other humans, even those who have done such horrific things, then why can't God?

These are very deep waters, and I'm not a certified theologian. So I'm not going to try to glibly summarize concepts with which much greater theologians than I have wrestled for centuries. But I will ask the reader to take a very close look at those aspects of the atonement which pertain to God himself requiring a death penalty or a blood sacrifice before he can fully forgive. Part of this collection of ideas includes scriptural passages or theological teachings which portray all of us as desperately wicked and deserving of a death sentence. I'm fully aware that some readers can immediately pull up passages from the Bible which they feel support this idea,[133] especially those who come from a Calvinist background (as I did). But again, just as with the creation account itself, those isolated passages sometimes seem to conflict with other scriptural

---

133. E.g., Jer 17:9; Isa 64:6.

passages, as well as with observations we can make all around us (such as people who just seem to be genuinely good at heart). For example, Enoch was said to "walk faithfully with God," so much so that he is said to have been spared the fate of all other humans.[134] It seems that Elijah also didn't experience death, but was taken up directly into heaven.[135] Enoch's great-grandson Noah was said to be "a righteous man" who "found favor in the eyes of the Lord."[136]

It is possible to find genuine good in people who aren't otherwise religious. The little old lady who lives next door, to the best of my knowledge, may have entertained a hint of petty jealousy or selfishness in her lifetime, but certainly hasn't done anything deserving of a death penalty, let alone one that involves eternal torment. The standard response to this is that her goodness pales in comparison to that of God, and in that sense she's no better than a psychopathic pedophile. But wouldn't any one of us still cry foul if a human judge condemned her to a month in jail for her petty faults, even though some might have no problem with an all-loving God consigning her to an eternity in hell for the same offence (and some would add eternal conscious torment on top of that).

I'm not at all saying that one can be morally perfect. Nor am I challenging the standard Christian doctrine that one can't work one's way into heaven based on one's own righteousness. I am asking, though, whether God is so abhorrent of sin that he would choose to, or is obligated to, destroy those he loves rather than "simply forgive"?

What about Christ's death on the cross then? This is not the book, and I am not the author, to get into the nuances of words and phrases like penal substitution, Christus victor, redemptive violence, propitiation, expiation, penal substitution, reconciliation, justification, and many others. I'm not sure the new data from genetics have anything to say about those concepts. But they do have something to say about the concepts of Adam and Eve and the fall in the garden. The fact that the new data call into question the very existence of Adam and Eve and the fall should prompt a reexamination of any theology that is based upon their existence. It wouldn't matter how much hand-waving was done, and argumentation given, by a historian trying to prove that Abraham Lincoln founded the Communist government in China in the twentieth century: as soon as it came to light that Abraham Lincoln didn't even live during the twentieth century and never left American soil, the story would need to be adjusted. In this case, if there was no fall in the garden, and if humans were never perfect to begin with, but have instead inherited a long list of flaws along the evolutionary path that brought them into being (for example, the selfishness, tribalism, agency detection, and sexual urges that were instilled into us during millennia of evolutionary processes), what needs to be said about Christ's death on the cross?

134. Gen 5:24.
135. 2 Kgs 2:11.
136. Gen 6:8–9.

First, could it be that Christ came not to be ripped apart, have his blood poured out, in order to satisfy the vengeful wrath of God against the little old lady who lives next door (and against everyone else), but rather to be an example? He modeled for us a life of complete self-sacrifice, giving up wealth or power or personal comforts like a wife or a house, even any claims on personal safety, security and independence, and instead gave us a different set of goals. Rather than follow that primal urge for self-preservation and personal gain, he demonstrated and taught us the complete opposite: to give food, water and clothing to those without, to care for the orphaned and widowed, to visit the sick and imprisoned, and to break the chains of the oppressed. It turned out that this example was just too radical for some people and so they killed him. The cross just happened to be the way the Romans did things back then in their occupied territories: thousands of other people were also crucified, but we don't attribute any theological significance to that.

I recognize that logic is not the best tool to use to distinguish between these two theological/philosophical perspectives. However, it may still be worth taking into consideration whether the timing of Christ's appearance makes more sense from one perspective or from the other. Let me put this thought on the table for readers to consider: I'm not claiming it as a bone fide argument, but simply something to think about. The Population Reference Bureau guestimate that 108 billion humans have been born over the past fifty thousand years (that estimate was made in mid-2011).[137] They came to this determination after making certain assumptions about life expectancies and infant mortality rates over the ages, as well as major stresses on the human population (the Plague, or Black Death; major wars; et cetera). Interestingly, between a third and a half of that number (forty-seven billion) had been born by the time Christ himself was born. If his primary mission was to satisfy God's vengeful wrath against sin, it might make more sense for him to carry that out at the very beginning or end of human history, rather than some arbitrary point in the middle of human history. On the other hand, if his primary mission was to be an example to us, it would make total sense for him to do so when we humans were ready to accept his challenge of being completely selfless, rather than at the very beginning of human history (which stretches back a hundred thousand years or more, when our ancestors wouldn't be ready for that kind of challenge) or at the end of our history (the final two thousand years of that hundred thousand-year history, when it would be too late for us to act on it). Again, I don't claim this to be a definitive argument, but perhaps some food for thought?

Second, could it be that *Homo sapiens* have always had this thing about sacrifice and the spilling of blood. Civilizations all around the world for thousands of years, even long before Moses set up the temple practices, have employed animal and human sacrifices in their religious rituals. The Bible itself records Noah offering sacrifices to YHWH thousands of years before Moses was born and the sacrificial system

137. Population Reference Bureau, "How Many People Have Ever Lived?"

instituted,[138] and also portrays Cain and Abel bringing offerings long before that.[139] Sacrifice brings out a visceral response in us: it adds a whole new level of meaning that simply beating on a drum or playing organ music in the background can't provide. Perhaps we humans just wrote this aspect into the script: the need to offer sacrifices was already there within the authors, and it resonated with the readers. In that sense, they developed those practices all on their own (which would certainly explain some of the more bizarre commands found in the Pentateuch). Or perhaps God said, "OK, this means something to your species. I can work with this," and decreed all the sacrificial practices found in the Pentateuch. Maybe that ritual killing carried on and escalated until God finally declared "Okay. Enough. I actually don't need these sacrifices and this blood spilling. If you really need something like that in order to be reassured of my forgiveness, then let me finish this off once and for all. I'll take care of the details. It'll be a perfect sacrifice that ends all need for any further sacrifice." Again, something to think about?

Third, could it be that God wants to just simply forgive and forget but that someone else cries foul, especially in the more egregious cases. The cry goes up: "That's not fair. A penalty needs to be paid. Hitler slaughters millions of people, ISIS lops the heads off others, and if they finally apologize and ask for forgiveness, you welcome them into Paradise with open arms. Letting them off the hook like that is not justice." And so to satisfy such an Accuser, God in Christ steps forward and says, "OK, you need justice to be done? Fine, let me pay the price myself."

I'm sure it's all those things and much more. It's a multidimensional thing. Like the seven blind men trying to describe an elephant, each one grasping a different part of the whole—the trunk, the tail, a leg, an ear, a tusk, the body, and so on—and attempting to come up with their explanation solely from their partial grasp of the data. But a sight-enabled observer watching the seven can see that the elephant is all of those things, plus more.

Why Is This a Genetic Thing?

How can one take questions which are so thoroughly theological in nature and conflate them with concepts that are so genetic in nature? Why is original sin only attributed to a certain genotype (human), and how is an abstract concept such as sin transmitted down a genetic line? This kind of thinking can even invite crazy questions like "can sin be treated using gene therapy?" and "if original sin is passed on through the act of sex, as some believe, what does this say about test-tube babies?"

---

138. Gen 8:20. As an aside, this passage is anachronistic in that it says Noah "*took of every clean beast, and of every clean fowl, and offered burnt offerings on the altar*"—the distinction between "clean" and "unclean" wouldn't be defined until Moses arrived two millennia later. This is one of many reasons that lead scholars to think that the Pentateuch appeared in its final form during the Babylonian exile.

139. Gen 4:3–4.

Our evolutionary history merges with that of Neanderthals. Their DNA is essentially identical to our own. They had a consciousness comparable to our own: they made stone tools, lived in social groups and may have buried their dead. Do we then need to start thinking about how original sin and Christ's atonement applies to them? Were they also in need of saving, and was salvation available to them? Lines of thought such as this are not unprecedented: again, they go as far back as Augustine's defense of the Plinians in the fourth century, to the Dominican philosopher Tommaso Campanella in his 1622 book *Defense of Galileo*, and as close to the present as Ken Ham's 2014 blog post, all addressing whether or not aliens needed salvation (see p. 215).

This theological concept of original sin (which is tied up so tightly with genetic descent) doesn't make it easy to deal with the moral problem of children inheriting a death penalty and all the incumbent curses from that first primal pair. In fact, that moral problem is worsened by the fact that the other hominids existing at the same time as that first primal pair, and all of their ancestors for millions of years before them, who shared all the same genetic and behavioral characteristics of that representative Adam and Eve, would have also been excluded from the redemption story. Can it really be that an eternal relationship with God, and access to a redemption purchased by the blood of Christ, is entirely withheld because of just a few base pair differences in a genome that sets an individual hominid apart from the genetic lineage of that first representative pair? In other words, are Neanderthals, Denisovans, and other pre-Adamites all going to hell? Do they have no spiritual/eternal hope whatsoever, even though genetically they are our brothers and sisters? Or were they intentionally killed off, either by God or by his prehuman ambassadors, before the actual date of the fall, so that they would predate that horrible crime and be excluded from its penalty, and therefore grand-fathered into the go-straight-to-Paradise policy? They could then be treated exactly the same as our house cats and dogs, which some theists believe "will meet us in heaven."

Or maybe it isn't a genetic thing. If it isn't, should we still be referring to "descendants of Adam"?

World Religions

Homo sapiens have been around for hundreds of thousands of years. Looking up into the sky, sensing the presence of a "Great Being," and coming up with a wide variety of images for the Divine.

Many atheists will relegate that to being merely some quirk of our neural wiring: part of our instinctual agency detection that I wrote about above (p. 222). That's a big topic, and not the focus of this particular book. For interested readers, I'd recommend Robert Bellah's contribution on this subject.[140]

---

140. Bellah, *Religion in Human Evolution*.

Theists, on the other hand, will attribute this sensing of the Divine to something real that exists deep within us. Blaise Pascal, the French mathematician, physicist and Christian philosopher (and the originator of the often quoted "Pascal's wager"), ascribed this to "a God-shaped vacuum in the heart of every man which cannot be filled by any created thing, but only by God, the Creator, made known through Jesus." I wondered above if it might have something to do with the *imago Dei* in which we were made (p. 210).

Either way, we humans have been devising all kind of religions by which the Divine should be worshipped, and writing down all those ideas on countless clay tablets, scrolls, animal skins and books. We and our now extinct cousins, the Neanderthals and Denisovans, have long been aware of an afterlife: all three have buried their dead with a variety of items—weapons, food, jewelry, religious icons, and other artefacts—to be used on the other side of the mortal curtain. Religion can be found in every human civilization past and present, in every part of the globe, in every socioeconomic slice of the human population pie. Irrespective of gender, age, race, wealth, IQ, and any other parameter by which *Homo sapiens* can be defined. Our existence can be described by our search for God. Even many of those who will emphatically distance themselves from any particular religion will still lay claim to certain aspects of spirituality. In one poll of 1,700 American scientists, one-fifth of those who identified themselves as atheists labelled themselves as "spiritual atheists."[141]

Recognizing this as a driving motive in what it means to be human is important. I've come across too many theists who are harshly judgmental of people who adhere to some religion other than their own. Instead, I would advocate using this as a starting point. A bridge to open dialogue.

Recall the story of Paul in Athens (Acts 17). He saw that "the city was full of idols" and that the people were "in every way . . . very religious." He pointed out to them that they even had an altar dedicated to "an unknown god." Why would the Greeks have done that? It seems they recognized that, although they had *many* deities, there was still something more, something different, but they couldn't quite pin down exactly what (or who) that was. Being rationalists, they weren't going to just give it any kind of name: the name had to make sense. "To an unknown god" was pretty all-encompassing and left nothing out. But, as Paul said, they were ignorant. He didn't rebuke them for their polytheism or their ignorance. He used it as a teaching tool, and explained to them the gospel.

The Bible gives us another example of tolerance toward other religions. When the Magi came to Herod to inquire about the birth of the King of the Jews, they explained how they had seen the new King's star in the east. These were astrologers; some say they were Zoroastrians, one of the world's oldest religions. Whether that's true or not, it can at least be said that these Magi believed the celestial bodies communicated to/with them somehow. In this case, they believed the star "told" them about the birth

---

141. Pew Research Center, "Scientists and Belief."

of a monarch of a particular nation (the Jews) far to their west. They took guidance from their reading of the cosmos, something which some would say is forbidden in the Bible. And yet there's no evidence that God rebuked them for this. Instead, he used something that was meaningful to them, and in the process he placed them squarely within the background scenery of one of the central stories in the Bible: the birth of Christ.

Jesus himself never spoke out against any religions at all, despite being surrounded by polytheistic Romans and Greeks. However, he frequently had very harsh and even condemning things to say about those within his own religion: the hypocrisy and criticism of the Jewish leaders.

I think many religions may give us some clues about who God is and how to find him. Many people believe they've found God through some of the most bizarre of spiritual journeys, and I can believe that even the parts of the journeys through very non-Christian beliefs actually led them to God.

If I'm lost while journeying through a forest, I can find my way out of a forest using a wide variety of strategies, some more precise or effective than others. I can get a sense of what direction to go if I use the sun and my watch to give me a crude sense of east-west directions. The proverbial "moss grows on the north side of a tree" or a night-time sighting of the north star can give me a sense of the north-south axis (or if I'm down under, the Southern Cross can point me to the South Celestial Pole). The direction of flow of a stream can tell me which way to find the valley (where civilization is more likely to be found than at higher elevations). Maybe I can listen for certain distant landmarks, like the ocean, a highway, an industrial center. Using those cues, and ignoring red herrings, and possibly a little luck, I may eventually find my way out of that forest, albeit with a lot of meandering and retracing of my steps. Or I may end up becoming worm food before getting out. But a map and a GPS unit will tell me exactly where I am, where I want to be, and the shortest and quickest route between the two.

Many religions in general, including Christianity, can provide rough clues for our journey through life to find God, but Jesus Christ the person provides the map, compass and GPS coordinates.

I'm not saying that all religions are the same. Clearly they are different, often contradictory, sometimes even at enmity. But I am saying that many of them have some ring of truth derived from earnest searches for God over many millennia; some have found more truth than others, mixed in with some wrong ideas.

This is not universalism, nor does it mean that all paths lead to the same God. Instead, it's a recognition that a theist trying to find the divine in Indonesia is highly likely to look for that in Islam. A theist in the high Arctic will find that in the Inuit religion. A theist in the Andes mountains of South America during pre-Inca times would have found that in Viracocha: today the latter would find it in Roman Catholicism. Will God hold people accountable for having been "geographically challenged" and doing the best they could with the information they had? C. S. Lewis asked the same question in *The Last Battle*, seventh in the *Chronicles of Narnia* series, when he wrote about Emeth, a worshipper of Tash (Aslan's opposite, who some would equate to Satan) who is forgiven by Aslan: Aslan explains that Tash didn't really know what he was doing, and credits the good and noble works done for Tash as having been done instead for Aslan.

Those searches for the ultimate truth are divinely inspired. I believe God created that urge within us to find him. And those doing the seeking in the ancient past may well have caught intellectual glimpses of him while they meditated and pondered the question: his power, or the fact that he created all things, or that he loves people, or hates certain actions. They may have alloyed their understanding of him with other human ideas and values. Christians have also been seeking the Divine, and have found him revealed in Jesus. But they too have likewise alloyed that understanding with human ideas and values: the Prosperity or Health-and-Wealth Gospel, gender inequality, the "mark of Cain" on black people, and the uber-importance of regular attendance in church on Sundays would be some examples of contaminations and distortions of the original message. We call ourselves Christians and claim to follow Christ's teachings, but find it very easy to tolerate poverty and injustice all around us, or to celebrate retaliatory strikes against our enemies.

The images of Christ talking to the Samaritan woman,[142] and referring to having "other sheep that are not of this sheep pen,"[143] Peter's vision of the blanket full of unclean animals,[144] and Paul speaking to the Athenians in the meeting on the Areopagus Hill[145] should dampen our enthusiasm to condemn our fellow humans who think and believe differently than us. Recall how Jesus rebuked his disciples when they were so quick to ask permission "to call fire down from heaven to destroy them," referring to certain non-Jews who didn't accept their message.[146]

The immediate rebuttal from many Christians at this point is to quote from the book of John: "Jesus answered, 'I am the way and the truth and the life. No one comes to the Father except through me.'"[147] This passage is often wielded as a weapon-of-mass destruction against all religions except the Christian one. This passage doesn't have to mean that unless one has a personal conscious encounter with Christ, then one is eternally damned. And it certainly does not mean that one has to self-identify with a particular human institution known as the Christian church. It can also mean that Christ has gone ahead and opened the door, paid the ticket, created the bridge, and made a way for people to come to the Father. An analogy might be people in a

---

142. John 4.
143. John 10:16.
144. Acts 10.
145. Acts 17:16–34.
146. Luke 9:51–56.
147. John 14:6.

coffee shop drive-through finding out that the customer in front of them already paid for their coffee and donut. Even if they have no idea who that customer was, and will likely never meet them again, they still derive the benefit. Could it not be that Christ is saying in John 14:6 that he's taken care of providing a way and simply invites everyone to not only accept the benefit but to also be part of God's solution for the world?

It seems that a common interpretation of John 14:6 is that unless one conforms to the catechisms of a particular human institution (notice the switch here: I didn't write "the teachings of Christ") then one is eternally damned. Sikhism arose out of a harsh environment of inequality and indifference to human suffering, and sought to address social injustice, to promote equality with respect to gender and social status, to share wealth and to be a positive force in the community: how is this different from the NT church practicing what Christ taught? Buddhism is in part dedicated to freeing self from unhealthy mental states and being the best person one can be: how is this different from the Christian goal of freeing self from sin or from the old nature, and living a good life? Likewise, Buddhism seeks to find joy, happiness and fulfillment (through suffering): how is this different from certain groups of Christians pursuing the Prosperity Message (otherwise referred to as the Health-and-Wealth Message, or the Name-It-Claim-It-Frame-It Message). With all the attention on Islam in the world today, there are reports of Muslims having a very Christian encounter with God in the complete absence of any church influence. Similar things can be said about many other religions.

Some will challenge these other religions by pointing out their differences from Christianity; in response, I would ask them to consider the differences between the ways they themselves deal with inequality (with respect to social status, wealth, gender, and so on) or suffering and how Jesus dealt with those.

Or the challenge will be raised that other religions concentrate incorrectly on actions and the things one has to do for salvation. Yet at the same time, so much of Christianity is busy condemning on the basis of things people do, especially when it comes to matters of a sexual nature (think how often attention is drawn to sex in the media, pornography, promiscuity, adultery, extramarital affairs, teen pregnancy, and homosexuality, even though other sins of the flesh like envy, fits of rage, and selfish ambition are given within the very same scriptural lists).[148] Or the attention to the things that Christians must do (attend church regularly, read the Bible, give tithes) or not do (smoke, drink, gamble).

But to avoid being misconstrued on this point, let me repeat once again: this is not universalism, nor does it mean that all paths lead to the same God. Even the modern version of the religion referred to as Christianity does not fully follow the life teaching of Jesus. Many times I've heard "What do they do with Jesus?" asked as a criticism against other world religions or even other Christian denominations, even though the ones asking don't themselves follow all of Jesus' teachings about: giving

---

148. Gal 5:19–21; Brauch, *Abusing Scripture*, 152–53.

away wealth; helping the poor and sick; visiting the imprisoned; "just war" in Middle East countries; hatred toward other people, often on the grounds of sexual orientation or race; lack of forgiveness; loving your enemy; and so forth. It's so easy to casually say, "Jesus is the Lord of *my* life," while still going on to live life according to one's own rules.

Jesus is a person, not a religion, and he gives the best example I know of to follow in our search to know God. Other individuals have been put on similar pedestals: Buddha (Siddhartha), Confucius, Mohammad, Ghandi, Mother Theresa, the Dalai Lama, and many others. When I take a close look at the life stories of even those highly revered individuals, I find that they fall short of the ideal. For some of them, it's a life of violence, a racist or misogynist attitude, or even activities that today would have them arrested. For others of those human heroes, it's an attitude of indifference to human suffering around them, or of haughty indignation to certain groups of people. This is not the case for Jesus Christ. In the end, I think that he gives us the best example to follow in getting to know God.

Inspiration and Authority of Scripture

The first criticism to be levelled against the geocentric theory was that it conflicted with Scripture. This is the exact same charge immediately brought up against the theory of evolution and the idea that humanity may have evolved from an ancestor that we share in common with primates. So this concept of the authority of the Bible is certainly one that needs to be addressed in detail. Closely related to that concept is its inspiration.

Yes, absolutely, Paul wrote that "all Scripture is God-breathed and is useful for teaching, rebuking, correcting and training in righteousness."[149] But what does this verse mean?

First, what did Paul mean by "all Scripture"? Most theists subconsciously substitute "all Scripture" with "the Bible." However, the Bible didn't exist at the time that Paul wrote that passage, and in fact many of the NT books that are now found in the Bible (including the gospels) hadn't yet even been written. "The Bible"—the canon of Scripture—came into being at the Council of Nicea almost three hundred years later. Prior to that, Scripture constituted a variety of Hebrew manuscripts which varied from synagogue to synagogue, and those manuscripts were almost exclusively ones that we call OT. Without a doubt Paul would have been thinking of the Pentateuch—the first five books of the OT—as being part of Scripture, as well as many (most?) of the other OT books.

Next, what would he have meant by "God-breathed" or "inspired"? Many authors and artists will speak about the idea or agent which inspired them to create their

---

149. 2 Tim 3:16.

works. The former don't mean that their novels were dictated to them by some external agent, and the latter aren't saying their art was a paint-by-number effort. In the same way, very few theists will say that God dictated the words to the OT writers, let alone handed them the original copy of the book in its complete form. Those that do think this have to then explain the various flaws we now find in those books. I won't take the time here to repeat the very long list of contradictions, errors and inconsistencies which Bible critics frequently point to, but most readers will already know of their existence, and they're very easily researched.

Instead, inspiration can refer to what motivates a writer or artist. That which gives them the core idea of an original thought. God certainly provided that to the authors of the books of Scripture. But the authors still had their preconceived ideas, culture, biases, and limited understanding which competed with those Divine influences on what they wrote. For me, the perfect example of this inner conflict in the authors of Scripture is Psalm 137:8-9, in which the writer refers to the joy to be obtained by smashing the heads of the children of their enemies against the rocks. I've tried my best to understand the hateful motive behind those two verses, and discussed them with others who have more theological training than I do: somehow the latter are able to reconcile these verses with a God of love, but I remain completely convinced that these words crystalize the angst and hate and vengefulness of the human heart, and do not reflect the God who loves and forgives and attributes guilt for sin to the one(s) who committed the sin. In the same way, many other passages in Scripture contain human misunderstandings and biases: chapters 2 and 3 summarize a whole host of other examples of this point.

Christians are quite comfortable with echoing John's ascription for Jesus as The Word of God[150] and believing that Jesus was both human and divine, and furthermore that his human nature limited him in many ways. For example, Jesus still needed to sleep,[151] still walked from one place to another[152] (a divine being could have flown, or transported telepathically), still became hungry,[153] didn't know everything but sometimes had to ask (he could have just stated how many loaves and fish they had[154] or how long the boy had been suffering with epileptic fits,[155] rather than asking), and ultimately could be killed. He still needed to learn obedience.[156]

If they are able to hold the divine and human natures of Jesus in tension in this way, why, then, do some Christians resist the idea that the Bible is also The Word of

---

150. John 1:1–18.
151. Matt 8:24; Mark 4:38.
152. Many dozens of gospel passages attest to this, the first of which is Matt 4:18.
153. Matt 4:2.
154. Matt 24:36; Mark 5:9; Mark 6:38.
155. Mark 9:21.
156. Heb 5:8–9.

God, is both human and divine, and is limited in some way(s) by its human authors, copyists, translators, and exegetists?[157]

If one accepts the idea that the writings of Scripture comprise divine inspiration as well as human limitations (our biases, culture, limited understanding), it becomes much easier to come up with alternative interpretations of difficult passages. Such as the ones which speak about slavery, the place of women in society, other religions, gay rights, dietary laws, how to discipline children, and scientific matters such as the origin of living things and of mankind, the nature of the cosmos, the evolution of language, and even the weather.

So when the charge is levelled against a particular point or concept that "it contradicts Scripture," one should keep in mind that this doesn't necessarily mean it contradicts a direct teaching of God on the matter. It might instead mean that it contradicts how the ancient writers saw things while grappling between divine inspiration and human limitation.

This clarification becomes important when we speak about the authority of Scripture. For some, this authority is absolute and final: "If the Bible says it, I believe it and that settles it." But we really need to be careful. God may well be the ultimate authority on a matter, but humans are not, and Scripture combines both inputs. When Scripture speaks about the dome or firmament which curves over the flat earth, does that constitute solely God's ultimate authority on the subject of cosmology, or does it also reflect human understanding on the matter? When Jesus taught that the mustard seed "is the smallest of all seeds on earth,"[158] was he speaking as an authority on horticulture? There are many plants with much smaller seeds. (Some translations try to mitigate or circumvent this problem by rewriting the text as "the smallest of *your* seeds," but that word "your" is not in the original Greek). Likewise, if he said that a seed must die before it can germinate and produce new seeds,[159] do we need to reject the biology books that tell us the seed is actually still very much alive, albeit in a dormant state?

Many former believers have given up their faith entirely because "a plain reading of Scripture" on this or that point conflicted with the plain evidence of science and/or a straightforward sense of morality and justice (by the latter, I'm referring to the slaughter of the Canaanites, or the subjugation and devaluing of women and children relative to men, for example). I nearly was one of them. But what rescued me from throwing the baby out with the bathwater was the realization that Scripture is not only inspired but also interpreted and responded to. Down through the ages, theists have regularly taken a given passage and interpreted it in quite different ways. The process of interpretation involves allowing one's worldview—one's "-ism"—to slant the meaning one way or the other. Sometimes in diametrically opposite directions. And this is

---

157. Enns, *Inspiration and Incarnation*; Ramm, *Special Revelation*, 33.

158. Mark 4:31.

159. John 12:24; Paul says the same kind of thing in 1 Cor 15:36-7.

where authority and reliability become vulnerable: it is not the authority and reliability of Scripture which one must question, but those of the interpreter of that Scripture.

I also realized the huge mistake that is frequently made by employing individual snippets of passages in one's arguments without keeping those passages in their broader proper context. Fundamentalists are often quick to toss out a single verse from the Bible, or even a fragment of a verse, like a hand-grenade when debating huge, complicated concepts. It should be obvious that this is a weak strategy. Consider if one were counseling another person about what life goal to set, and gave them this scriptural advice from the author of Ecclesiastes: "money is the answer for everything,"[160] or "I commend the enjoyment of life, because there is nothing better for a person under the sun than to eat and drink and be glad."[161] Telling a friend who is despairing over their family relationship difficulties to simply look at the verses which teach us to drink our problems away.[162] Coming to a suicidal friend despairing over the meaning of life, and sharing with them the third chapter of Job. Or counseling someone about a sexual hang-up they're struggling with and gave them this from Paul: "I wish they would go the whole way and emasculate themselves."[163] Obviously that would be taking those verses out of their proper context.

So as we work our way through these new revelations of knowledge coming out of the sciences of paleontology and genetics, let's be careful to keep in clear view the concepts of Scripture, inspiration and interpretation and be careful about muddying the waters with our preconceptions.

Reframing Our View about the Bible

I'll admit that my own book (and a previous one)[164] are both at odds with some traditional church thinking which we've inherited for many millennia. They seem to deviate considerably from "a plain reading of Scripture." They're motivated by the struggle I had as I came to grips with my own faith-versus-science dilemma. Holding up the new findings from genetics against my theological framework was massively disruptive for me, and so I can fully understand on a personal level how disorienting this kind of discussion can be for some readers.

But I think the main reason this was so damaging for my belief, and likely also for that of many readers, is that I previously held the view that the Bible was God's word. Not that it *contained* God's word, but that it *is* God's very word. That view was instilled into me for the first several decades of my life. And that colored how I viewed everything in life around me, including the findings of science. Actually, what that

160. Eccl 10:19.
161. Eccl 8:15.
162. Prov 31:6–8.
163. Gal 5:12.
164. Janssen, *Plato's Cave*.

view often required me to do was to deny the findings of science. Just file them away and ignore them.

But I can't do that anymore. There shouldn't be such a conflict between what I read/believe and what I see. Sir Francis Bacon referred to God's word and God's work (or God's world) needing to say the same thing. The Bible just isn't a science textbook for the modern world. It once might have been construed to be a science textbook to an ANE or ancient Hebrew audience, but all science textbooks eventually become outdated as science moves forward.

No credible theologian or biblical scholar believes that God dictated—in a word-for-word fashion—the scriptural texts to some scribe, be that Moses or any other. The authors of Scripture were indeed inspired to write what they did, but they also had their preconceived biases, their cultural and scientific limitations, their inner struggles against sinful thoughts and desires (see the preceding section entitled "*Inspiration and authority of Scripture*," p. 240). In fact, *Homo sapiens* have been struggling all through history with their theistic/deistic questions about the Great Being, and we've elaborated all kinds of religions out of that (see the preceding section entitled "*World religions*," p. 235). Ever since we were able to speak, we shared our thoughts on that big question. Ever since we've been able to write, we wrote books. Innumerable books containing all kinds of theologies. Different societies and civilizations have gathered around them certain of these books which they have declared to be distinct in some respect. Daniel Jeremy Silver—a now deceased Hebrew scholar, Jewish rabbi, and adjunct professor of religion at Case Western Reserve University—wrote, "Judaism, Christianity, Islam, Zoroastrianism, Buddhism, Jainism, Hinduism, Confucianism, Taoism all developed sacred books to which was ascribed a high degree of authority and infallibility."[165] He did not include within this list the ANE writings which we now have, particularly the Akkadian, Sumerian and Egyptian texts.

It isn't necessary to emphasize which of these texts preceded which others, as if to suggest that the "older" ones might be closer to a truer understanding of God, or that the "more recent" ones represent an improvement on the others. In a superficial sense, some are "older" and others "more recent." But from the perspective of human evolution occurring over several hundred thousand years, all those texts are relatively concurrent and recent, having appeared in the last few percent of our history as a species. In this sense, they all represent a relatively simultaneous crystallization of our meditation on the divine once our species reached a certain level of maturity and sophistication to not only ponder the divine, but to write those thoughts down. Of course, divine input—inspiration—would certainly separate one text from the other. However, all of them contain at least some degree of human input. The Holy Spirit is a powerful ally in helping us to sift through the ideas, separating out "the wheat from the chaff." When you read any given writing, be it Christian or not, be it contemporary or ancient, you need to ask the Holy Spirit to open your eyes to the truths that might

---

165. Silver, *Story of Scripture*, 4.

be contained in them, and to help spot the errors that are in them. We don't have to hide ourselves or run in fear from those other texts.

I do not at all suggest that these sacred books are equivalent in their content. But could they be complementary? Could it be that certain of these texts raise questions that another text answers, or perhaps the latter raises questions that the former try to answer but fail miserably, thereby highlighting the worth of the other text? Might they do so in the same way that the OT laws underscore Christ's teachings, and raise questions that are answered by the NT and the ministry of Jesus Christ?

In one sense, one could view that diverse collection of religious writings as a human diary, into which we scrawled our innermost musings, questions, experiences and conclusions on the subject.

Just like a child who maintains a lifelong diary that captures these kinds of details as they grow up, and live life, and learn about who they are. The earliest entries reflect their immaturity and lack of experience. But as one turns the pages, one can see the writing becoming more mature, insightful, and nuanced. Closer to reality (or at least to one perception of reality).

Or like a student keeping a detailed notebook summarizing what they learn about an exceptionally complex subject, such as biology or computer programming. Again, the first few pages are quite simplistic in their content. They just brush over the surface of the topic. They lay out the scope of what will be studied. Provide basic definitions of what will become important and central concepts. And again, as one turns the pages, the content will become increasingly detailed, sophisticated and "accurate." Closer to the whole truth.

In that sense, I see the Bible as part of the human diary or notebook on the subject of God. Recording our successes and failures in trying to understand the Divine.

Subconsciously, many Christians see the Bible as a single document which entered the human experience at a certain point in our history. If pressed hard on their understanding of the Bible's origin, they might acknowledge that there were a number of authors separated by long periods of time. But by and large, in the course of their daily lives when the Bible enters their thought stream, they will see it as one single document rather than as a library of books (which is the meaning of the root word from which we get the name "Bible") written by many different authors coming from very different perspectives. Moreover, they will acknowledge that portions of the Bible will have been written some two to four thousand years ago, but will nonetheless believe that those ancient texts crystalize the final version of our understanding of God, or of what he wanted us to know about him (and about life and the universe, for that matter). They will not appreciate that the texts reflect a growing understanding of the Divine. That our theology has been evolving.

Millennia ago, the Jews collected texts which they called Scripture, and kept them in their temples and synagogues. The priests could access them, but, generally speaking, the average Jew could not (and certainly non-Jews couldn't). It wasn't until hundreds of years after the events described in the gospels, and in Acts, and in the letters and epistles also contained in the NT, which the Council of Nicea distilled from the many texts that had been written at the time into a book which we now call the canon of Scripture: the Bible. We know that there were many, many other texts which had been written, many of which are even referred to within the Bible itself but which we've long since lost. Many are cited by OT writers: the Annals of Samuel the Seer;[166] the Records of Nathan the prophet;[167] the Records of Gad the Seer;[168] the Visions of Iddo the Seer;[169] the Records of Shemaiah the prophet;[170] the Book of the Wars of the Lord;[171] the Book of Jashar;[172] the Annals of the Acts of Solomon;[173] the Annals of the Kings of Judah.[174]

In Luke 24:45–47, Jesus refers to writings which don't match up easily with any text in the various Bibles I've looked into, which might suggest there was at that time another OT book that we've since lost, or that he was summarizing and paraphrasing several different writings.

Other texts from Judeo-Christian writers which are not included in my Protestant Bible include: the book of Enoch; Ecclesiasticus; the Wisdom of Solomon; the Acts of Thomas; the Revelation of Paul; the Apocalypse of Peter; the Gospel of Peter; the Gospel of Philip; the Gospel of Judas; the Gospel of the Nazarenes; the Gospel of Truth; the Gospel of the Ebionites; the Epistle of Barnabas; Paul's letter to the Laodiceans; Third Corinthians; as well as many other texts which have been retained within the Roman Catholic Apocrypha.

Depending on your Christian roots, some texts have been added or discarded, such that the Protestant Bible has sixty-six books, the Catholic Bible has seventy-three, the Greek Orthodox Bible has seventy-six, and the Ethiopian Orthodox canon has eighty-one books.

How does this knowledge of the origin of the Bible help me in my struggle to accommodate the findings of science with my personal beliefs?

From the earliest writings of humanity (the oldest one that we have at this time is the *Epic of Gilgamesh*) until the present, our thinking about the Great Being has ranged all over the map. Some would dismiss those as being merely pagan religions. I believe those were motivated by the Spirit within humans, creating a curiosity or

---

166. 1 Chr 29:29.
167. 1 Chr 29:29; 2 Chr 9:29.
168. 1 Chr 29:29.
169. 2 Chr 9:29 and 12:15.
170. 2 Chr 12:15.
171. Num 21:14.
172. Josh 10:13.
173. 1 Kgs 11:41.
174. 1 Kgs 14:29.

even a hunger to know the Great Being. And we've been busy taking down notes. Comparing ideas.

Even within the Bible, I can see a progression of thought.

The first few chapters describe an obviously incorrect understanding of cosmology, and an arguably incorrect understanding of the origin of species, of language, and even of the rainbow.

The opening chapters of the Bible present an image of God who isn't always fully aware of what is going on. He called out, "Where are you?" when Adam was hiding in the garden (although I understand this could have been just as much a rhetorical question, intended to make a point to Adam rather than gather information). The *elohim* needed to check out what was going on with the Tower of Babel.[175] They also needed to see if Sodom and Gomorrah were as wicked as rumors claimed.[176] That latter passage—which quotes the Lord as saying "If not, I will know"—certainly calls into question God's omniscience.

The OT texts also present God as too often angry and vengeful and destroying things: casting mankind out of the garden of Eden; putting curses on Adam, Eve, the serpent, and Cain; Noah's flood; Sodom and Gomorrah; the Tower of Babel; the Canaanites. (Reader: please note, I'm not contending for or against the actual historicity of any of these things; instead, I'm just pointing out the nature of how the authors seem to have perceived YHWH at the time.)

Isn't this much like the way children describe the world and their parents in the earliest pages of their own diaries? "Dad doesn't know anything. He's always checking up on me. He's always mad, always yelling at me."

Those earlier biblical texts contain within them the idea that if God's favour rests on us, we will enjoy protection, health and wealth (Enoch, Noah, Abraham), but then it begins to explore the unsettling possibility that one can please God greatly and yet suffer great loss and suffering (Job).

Turn the pages on this human diary and we learn about the reflections of Elijah, of David, of Solomon, and of the various prophets, on the subject of who God is, especially within the context of a covenant relationship with the nation of Israel. These include images of a God who protects and loves. And eventually we come to the pages which present Jesus and a whole new image of God: one of love, compassion and forgiveness.

Likewise when it comes to defining morality. Do the laws in the Pentateuch sound like they came from an omnipotent, omniscient, omnipresent, omnibenevolent Being who is concerned with ultimate justice, or do they sound more like the biases of an ancient patriarchal Semitic society? For example, I can understand why men from such a society would be concerned about whether a man was hunch-backed or

---

175. Gen 11:5
176. Gen 18:20–21.

a dwarf,[177] or had crushed testicles,[178] or about the menstrual state of a woman,[179] or how to go about claiming a beautiful woman captured in war as one's sex slave, and how to dismiss her if she doesn't prove to be satisfactory,[180] or prescribing a death penalty by stoning to a stubborn, rebellious, disobedient son.[181] But I do indeed find it very hard to see these values coming from God. In the same way, I don't see why a God of infinite proportions would care about whether one ate shrimp or lobster,[182] or about trimming one's beard.[183] Laws like these just sound too petty and human. Nothing like the morality taught in the NT: "Love your enemies, do good to those who hate you, bless those who curse you, pray for those who mistreat you. If someone slaps you on one cheek, turn to them the other also. If someone takes your coat, do not withhold your shirt from them. Give to everyone who asks you, and if anyone takes what belongs to you, do not demand it back. Do to others as you would have them do to you."[184]

Why do we have to put equal weighting on each of the pages of our diary/notebook? Why insist that the first few pages are just as accurate and mature as the later passages of the book? And for that matter, why consider that the last pages of that diary/notebook constitute the end-of-discussion on any matter.

Jesus repeatedly taught "You have heard it said . . . but now I say . . ."[185] Despite the many dietary laws in the Pentateuch (and there were indeed many), he said: "What goes into someone's mouth does not defile them, but what comes out of their mouth, that is what defiles them."[186] Concerning divorce: "Moses permitted you to divorce your wives because your hearts were hard. But it was not this way from the beginning. I tell you that anyone who divorces his wife, except for sexual immorality, and marries another woman commits adultery."[187] Each time he does this, he's emphasizing that the Scriptures may be a good starting point, but we need to take them to a higher, more complete and more accurate level of understanding.

The NT church took the Scriptures and bent them backwards against what the text actually said. Contrary to what was explicitly stated in the OT, they taught that it was OK not to be circumcised, that we no longer need to bring sacrifices for sin,

---

177. Lev 21:18.
178. Lev 21:20.
179. Lev 15:19–22.
180. Deut 21:10–14.
181. Deut 21:18–21.
182. Lev 11:10–12.
183. Lev 19:27.
184. Luke 6:27–31.
185. Matt 5:27–28, 31–32, 33–34, 38–39, and 43–44.
186. Matt 15:11; also see vv. 17–20.
187. Matt 19:8–9.

and that it's OK to eat meat offered to idols, just to name a few reversals on scriptural teaching.

And then there's the universality of the gospel message. It's easy to see why Jews at that time thought that YHWH was only interested in them. YHWH chose their forefather Abraham and created a nation out of them, giving them a homeland and a monarchy. YHWH's prophets spoke to the Jewish people. Jesus Himself said he was "sent only to the lost sheep of Israel."[188] But now the apostles were saying that the good news is for all people, not just those who are genetically related to Abraham. "Through the gospel the Gentiles are heirs together with Israel, members together of one body, and sharers together in the promise in Christ Jesus."[189]

And we've since moved even further beyond the thinking found within the biblical texts.

The subjugation of women and their role in the church, for example. Despite some claims today that even Jesus did away with gender inequality, he still only called male disciples. Perhaps that wasn't so much a result of him viewing women as less worthy, and more because he came to confront the Jewish religious establishment, and including women in his inner circle would only bring public controversy and thereby compromise that mission.[190] And even Paul the Apostle, who said there is no male nor female,[191] also wrote things with which many theologians and female believers have long since struggled.

Another change in our thinking that is ongoing today pertains to homosexuality. Yes, this is indeed spoken against within the Bible. So is eating shrimp or lobster[192] and women having short hair or men having long hair.[193] Every theist that I know of is able to dismiss the latter as simply vestiges of the culture of that time, but some will be adamant that the same cannot be said about homosexuality. I'm not convinced. But this book is not the place to wage that exceedingly contentious battle.

And of course, what *is* central to the point of this book is this matter of science. In the same way that we've long since moved beyond our earliest conception of the universe—a three-tiered heaven, including a dome, over a three-tiered earth, with the waters above and the waters below and a sun circling the earth—we now should move beyond a conception of the origin of mankind that involves a single couple living in the garden of Eden six thousand years ago.

188. Matt 15:24.

189. Eph 3:6.

190. There are other examples in which he makes astute strategic decisions regarding his mission . . . healing a leper but instructing the latter to not tell anyone who did the healing (Matt 8:4, Mark 1:44, and Luke 5:14) . . . foregoing on a suggestion from Mary because his time had not yet come (John 2:4) . . . claiming he couldn't go publicly to a festival because his time was not yet, but then shortly thereafter going in secret (John 7:6, 10).

191. Gal 3:28.

192. Lev 11:10 and 12.

193. 1 Cor 11:13–16.

All of these things involve interpretation and reinterpretation of the texts. It's not enough to simply have the texts and to be able to quote chapter and verse for any given situation. We need to recognize that not all passages were meant to be taken literally: we need to mine the text, and its context, for the deeper meaning. I've already mentioned above what happens when you take certain verses from Ecclesiastes at face value, in isolation and out of context: you end up searching after money and partying it up (see p. 243). But there are many other passages which also need to be digested slowly, and compared with other passages. Imagine if we all took literally Jesus's recommendation about chopping off our hands or popping out our eyes whenever we think they cause us to sin (I've learned that it's not one's hands or eyes which cause one to sin, but rather one's mind and inner impulses, and there are many other biblical passages which talk about doing radical surgery on thoughts and impulses).

Interpretation is a critical skill to learn if one wants to properly understand Scripture. It takes maturity and long, hard practice. We need to interpret Scripture in light of other Scripture and ANE literature. Whole books have been written on that skill, and it's too big of a subject to cover here. But suffice to say that "a plain reading of Scripture" is far from sufficient, and restricting oneself to that level is not something to boast about.

Some won't like this idea of holding less tightly to the wording of the Bible, and allowing more fluid interpretations of its texts. They much prefer the security in being anchored firmly to something immoveable. Their fear is that by letting loose from that anchorage they will drift; and that as long as they remain tightly moored, they will be safe. But there is a danger in this security. Sometimes it's necessary to take the boat to another dock, or into slightly deeper water when a storm comes in: otherwise, if the lines aren't removed, they will cause the boat to begin to crash against the dock, or hold them in the surf where they are dashed by the other rocks. This is exactly what happens on academic campuses all across the country. The children of the biblical literalists leave home and the sheltered environment of their local church and Christian community to go to university or take their place in the public marketplace. There, the children encounter the overwhelming findings of science for which they weren't prepared, and the beliefs that they were taught from the Bible will be shattered on the rocks of modern science. In the end, we have a scene of some boats securely anchored to a dock, surrounded by shipwrecked boats, broken and rolling in the surf.

Why Then Are the Scriptures Given to Us in Their Present Form?

It's quite evident that the original writers of the Genesis text, and their readers at the time, certainly believed these texts to be true. But their sincere belief doesn't mean then that the texts therefore must necessarily be taken literally today. It's worth pointing out that many great Christian leaders didn't require a literal interpretation of those texts, including Augustine of Hippo and Origen. Nor did many leaders during the centuries which followed. The discussions during the sixteenth century around the heliocentric theory (chapter 2) featured proponents from both camps. Even as recently as 1880 (or as far back as that time, depending on one's perspective), it was estimated that perhaps half of the ministers in major evangelical denominations did not hold to a historical Adam.[194]

There are several explanations for why those problematic passages are there, or how they got there.

The first obvious explanation is to attribute it to the human authors. As I've already stated several times, humans were involved at every stage in getting the Scriptures into our hands: the writing, copying, translating, interpretation and distribution.

This helps me understand passages which just don't sound like God to me. Again, as I pointed out on page 241, the perfect example of this for me is Psalm 137.

It also helps me understand passages such as those in which Paul is absolutely explicit about Adam being the first human,[195] or Peter referring to the Noah flood.[196] The biblical authors were products of their time. They were thoroughly steeped in Jewish history, Jewish culture and Jewish thinking. And they didn't have our level of scientific understanding. It's highly likely that they did in fact thoroughly believe in the historicity of Adam and of Noah and the various anecdotes written about in their cherished Scriptures. They also believed in the third heaven,[197] and in hell being somewhere down in the depths of the earth. That the earth was flat and the sun revolved around the earth. They believed in the existence of Leviathan, which they took to be either a great sea monster[198] or a fire-breathing dragon.[199] They held that dietary laws and circumcision were of paramount importance, that animal sacrifices needed to be offered at various times, and that people and things were often susceptible to becoming religiously unclean (for example: men having a wet dream; women menstruating; people touching a dead body), and sometimes needed to undergo elaborate purification rites to become religiously clean or to atone for their sins. These are all legitimate premodern Jewish beliefs and values that most of us don't hold anymore.

The next obvious explanation is to attribute it to the human audience. The authors wrote what they did for the sake of their audience: to meet them where they were at. Everybody in the Middle East at the time of the writing of the OT was talking about the primeval waters, and the dome over the earth, and various other aspects of the cosmology I presented in chapter 2. There was no way for that kind of audience to accept an entirely new story involving a Big Bang and the process of evolution, and genetic cousins which had long before gone extinct. Likewise, Paul showed an

---

194. Barrett and Caneday, "Adam, to Be or Not to Be," 15.
195. Rom 5, and 1 Cor 15.
196. 2 Pet 2:5.
197. 2 Cor 12:2.
198. Job 41:1; Pss 74:13–14, 104:26; Isa 27:1.
199. Job 41:18–21.

acute sensitivity to his own audiences, and modified his strategies accordingly: for example, standing up in the meeting of the Areopagus[200] or defending himself before the Sanhedrin.[201] And so, as I pointed out on page 173, perhaps Paul referred to Adam in order to bridge the gap between the Jews and the Gentiles and thereby bring a very tribal message to a non-Jewish audience.

Likewise, Peter's reference to the flood story,[202] and Jesus's to the same event,[203] to the blood of Abel,[204] and even to the creation anecdote itself[205] could have been for the sake of their Jewish audiences. They were using something familiar and comfortable. Mental pictures which they could use to paint a thousand words.

Jesus was also using word-play when he referred to the blood of Abel, and linking it to the "blood of Zechariah." (As an aside, much debate has been raised about Jesus referring to this Zechariah as being the son of Berekiah, while the passage in Chronicles indicates this Zechariah to being the son of Jehoiada). The murder of Abel[206] is recorded in the first book of the Jewish Bible (their book of Bereshit is our Genesis), while the murder of Zechariah the son of Jehoiada[207] is recorded in the last book of the Jewish Bible (our 2 Chronicles is part of their book of Divrei ha-Yamim). In both stories, the underlying motive for the murder pertained to tensions in working out the human-divine relationship: Abel's sacrifice to YHWH drove Cain to extreme jealousy or embarrassment, and Jehoida was calling attention to the apostasy of his contemporaries. According to the footnote in my NIV Bible to this reference to the murders of Abel and Jehoida: "The expression was somewhat like our 'from Genesis to Revelation.' Jesus was summing up the history of martyrdom in the Old Testament." His audience certainly believed Abel and Jehoida to be real historical people, but this doesn't necessarily require that Abel therefore was a real historical person. A modern-day historian could write, "From the days of King Arthur to the days of Queen Elizabeth II, the monarchy of Britain has been . . . ," and only the most unreasonable readers would then criticize that historian's understanding of history, or might on the other hand conclude that King Arthur must necessarily have been an actual historical figure "because this noted historian said so."

For some readers, this idea that the authors of Scripture—or the Apostles, or even Jesus himself—were limited by their audience shouldn't be too provocative: I've heard this kind of argument used to defend God himself, the one who provided the Divine inspiration. That is, I've heard similar defences against the accusation that

200. Acts 17:16–34.
201. Acts 23:6.
202. 2 Pet 2:5.
203. Matt 24:38–39.
204. Matt 23:35.
205. Mark 10:6.
206. Gen 4:8.
207. 2 Chr 24:20.

many passages in the Bible are misogynistic or racist; the rebuttal to this is that the Bible in fact advanced gender and racial equality far ahead of its time, but couldn't push further because the people of that time were not ready for too much radical change. In other words, the implication is that God could push only so far, but not further. The Bible already shows itself to be able to deal with much bigger issues, like the divide between the human and the Divine, and the problems of evil and death. It is prepared to challenge the ANE cosmology of its day, dismissing the gods of the other civilizations as impotent or even non-existent, and to even declare the peoples of heathen nations as worthy of extermination. If it is able to be that radical, why could it not also dismiss the social values which prevailed at the time? Why couldn't it just say that a woman is of equal worth, rather than stating a monetary value on females which is always less than that of a male?[208] Why not command an outright abolition of slavery, rather than describe how badly one could treat a slave[209] or how long one could own them depending on whether they were Hebrew or Gentile?[210]

And as far as the origin of the cosmos, of life, of species, of language and even of rainbows is concerned, it certainly would have been possible to give the ancient Jewish readers a story which is closer to the truth. Or at least been a little more vague, using wording such as "a long, long time ago," rather than fabricating a very detailed story which only goes back six thousand years and which contradicts history and science in untold numbers of ways. Why not let them know that, at a time when they thought they were the only people on earth and that "the world" only extended about as far as they could see, there were actually people scattered all across the globe?

Instead, passages like the ones which I've highlighted throughout this book as being problematic sound more like the paternalistic thinking of an ancient tribal society rather than the loving directions of a God who isn't able to push the redemption/restoration agenda too quickly.

Adjustments to Christian Perspectives of Biology

So far, this chapter has focused on how Christians might need to adjust their theology based on biological evidence. Before closing the chapter, I felt it was necessary to include in this chapter a short section on how Christian's might need to also change their interpretations of biology, based on their theology. I've already touched on this in chapter 7, when confronting objection #6.

I've often heard Christians refer to whole segments of biological life as products of the fall: viruses; predation; thorns; thistles; snakes; mosquitoes; the horrible smell of a skunk. Generally speaking, anything that has an undesirable aspect to it is labelled

---

208. Lev 27:3–7.
209. Exod 21:20–21.
210. Deut 15:12–18; Lev 25:44–46.

a product of evil, or even evil itself: a part of God's curse on creation. The underlying rationale is that nothing that we might find undesirable was present prior to the fall.

When it comes to death and predation, they will point to God's instruction in Genesis 1:29-30: "I give you every seed-bearing plant on the face of the whole earth and every tree that has fruit with seed in it. They will be yours for food. And to all the beasts of the earth and all the birds of the air and all the creatures that move on the ground—everything that has the breath of life in it—I give every green plant for food." This, they say, implies that all animals and people were only herbivores prior to the fall.

But we need to be careful about taking this too literally. The text emphasizes "every seed-bearing plant" and "every green plant." This would include poison ivy, poison oak and hemlock, which most people recognize as plants best to be avoided. It would also include *Abrus precatorius* (or Indian licorice), which causes nausea, vomiting, convulsions, liver failure and death, even after eating a single seed. And *Atropa belladonna* (or deadly nightshade plant), which causes a very long list of undesirable symptoms that include convulsions and death. I could go on, but suffice to say there are hundreds of different plants, many of which feature beautiful flowers and attractive brightly colored berries, which cause all kinds of suffering and death in humans and animals. One could propose that none of these plants existed prior to the fall, or that they did not have such poisonous properties until after that event. But there's no scriptural support for that idea (and a great deal of scientific evidence against it), and that's a lot of biology that God would have to go out of his way to spoil.

Besides, as already noted above, we have an abundance of fossilized evidence of predation going back millions of years. Bones with teeth marks on them. Dinosaurs which appear fully adapted for a life of predation, complete with fearsome teeth and claws. Other dinosaurs fully adapted for protection against predation, including horns and massive boney plates on their neck and back which effectively give them a suit-of-armor. We have fossils of predators with the skeletal remains of their prey in their digestive tract. That too is a lot of false evidence for God to have to plant all around the world.

Or one just accepts that these unsavory elements were all present for millennia as the data attest.

Even many of the most pleasant aspects of creation, ones that many creationists would insist were part of God's original design, are also a direct result of predation, and are therefore equally deserving of being labelled part of God's curse on the world. Camouflage is an adaptive response to predation. We marvel at the ability of a chameleon to quickly change its colors in order to blend in with its surroundings. The only reason it would want or need to do so is to evade predators. Many different kinds of animals and insects look almost identical to parts of plants. *Phycodurus eques* is a beautiful sea horse that looks almost indistinguishable from sea-weed. The walking stick (*Phasmatodea*) looks like, well, like a walking stick. *Phyllium celebicum* (leaf insect) looks like a decaying green leaf. *Membracidae*, or thorn bugs, look very much like a thorn. Many brightly colored caterpillars, butterflies and moths feature a pair of spots which look like big eyes, intended to scare off a hungry bird. If there were no such thing as predation, there would be no need for any of these astoundingly beautiful kinds of deception.

Other Christians will point to passages in Scripture which describe a future paradise in which "the wolf will live with the lamb, the leopard will lie down with the goat, the calf and the lion and the yearling together; and a little child will lead them. The cow will feed with the bear, and their young will lie down together, and the lion will eat straw like the ox,"[211] and interpret this to indicate God's displeasure with predation. I'm not convinced that these passages should be taken too literally, because Isaiah writes in these passages of lions eating straw, but later writes "no lion will be there."[212] Instead, I think one needs to read between the lines. The context around all those passages speak of the conflicts and suffering the Israelites experienced at the hands of the Assyrians and Babylonians, and judgements on those nations as well as on Moab and Egypt and Cush. The passages I've quoted above convey restoration and the ending of hostilities: enemies will set aside their differences, sit down and eat together.

The words on the pages of Scripture—the ones speaking of creation, and God's provision and lion's lying down with lambs—may be factually incorrect, but that doesn't mean one must therefore reject Scripture wholesale. Instead, we need to dig deeper into it as something conveying far greater truths: of God's provision, and care and restoration. It's all about interpretation.

Conclusion

A great deal of theology hangs on the story of Adam and Eve in the garden, created in perfection six thousand years ago, and choosing to cross a line drawn in the sand by God. But a mountain of evidence speaks against humanity arising from a single couple such a short time ago. In fact, that mountain of evidence, created by God, says we never numbered less than a few thousand, and arose over millions of years through a gradual process of evolution. Accepting that evidence as fact requires adjustments to our interpretation of the creation accounts, the place of Adam and Eve in our history, the fall and original sin, Christ's atonement on the cross, and even how we view scriptural texts (with respect to inerrancy, infallibility, and authority).

But simply denying the existence of that mountain of evidence because it requires too much change is not intellectually honest.

A variety of responses on the part of theists to this are considered in the next chapter.

---

211. Isa 11:6–7; similar passage found in Isa 65:25.
212. Isa 35:9.

9

Various Responses from the Church

THERE'S JUST *TOO MUCH* evidence that humans evolved over the course of millions of years to simply ignore or deny them. The church needs to come to grips with those facts, just like it did with conflicting astronomical data that overturned their biblical understanding of the universe.

I have read countless articles and blog posts online, and often take the time to read the comments which follow, in part because of additional insight or differing points that are sometimes made, but also to gauge the response of the reading audience to the views of the authors. Out of that experience, I fully expect that the views I promote in this book will elicit a variety of very negative responses from Christians, theists and atheists alike. That expectation is reinforced by the experiences of many actual face-to-face dialogues with people from all these groups.

In this chapter, I want to speak to the various groups of readers, anticipating their responses to my thesis and rebutting any potential misperceptions or challenges.

Apostasy? Defeat?

One response I have frequently witnessed, and always lamented, is a former believer feeling they have to throw out the baby with the bath-water when they find reason to question certain interpretations of Scripture or certain core theological tenets.

It's one thing if they choose to give up because they don't like the demands that their belief places on their lifestyle choices, or because they feel it's just too hard a path for them: both of these are represented in Jesus's analogy of the seeds scattered on the path.[1]

There's no need to give up the whole package when only a part of it needs to be thrown out or adjusted. This is a dilemma that is too often forced particularly upon believers immersed in a Fundamentalist worldview. From that rigid perspective, the

---

1. Matt 13:1–9.

Bible is God's word and therefore has to be all completely and absolutely true. It draws an unnecessary line in the sand which basically says: "If you can show me convincingly that even a part of it isn't true, then the whole thing can't possibly be true." But discarding part of one's belief doesn't mean one has to throw it out in entirety. Many people have been lured into the investment world by certain ideas of how to become wealthy, only to find out that some of those ideas might have been misplaced or mistaken: but this doesn't mean the rational response is to quite a paying job, close all one's bank accounts, liquidate one's assets, and "live off the land." Scientists have sometimes found a certain pet theory turned out to be a dead end, possibly because it was based on certain erroneous observations or biased interpretations: but they shouldn't then smash their computers and equipment, throw out all knowledge, and stop pursuing scientific questions.

Fundamentalists will often refer to a slippery slope, and put the blame on believers who are experiencing doubt or asking questions for creating that slippery slope. However, could it be that the Fundamentalist viewpoint itself is primarily responsible for that slippery slope? There is no slope if there is no elevation: and Fundamentalists are the ones who have elevated Scripture to a very high level. It is deemed to be beyond questioning, inerrant and infallible. These are qualities of God, but not of a text which we humans have participated in writing. We've been centrally involved vis-a-vis the writing, copying, translating, interpreting and distribution right from the very beginning. Our fingerprints are all over those manuscripts.

The texts may not be perfect, but they're nonetheless incredibly valuable. As cherished to us as a diary we might have written during the course of our entire life. We can learn so much about ourselves by flipping back through the pages of that diary or notebook, observing how much we've learned or matured, noticing character flaws develop through our life's experiences, relive our triumphs and learn from our failures. It's as valuable to us as the notebook into which we've recorded much of what we know about an important subject; something that we've spent a lifetime trying to understand, and haven't yet finished.

My own personal experience has been that it isn't necessary to jettison faith upon realizing and accepting the conflicts between science and Scripture. I know of others who have found the same.

Avoidance and Fear

Certain literalists will respond with fear. They will choose to not expose themselves to conversations involving evolution and OEC viewpoints, for fear of "poisoning their minds." When they come across a documentary on the TV or radio which addresses topics in these areas, they will turn it off or change the channel. This is really no different than the proverbial ostrich burying its head in the sand at the first sign of danger, or putting one's hands over their ears and making noise to block out the offensive noise. Not a very dignified way to deal with a threat. It's unfortunate that people who choose to hide themselves from ideas which challenge their faith obviously perceive the latter to be too delicate to stand up to scrutiny. "Always be prepared to give an answer to everyone who asks you to give the reason for the hope that you have."[2]

Any reader who plays a role of spiritual leadership in the lives of other believers—either as pastors, youth group leaders, or parents, for example—needs to inform themselves of the new scientific findings and how they impact theology. The believers that they are leading are going to encounter these ideas on campuses and in the marketplaces all over. If they're not properly prepared for this, it may overtake their faith.

Deny the Existence of the Data

AiG are very up-front in stating on their website that: "When teaching children, we tell them they should politely ask the question 'Were you there?' when talking to someone who believes in millions of years and molecules-to-man evolution."[3] The strategy is that it doesn't matter what the evidence seems to indicate: those who were not present to actually witness an event are declared to be in no position to comment on what happened. In the minds of those who take this approach, the only credible witness is the one who observed the event. And in the case of the creation account, we have only one direct eye-witness and sole participant: God himself. Who could be a more credible witness? Those who employ this strategy say we have the direct testimony of God, and "a plain reading" of that testimony informs us as to exactly how it happened.

On page 83, I related coming across a comment made by someone with clearly YEC views on an online article pertaining to evolution: the commenter wrote that science must give way to a plain reading of Scripture, that evolutionary theory must therefore be rejected and that no matter how much evidence that kind of science collects, it must be seen as misleading and bad science. Unfortunately, that comment has since been deleted from the website, so I can't refer the reader to it. But in searching for it, I did find what appears to be the original source of a comment like that: the opening page of the controversial textbook *Biology for Christian Schools*,[4] which states, "If [scientific] conclusions contradict the Word of God, the conclusions are wrong, no matter how many scientific facts may appear to back them."[5]

Collectively, Fundamentalists will insist that the Bible must be taken literally, and emphasize that it says nothing about Big Bangs or evolution but instead refers to "six

---

2. 1 Pet 3:15.
3. Ham, "Were You There? Evidence"; also see Ham, "Were You There? How Can Anyone Know."
4. Pinkston, *Biology for Christian Schools*, 1.
5. In the process of trying to relocate that blogger's comment, I found the same sort of defiant and blinkered denial being pronounced against subjects as diverse as global warming, the dangers of cholesterol in our diet, fluoride in our water, vaccines, and even 9/11. It seems that it isn't only theists who prefer their original biased beliefs "no matter how much evidence that science collects to the contrary."

days" and God creating the living things "each according to their kind." This viewpoint has never been a majority view within the church as a whole over the ages; Augustine of Hippo and Origen certainly didn't take Genesis literally, nor did Sir Isaac Newton. Instead, it is a very modern idea, arising in the United States less than two hundred years ago, and is exclusively Protestant (although it does have close parallels with the strict interpretation of Islamic scriptural texts describing the creation of everything by the God of Abraham). It was formulated in response to what some people saw as attacks against the faith. In other words, it's a defense. A circling of the wagons. Digging a moat and pulling up the drawbridge. And unfortunately, it has been getting us in trouble. Encouraging nonexperts to reinterpret fossils and archaeological data in ways that make experts shake their heads. To ignore mountains of genetic data which point to gradual changes over long periods of time. To create new physical laws, or bend or negate existing ones, in order to explain observations made by astronomers and quantum physicists.

Do we really need to do this? Denying the obvious and experimentally proven because it doesn't fit with our worldview? Should we put the Bible in authority over science and actual hard data solely in the name of inerrancy and infallibility (neither of which the Bible claims for itself)? Do we really have to insist that God was handing us a science book, history book and theology book all wrapped up in one? Or was God instead using words and imagery of that time—when the Babylonians, Sumerians, Akkadians and Egyptians were the intellectual powerbrokers—to get us thinking differently from the way we humans had been thinking for millennia. Get us away from thinking that the cosmos was filled with many gods. Away from the idea that people were created to do labor for the gods reclining in their temples, and toward the idea that we were meant to enjoy life to the fullest, as if walking with God and our partner through a garden full of food for the taking. Away from the idea that the sun, moon, stars and animals were gods to be feared, but were instead mere objects created for our use and benefit (marking off time; providing light, warmth and food). And many other theological ideas.

Accept the Data, but Attribute Their Authorship Differently

Instead of actually denying the existence of those data, another strategy is to reinterpret them. In one sense, this is a role-reversal on the part of YECists: rather than accept "a plain reading" of the data, they look for ways to interpret the data differently, or coming up with alternative explanations for how the data came to be where they were found, rather than being products of actual historical events.

One explanation that the YECists have given for the discrepancies between the biblical text and scientific observations is that God planted all that evidence intentionally. He placed fossil bones and stone tools where we would find them six thousand years later, and gave them a molecular structure which made them look much older (that is, God carefully calibrated the ratios of the various radioactive isotopes to indicate a much older dating). He actually created the human, Neanderthal, Denisovan and great ape genomes to look similar, including the genetic mistakes they have in common.

The explanation given for why he would do this is to test our faith: to see if we would believe him or the scientists. Faith versus fact. But that turns Him into a liar, making things appear to be what they aren't. That's like me leaving all kinds of false evidence that I'm having an extramarital affair in order to test my wife's love for me: my reasoning would be that if she can't overlook the overwhelming mountain of data saying that I was being unfaithful, despite my insistence otherwise, then she isn't good enough for me. To do that would make me a horrible husband. Is the deception supposedly carried out by God any different? Isn't that made worse by the fact that he is the one who created some of us with a scientific or logical mind: ones that accept observations at face value and are skeptical of conclusions that go against the data?

Others will say it wasn't God who planted those deceptive clues, but the devil. To mislead people. Sure, it's easier to accept that such a grand deception was orchestrated by the Great Deceiver himself, rather than by God. However, that kind of explanation then makes me wonder why God would stand idly by while so many innocent people are misled by the ruse. The analogy that comes to mind here is of a line of trusting, uncritical people stepping through an open door into a yawning abyss and falling to their deaths, with the devil on one side beckoning them to pass through and God on the other side standing silently with a small finger pointing meekly to a tiny little sign posted at knee-cap level which says, in letters as small as those found on the back of a package of prescription pills, "Stop. You're making a mistake. Turn around." How/why would he not intervene?

Distort the Laws of Nature

Alternatively, one can accept the existence of the data which challenge the traditional teaching from the church, and also accept that those data are a product of natural processes and actual historical events, but distort the laws of nature which govern those processes. Claim that the physical constants of various parameters are actually not so constant after all. With little or no basis for such extraordinary claims.

For example, some take scientific reports looking at changes in the speed of light, and then distort those wildly to fit onto a YEC viewpoint (not realizing that squeezing all the light energy which would otherwise fall on earth over the course of billions of years into just a few thousand years would simply vaporize everything on earth). Others try to explain away the long ages of time indicated from radioisotope dating by claiming that the rates of radioactive decay were different around the time of creation, without any underlying evidence for such a claim.

More to the central point of this book, though, are claims that the rates of genetic mutation were fantastically increased around the time described in the book of Genesis in order to account for the large genetic changes found in our genome or that of other animals. Proponents of this idea find the flood described in Genesis chapter 7 provides a perfect mechanism to explain a sudden change in mutation rates. They would hold that the firmament described in Genesis chapter 2 (whatever that was) collapsed, providing not only the water for the flood itself, but also exposing life on earth to massive amounts of cosmic radiation. At the same time, the flood was accompanied by massive geological events—earthquakes, volcanoes, continental shifts—which not only opened up the floodgates under the earth but also released all kinds of radioactive, mutagenic and toxic substances from deep under the earth. As a result, the background rate of mutations went off the scale, evolution went into overdrive, and all kinds of novel life forms appeared. To such proponents, these massive ecological disasters which followed after the flood also explain why the life spans of humans decreased from many hundreds of years to only "three score and ten."[6] This mechanism fails on so many counts. It's absolutely inconceivable to condense many millions of year's worth of evolution, spread out over millions of generations, down to just a couple thousand years spread out over a handful of generations. It just doesn't work that way.

Repeat a Lie till It Becomes Received Truth

I've already given one example of this on page 183: the frequently parroted claim that there are too many gaps and "missing links." But there are many other examples. Often, these misperceptions begin with the phrase "I heard that . . . " Nonetheless, the propaganda that there are too many "missing links" continues to be repeated and promoted. Likewise, the validity of radioisotope dating continues to be questioned and challenged by people who are not experts in that area. The arguments haven't changed in decades, despite the fact that all the concerns have been answered again and again.

Too often I've heard Neanderthals dismissed as representing fully human *Homo sapiens* with skeletal deficiencies owing to old age, disease and/or malnutrition. This seems to be in part due to an exhibit in the ICR Museum, which stated: "Many Neanderthal features are similar to those in elderly humans today. Since humans lived to great ages in the initial generations after the flood and Babel, perhaps the features are primarily due to advanced age." A popular book published in 1998 by Jack Cuozzo, entitled *Buried Alive*, added strength to this distortion by promoting the same thesis, despite the fact that he did not have the proper scientific credentials (he was an orthodontist). This misperception (some might call it deception) is becoming fixed in the thinking of some groups, even though we have many skeletons of very young healthy

6. Ps 90:10.

Neanderthals, and that the unusually stronger bones and the muscular attachment sites on those bones suggest that they were exceptionally powerful individuals, much more than the average *Homo sapiens*; in other words, they were not weak, unhealthy specimens.

Many YECs believe that humans walked with the dinosaurs because they've heard many times that fossilized footprints of both have been found which confirm this. They may have heard second or third hand about such sets of footprints found in the limestone beds of the Paluxy River, near Glen Rose, Texas. However, careful examination of those tracks by expert paleontologists have shown the "human" tracks have in fact been made by another species of dinosaur. Once again, the deception is perpetuated by museum exhibits, this time ones in the Creation Museum run by AiG in Petersburg, Kentucky, portraying humans and dinosaurs coexisting. In so doing, they continue to put an image in people's minds which then becomes truth to them. The fact is, no expert paleontologist—someone who has spent decades studying things like this in great detail and who understand the mechanisms—will agree that we have evidence of any kind for humans and dinosaurs coexisting.

Unfortunately, too many churches, youth groups and other gatherings of believers will invite speakers who do not have the proper accreditation to make the claims that they make. And too many books have been written by authors lacking the proper credentials, and too many readers of those books don't think to check out whether the author is adequately trained to make the claims that they do. Too much authority is placed in people who are able to portray themselves as knowledgeable, but who in fact have only a cursory exposure to the material. They should be properly recognized as nonexperts, and their testimony should not be placed over that of true experts simply because the listeners prefer the one message over the other.

Every time someone repeats something that they heard but didn't check out for themselves, they reinforce that statement in their own mind, making it that much more true to themselves, while at the same time they trigger the same sequence of events in the minds of their listeners. Those listeners in turn repeat the claims to each other, to their families at home, to their friends and neighbors, and thereby perpetuate the cycle. These people should practice the habit of catching themselves every time they begin speaking the phrase "I heard that . . . ," and commit to first checking out their facts: once they've done the latter, they'll be at liberty to speak as someone who's done their homework (or will find that what they'd heard was actually incorrect).

Changing the Definitions

Richard Dawkins has famously stacked the table in his favor by redefining biology as "the study of complicated things that have the appearance of having been designed with a purpose."[7] How can he not lose in any discussion around evolution if that definition is left on the table?

But theists also play loose with definitions, either intentionally or unwittingly.

Whenever a new hominid species is discovered, a common response from the YEC community is to label it as either a human or an ape. Never as something in between—a transitional form—because this conflicts with their worldview. By labelling it as "a human with certain physical deformities," or as "an ape with unusual features," they win the debate (in their minds) because they can point to the scientific record and show that it only ever has groups of apes on the one side, and groups of humans on the other side: that is, no transitional forms. It's circular argumentation.

The best example of this is given by AiG, the go-to source for many Christians; they maintain an ever growing website which expounds their position on a wide variety of topics and questions. In this online database, they describe Neanderthals,[8] Denisovans[9], *Homo floresiensis*,[10] and *Homo erectus*[11] all as "fully human" and "all descended from Adam through Noah's family."[12] *Australopithecus*, on the other hand, is said to be "an extinct kind of arboreal ape,"[13] while the more recently discovered *Homo naledi* is "most likely an ape."[14]

Not only is their motivation for relabelling each fossil discovery as either ape or human questionable, it is also fundamentally flawed: scientists do not claim that there is a link or transition between apes and humans. Instead, the consensus is that there was a series of branching events from common ancestors which produced a wide variety of hominids in various directions with widely varying degrees of lineage. I've already belabored this point above (page 184).

Whereas AiG view Neanderthals as descendants of Adam and fully human, RTB "identifies hominids, Neanderthals, *Homo erectus* and others, as animals created by God."[15] The justification for this distinction is that, in their opinion, those other hominids lack behavior which is consistent with the *imago Dei*; I have already discussed above the arbitrary nature of this distinction (p. 214). The definition which they have created carries with it various theological implications, one of which includes the conclusion that Neanderthals could not have had a soul.

AiG have also recently taken to redefining the process of Darwinian evolution and, ironically, incorporating this into their latest model of creation. In particular,

---

7. Dawkins, *Blind Watchmaker*, 1.
8. Lubenow, "Neanderthals: Worthy Ancestors"; Mitchell, "Does Hugh Ross Believe."
9. Mitchell, "Denisovan Gene."
10. Mitchell, "*Homo floresiensis*."
11. Mitchell, "*Homo erectus*."
12. Menton and UpChurch, "Who Were Cavemen?"
13. DeWitt, "It's an Ape."
14. Mitchell, "*Homo naledi*."
15. Reasons to Believe, "Hominids."

they are promoting a model in which Noah only took onto the ark "representative species"—one type of dog, one type of cat, one type of cow, and so on—which then gave rise to all the diverse forms of animals which we have today. To achieve this, AiG propose a period of hyperaccelerated change and adaptation (due to environmental stresses, genetic drift, founder effects, and even selection by humans) such that in only a small number of generations, each representative species morphed into the various unique species. On every level but one, this mechanism is absolutely no different than the standard process of Darwinian evolution. The one key difference is the timeline: the former is proposed (by nonexperts) to have taken place over only a few centuries, whereas the latter is widely accepted (by experts) to have taken millions of years. Previously (p. 181), I've elaborated on the inadequate academic credentials of the proponents of this very unorthodox YEC model.

Another redefinition of terms is to distinguish between micro-evolution and macro-evolution, as I've already pointed out on page 185. This allows them to appear to be open-minded and receptive to certain evidence for evolution while at the same time closing the door on a mountain of other evidence which also speaks to evolution on a grander scale.

Employ Rhetorical Distortions

All too frequently, when a new fossil find is made, some Christians react against it with flourishes like "this forces a complete rewrite of the whole story," or "turns evolution upside-down."

In truth, the revisions to the evolutionary story are rarely that dramatic. Each fossil find merely adds another piece to the puzzle and helps us bring the final picture into better focus. Just because scientists find another fossil that is older than previous ones doesn't mean the previous story was all wrong. Fossils continue to be found, and if they're younger, the announcement is low key; if they're older, the excitement level is raised in proportion. The same thing happens in other areas of our lives. Millions of athletes are always pushing their bodies to the limit: but it's only when one beats a world record that we hear about it. People are always doing strange things, but it's only when they exceed previous attempts at doing such a thing that Guinness write it down in their infamous book.

The motive for this kind of strategy is evident. By distorting a given claim to absurdity, many gullible or non-critical listeners will unconsciously discard that claim.

Condemnation and Discrediting

A large number will respond with harsh condemnation of those who hold, let alone teach, anything other than the standard YEC worldview. They will discredit the latter, and thereby give their fellow YEC peers license to reject or ignore anything that "the heretics" have to say.

Discrediting can include a judgement that the latter have lost the faith. I came from a very Fundamentalist background, and nearly gave up my faith because of that: the cognitive dissonance between my midweek worldview and my weekend worldview was just too great. I've since found that letting go of literalism and instead reinterpreting the Bible through the teachings of Jesus has allowed me to keep my faith.

Discrediting can come in the form of *ad hominem* comments using emotionally laden terms. They will refer to themselves having "a high view of Scripture," and/or to others having "a low view of Scripture," which I find to be quite condescending. It is possible to place a high value on Scripture as representing the collective insights and experiences of fellow believers in our passionate search to understand God, while at the same time feel no need to read it one-dimensionally or superficially: no need to see it as a modern science textbook.

Another *ad hominem* attack involves referring to the others as liberal Christians. It taps into the tribalism I referred to above (p. 220), and segregates them as "other" or as "the enemy." I find it very interesting how often it happens that when a believer commits to carefully scrutinizing what they believe, digging into the original documents and primary data, and especially going so far as to go to a seminary or divinity college, they end up softening their dogmatic or fundamentalist stance and becoming more liberal in their thinking. Why should the outcome of an earnest seeking for truth so often leave the seeker in a position that certain other fellow believers condemn as apostasy?

Or an accusation is made that the "other" has an agenda, and therefore their testimony is discredited accordingly. Scientists are said to have an agenda to destroy faith; despite the fact that a substantial proportion of scientists are theists themselves. As I wrote earlier, the agenda most scientists have is to reveal further truth and to oppose un-truths.

Discrediting may even mean declaring these ideas to be constructed and orchestrated by the devil himself, intended to destroy the faith of others. I've seen accusations such as these in many comments sections of online blogs. These people fail to realize that there are faithful Christians on both sides of the divide, just as there were on either side of the debate over the heliocentric theory. Christians on both sides sincerely seeking truth, and asking God to guide them to it.

Forcing a Round Peg to Fit into a Square Hole

Some Christians have tried to have it both ways: to accommodate science with a literal reading of Scripture. There have been at least three broad attempts at this.

When findings from geological studies suggested that the age of the earth was very much more than ten thousand years old, and later radioisotope dating put the estimate into the billions of years, some suggested that God in fact created the universe and an earth full of animals, including dinosaurs, approximately fifteen billion years ago but then had to destroy it all because of a spiritual rebellion led by the archangel Lucifer. The proponents of this idea say that Genesis 1:1 should actually be translated as: "Now the earth *became* chaotic and in confusion" (emphasis added), rather than the current translation: "Now the earth was formless and empty". God then presumably left the rubble and debris and dinosaur bones lying around until roughly six thousand years ago when he started all over again with the six-day creation event described in the rest of the first chapter of Genesis. In brief, then, this accommodative model inserts a fifteen billion year gap between the first creation event (which precedes Genesis 1:1, and therefore isn't recorded in the Bible) and the second creation event six thousand years ago (Genesis 1:2—2:3), and is therefore referred to as the gap theory. This model can be traced back to the second half of the nineteenth century within the writings of Herbert William Morris (professor of mathematics at Newington Collegiate Institute in London) and Isabelle Duncan (British author with unclear scientific credentials) in the late nineteenth century.[16]

Although the gap theory may have once satisfied some people as an explanation for the old age of the universe, the geological evidence and the fossil record, it does nothing to explain the genetic evidence of a long and gradual evolution, nor the fossil evidence of humans going back hundreds of thousands of years, nor many other lines of scientific observation. I'm not aware that any subscribe to this notion anymore.

A second accommodative model has been given more recently which attempts to account for the abundant evidence for evolution at all levels, including certain data pertaining to human evolution, but still remain faithful to the Genesis text. This explanation takes the first chapter of Genesis describing the origin of everything at the hands of God, and allows Big Bang cosmology and naturalistic evolution to reinterpret the verses in that chapter, complete with death and predation. Each day in Genesis chapter 1 is said to refer to an age or era in universal history, which themselves might last millions or billions of years: hence, this model is often referred to as the day-age theory. This proposal that the Genesis days correlate with geological epochs is elaborated in detail in Isabelle Duncan's book *Pre-Adamite Man; or, The Story of Our Old Planet and Its Inhabitants* (first published anonymously in 1860; reprinted and edited several times in response to favorable reception by the general public),[17] but has more recently been updated, modified and greatly popularized by Dr. Hugh Ross from RTB.

In this model, the Big Bang itself is to be found in the first two verses in Genesis: all the matter in the universe ("the heavens and the earth") being created but being completely disorganized and in a high-energy state ("formless and empty"). Cosmological evolution proceeds: matter cools and condenses to form stars and planets. Earth itself solidifies, but has a dense, cloudy, dusty atmosphere. Eventually,

16. Livingstone, *Adam's Ancestors*, 85–90.
17. Ibid., 80.

the atmosphere clarifies sufficiently to allow diffuse sunlight to penetrate to the surface ("Let there be light"). Earth eventually cools sufficiently to allow a water cycle to become established: water condenses into clouds which drop rain to form large bodies of water which in turn rise back up again by evaporation (the "waters above" and the "waters below"). Chemical and biological evolution proceed and give rise to life forms, beginning with microscopic ones and then later plant life ("Let the land produce vegetation"). The atmosphere continues to clarify to the point that finally the cosmological bodies themselves can be seen clearly ("Let there be lights in the expanse of the sky"). Biological evolution proceeds and we see the appearance of animal life and finally humans (Let the land produce living creatures . . . Let us make man . . .).

Although this model appears to reconcile the scriptural text with scientific findings, it isn't perfect. In particular, the Genesis text has birds appearing on day five, before land-dwelling animals on day six, while the fossil evidence has the ordering of their appearances the other way around (the latter says birds evolved from reptiles). This may seem like a very minor picky point. But the day-age model is rooted in the idea that Genesis describes God's own eye-witness testimony, given to the author(s) of the scriptural text, and therefore it would have to be correct in every detail. If, on the other hand, it isn't completely correct, is that because God's memory is defective? Is it because the human authors got it wrong? Either way, this relatively minor discrepancy between the scriptural account and scientific data weakens the authority of the scriptural account, and calls into question the justification for trying to make the two fit together.

A third accommodative model was proposed to account for the ever accumulating paleontological evidence of human evolution. In this case, the day-age model is modified further by claiming that the second chapter of Genesis describes God creating, in the relatively recent past, a separate, protected garden in which death, predation, disease, weeds and pain were physically excluded, and God placing two of the most evolved animals—hominids—into that garden after breathing into them a whole new kind of life. The in-breathing included an eternal spirit, but also a knowledge of his existence, and aspects of his own nature—the *imago Dei*—which were never defined in the scriptural texts and have been debated ever since. Was this free will? A sense of right and wrong? Other-centeredness (in contrast to self-centeredness)? Creativity? Reasoning? Language?

But this day-age interpretation of the first two chapters of Genesis also fails to explain much of the scientific evidence, and so it continues to require dilution and modification and a loosening of the "plain reading" of the text.

For example, to account for the much longer timeline, the garden experience is pushed back in time, even to fifty or a hundred thousand years ago rather than six thousand. But even that isn't sufficient to account for data which speak of many hundreds of thousands or even millions of years of human evolution. To account for our genetic diversity, it is said that Adam and Eve are only representative of a subset of humans existing at that time which were given the special soul-awakening experience which included eternal life. This makes God capricious, and calls into question his justice: would this be fair to those other humans around at that time and who would likely have descendants living today. In this context, Davis A. Young raised a particularly troubling conundrum: "Missionary strategists would be put in the very uncomfortable position of identifying those groups of anatomically modern 'people' who are not descendants of Adam and Eve and thus not really human. . . . As non-image-bearers, such 'peoples' are therefore not sinners and are ineligible for salvation. They do not need it. Missionary activity among such groups is unnecessary. We do not evangelize non-humans."[18] Davis also rightly asks what might be the status of the descendants of the half Adamites if interbreeding occurred between the chosen and not-chosen tribes. If one tries to resolve both problems by declaring that the unique pair—Adam and Eve—represented all modern humans existing at the time, what about the Neanderthals and Denisovans also living at that time, and which differ from humans by some relatively small finite number of genetic base pairs, or of the progeny produced by interbreeding between humans and Neanderthals in the relatively recent past (for example, we now have a specimen from a human who lived forty thousand years ago and whose ancestors included a Neanderthal only four to six generations prior to their own birth[19]). How could something as colossal as a creature-divine relationship and eternal salvation be dependent upon the orientation of a small finite number of pyrimidine and purine base pairs?

From a scientist's point of view, these latest developments in trying to reconcile the new genetic data with the original theological principles, summarized by the theologian Dr. C. John Collins (see p. 191 above), suffer from one central flaw: presupposition. That is, before trying to understand what the data are trying to tell us, they first require that "the data must say 'this,' 'this' and 'this' before we'll work with them." Recall the admonition of a dear friend: "I don't care what you do with all that as long as you leave me with Jesus dying on the cross for my sins."

As I've already mentioned above, Dr. Collins listed a number of provisos or initial criteria which must be met before a given model could be taken seriously. These include the presupposition that there must be a primal couple—Adam and Eve—or at least a tribe over which such an Adam ruled. But what if the data tell us that mankind as a species never numbered less than a few thousand? To solve this problem, adherents suggest that Adam and Eve were plucked out of the existing population of hominids and "refurbished," which only leads to the other problems raised above.

Another presupposition is that mankind was at one point perfect, then fell from perfection in some way, and must now be restored back to perfection. But what if we were never perfect? What if we were never much different than the animals? We have an abundance of anthropological and genetic data which place us directly within a

---

18. Young, "Antiquity and Unity," 380–96.
19. Fu et al., "Early Modern Human," 216–19.

Tree of Life which includes primates, dogs, cats and mushrooms. And in that data set we find a lot of evidence for mistakes, and diseases, and even death itself. Not to mention evidence that we killed and apparently ate each other at various times in our history. At what point can we be said to have been perfect? One can still believe that God wants to eventually make us perfect, but that doesn't require us to believe that we as a species had at one time been perfect.

So after addressing in previous sections the problems with taking the Genesis account too literally (Fundamentalism), and now in this section the problems with three newer approaches which take the texts less and less literally, I feel a need to ask why not dispense with half-measures and just go the whole way. Why not just accept that the creation account is not meant to be taken literally at all by us in the twenty-first century? That it reflects the understanding of the Jewish people thousands of years ago with respect to the origin of the universe, of mankind, and of their own nation? That it is indeed rich with metaphor, allegory and a deep mythical value that transcends merely a scientific description? Even for us Gentiles living today.

Intelligent Design (ID)

ID is a relatively nebulous term. It means many things to different people. For some, it's just YEC masquerading under a different name. But in the strictest sense of the idea, any explanation that involves another being or beings bringing us into existence must also hold the latter to be "intelligent" in some way. As such, even Panspermia—the idea that life was planted here by some extraterrestrial alien civilization, as portrayed in Hollywood movies like *Prometheus, 2001: A Space Odyssey*, and *Contact*, and vaguely alluded to in other ones like *Star Wars* and *Star Trek*—is also ID. The same could be said about the Babylonian version of the origin of the world and of mankind. Or any form of creation mythology in which a deity created things to serve a function or purpose. So it's not purely a Christian explanation.

The critics of ID label it a "science-stopper." However, many famous scientists past and present who have made substantial contributions to science also have believed that we were created by God, and sought to pursue how he did so. They saw something in nature—elucidated some of the mechanistic details of a new biochemical pathway, astrophysical phenomenon, or ecological wonder—and were amazed: in response they became excited to elucidate *how* God accomplished it. In that sense, ID is not so much based on ignorance or lack of knowledge, but is rather based on revelation and understanding. Sir Isaac Newton apparently wrote more about Christian religious matters than about science, but nonetheless is noted for discovering gravity and Newtonian mechanics, and for contributing to the development of modern calculus.

Admittedly, though, ID has indeed been horribly misused by people who use it as a catch-all explanation for anything that otherwise takes a bit too much effort to discover or explain. Sometimes ID is indeed simply Creationism repackaged. Equally embarrassing are ID claims which haven't been well researched.

For example, many nonexperts will remark about the awesome design of our human bodies—how much we are "fearfully and wonderfully made"[20]—without being aware of the many flaws in our design. I've elaborated on several of these already on pages 163–167. The recurrent laryngeal nerve. The retina of the mammalian eye. The erector pili (muscularized hairs on our skin). The muscle loosely attached to our eyes. The convergence of the trachea and esophagus on the one hand, and the convergence of our digestive tract and reproductive system on the other hand. And childbirth when it goes wrong.

A similar point could be made about design in the natural world. The symbiosis observed between flowers and bees is amazing. The annual migration of the Monarch butterfly. The biodiversity and harmony of a rain-forest. There are many such examples which would seem to point to an amazing designer who created these intricate webs of life. But that same designer would be the one who also created parasites and predators and the web of life in which they play crucial roles. The zombie wasp, and the Loa loa filarial worm which invades the human eye. The boa constrictor which squeezes the breath out of its prey. It's also the same designer who made sea turtles to lay their eggs high up on the beach, requiring the baby turtles to have to run a gauntlet of sea gulls and other predators before they can reach the relative safety of the ocean.

Many examples of apparently bad or poor design are put forward frequently and enthusiastically by passionate critics of the ID movement.

This careless use of the phrase "Intelligent Designer" has therefore led to references to the Malevolent Designer, who is responsible for those examples of design which we find morally reprehensible (the solution to human childbirth, and the eye-invading worms, for example).

Or to an Optimal Designer, to explain the examples of questionable design (the retina in our eyes). Such designs are cast as the best that God could possibly achieve, in lieu of a perfect outcome. Something that we otherwise would call a design flaw is in fact the best balance which can be reached between various competing factors. Certainly human engineers and inventors have to constantly struggle against competing factors and look for the best balance between them. But is this an acceptable explanation if one wants to believe that God is the designer and is omnipotent and omniscient, and was present and intervening at every step in creation? Or has this become a convenient way to dodge criticisms against the actual intelligence of the design?

One counterresponse from Christians against these criticisms of evil or poor design is to redefine them as a consequence of the fall. It is asserted that all was wonderful, harmonious and perfect before the fall, but that the flaws and problems were introduced when Adam and Eve made their fateful choice. However, setting aside

---

20. Ps 139:14.

the central point of this book that Genesis chapters one, two, and three are likely not historically accurate accounts, this assertion is based solely on an interpretation of Scripture and not justified by any concrete evidence.

Altogether, we need to be very cautious about ID claims. There is indeed merit in the idea, but we have to be fully aware that ID claims can easily explode in one's face.

Cautious Acceptance

Although not Fundamentalist at heart, nor requiring a literal reading of Scripture, there will nonetheless be some who are wary about changing long-held theological views. They will advise caution whenever we begin to move away from what the church taught for millennia.

Nonetheless we've done exactly that more than once before. For millennia, the church held to the three-tiered earth, three-tiered heaven view of the cosmos. But we don't think that way anymore. Very few Christians today will deny the findings of modern astronomy which refute this ancient cosmology. We fully accept that it was a naïve understanding based on simple observations and an unnecessarily literal reading of Scripture.

Theological conservatives would prefer to reject human evolution and insist on the historicity of Adam/Eve in the garden of Eden, not for any scientific or logical reasons, but because to do otherwise would mean having to re think the concepts of original sin, death and Christ's atonement.

But what if a mountain of scientific evidence supports human evolution occurring over hundreds of thousands of years? What if that mountain in fact argues directly against the historicity of an Adam and Eve? What if there's evidence of death, disease and predation going back for millions of years: fossilized dinosaurs with cancers, and fossilized evidence of a carnivorous attack on some prey animal? What if there's absolutely no evidence either way for a garden of Eden?

A very long list of observable facts now require us to adjust any theology based on a direct reading of the first few chapters of Genesis. This includes Paul's teachings in Romans 5 and 1 Corinthians 15 when he talks about the origin of sin and death.

I know fully well that even asking these questions will be disturbing for many readers. But I think we're allowed to ask questions like these. Many prominent figures in the Bible asked challenging questions. Abraham asked God "Do you really have to kill all those people?"[21] Moses tried to embarrass God out of wiping out the Israelite nation ("What will the surrounding nations say?"[22]) Job demanded an audience with his accuser, and with God himself, to explain the injustice of his situation.[23] Jesus

---

21. Gen 18:23–33.
22. Exod 32:11–13.
23. Job 31:35.

asked "Do I really have to go through this . . . to drink this cup?"[24] Thomas demanded proof before he could believe.[25] So I get the feeling that God's OK with us asking questions. It isn't sacrilege. To run from questions like these—with hands on ears, humming and babbling loudly to drown out the noise—simply because the answers might be challenging, or require some major renovations to our theological cathedrals, is not a sign of healthy pursuit of truth.

A Few Words from St. Augustine

Some theists think it is wrong to allow or even require theology to yield in front of the advances of science. It's a compromise. Moreover, they see this kind of confrontation as a relatively recent phenomenon: certainly not preceding the Renaissance, probably not older than just a couple centuries, certainly evidence of apostasy, and possibly even a fulfillment of the prophecy that in the end times the church will grow cold. To those who hold views anything like this, I would simply produce the following words of Augustine, one of the great church fathers of the fourth century:

> Usually, even a non-Christian knows something about the Earth, the heavens, and the other elements of this world, about the motion and orbit of the stars and even their size and relative positions, about the predictable eclipses of the Sun and Moon, the cycles of the years and the seasons, about the kinds of animals, shrubs, stones, and so forth, and this knowledge he holds to as being certain from reason and experience.
>
> Now, it is a disgraceful and dangerous thing for an infidel to hear a Christian, presumably giving the meaning of Holy Scripture, talking nonsense on these topics; and we should take all means to prevent such an embarrassing situation, in which people show up vast ignorance in a Christian and laugh it to scorn. The shame is not so much that an ignorant individual is derided, but that people outside the household of faith think our sacred writers held such opinions, and, to the great loss of those for whose salvation we toil, the writers of our Scripture are criticized and rejected as unlearned men.
>
> If they find a Christian mistaken in a field which they themselves know well and hear him maintaining his foolish opinions about our books, how are they going to believe those books in matters concerning the resurrection of the dead, the hope of eternal life, and the kingdom of heaven, when they think their pages are full of falsehoods and on facts which they themselves have learnt from experience and the light of reason? Reckless and incompetent expounders of Holy Scripture bring untold trouble and sorrow on their wiser brethren when they are caught in one of their mischievous false opinions and are taken to task by those who are not bound by the authority of our sacred

---

24. Mark 14:36.
25. John 20:25.

books. For then, to defend their utterly foolish and obviously untrue statements, they will try to call upon Holy Scripture for proof and even recite from memory many passages which they think support their position, although they understand neither what they say nor the things about which they make assertions.[26]

Later, he also wrote:

In matters that are obscure and far beyond our vision, even such as we may find treated in Holy Scripture, different interpretations are sometimes possible without prejudice to the faith we have received. In such a case, we should not rush in headlong and so firmly take our stand on one side that, if further progress in the search for truth justly undermines this position, we too fall with it. That would be to battle not for the teaching of Holy Scripture but for our own, wishing its teaching to conform to ours, whereas we ought to wish ours to conform to that of Holy Scripture.

Conclusion

We need to change our thinking. The few people who remain ardent Flat Earthers are no longer taken seriously (do an internet search: you'll find they do still indeed exist). They are dismissed as irrelevant, and anything they might have to say on a given topic is lampooned. In other words, they've lost their ability to be witnesses for the gospel. They're like the servant who took his talent (an ancient form of money) from the master and buried it in the sand (the metaphorical investment being their own heads), and returned it to the master without any gain, unlike the other two servants, who increased their allotment tenfold and fivefold respectively. That overly cautious servant was condemned.

Likewise, Christian apologetics has long ago moved beyond disputes over radioisotope dating, fossilized dinosaurs and geological formations. We can no longer maintain unbending support for a six thousand–year-old earth which bears the scars of a global flood. Those that do are just not taken seriously anymore. The "persecution" that one faces for holding to this worldview may be applauded within the walls of the church, but it too, like holding on to a Flat Earth worldview, seriously compromises one's witness to the world outside the church walls. Those battles were fought in the previous century and it's time to move on to a newer conflict.

Likewise, the story of Adam and Eve blissfully living in the garden of Eden until that fateful day when a talking serpent entered the scene is becoming indefensible in today's society. There is a tsunami of genetic data which directly challenge the ideas that humanity originated from one single pair, or have been around for less than ten thousand years. There is an irrefutable abundance of evidence for death, disease and

---

26. Augustine of Hippo, *Literal Meaning*.

predation going back millions of years, including fossilized forms of cancer. All of this, and much more, is entirely inconsistent with the fall in the garden being the origin of death and disease on the planet. It's irrational to deny or ignore this, and deplorable to promote a theology in which God stages an overwhelmingly persuasive deception designed to mislead thinking observant people in order to test their faith. Theology should resonate with both spiritual truth and natural truth. We can't simply discard science just because it isn't consistent with our theology. This is called presuppositional apologetics: allowing the outcome we're looking for to be the starting point for interpreting the facts in front of us.

I'm now more acutely aware of theistic attempts to maintain certain tenets simply because they support other theological tenets, and to avoid certain conclusions because those undermine other important theological tenets. The image it brings to my mind is of a house made of cards. One card in this image represents Adam's federal headship over the entire human race. Other cards represent: special creation; the *imago Dei*; the perfect serenity of the garden of Eden; the fall from grace; the universality of original sin; the atonement; inspiration and the authority of Scripture; the exclusive claims of Christianity; world religions. Removing certain cards threatens to bring the entire house down; nudging other cards causes the ones they support to shift and possibly cause whole sections to collapse, again threatening the integrity of the entire structure. According to Edward Stillingfleet, the orthodox Anglican bishop of Worcester, "a range of fundamental Christian doctrines rested on the assumption of worldwide human descent from Adam, and any tampering with that foundation would result in the collapse of the entire edifice of biblical anthropology."[27] Why try to maintain such a delicate structure? Why not remove the cards which can no longer be trusted to bear significant weight—particularly the one picturing an Adam (and Eve) as the federal head over the human race, being tempted in the garden of Eden—and begin the process of rebuilding? Peter Enns, a modern liberal theologian, recently put it this way: "Theological needs—better, perceived theological needs—do not determine historical truth. Evangelicals do not tolerate such self-referential logic from defenders of other faiths, and they should not tolerate it in themselves."[28]

---

27. Livingstone, *Adam's Ancestors*, 39.
28. Enns, "7 problems."

10

Atheist Worldviews Also Color Their Belief Systems

THE PRIMARY GOAL OF this book has been to address the response on the part of Christians to the new data coming out from the field of genetics.

However, a lesser goal has been to challenge everyone's tendency to allow one's worldview to color one's interpretation of the world around us, rather than to let the facts speak for themselves.

This applies equally to atheists.

The Faith/Belief Systems of Atheists

In my previous book, I devoted a chapter to exploring two articles of faith that atheists hold. Both pertain to the questions of origins: the first pertaining to the origin of all the "stuff" in the universe, the second to the origin of life.

The current, widely accepted understanding of the origin of all the matter and energy in the universe is the Big Bang.[1] To be precise, though, this theory doesn't actually explain the origin of everything in the universe. Rather, it explains what happened after the "stuff of everything" came into being: the unfurling of the many dimensions of our existence and the transmutation of an inconceivable amount of energy into different particles, and then those different particles into different forms of matter, as well as the expansion of all that material. But we still have no idea of how we went from nothing to everything in that first blink of existence.

Stephen Hawking is the one often credited for arriving at the equations describing the Big Bang: and the story goes that he did so in part by reconsidering the observed expansion of the universe in reverse. It is indeed an amazing theory, and provides a

---

1. As an aside, I always like to point out that the term "Big Bang" is a bit of a misnomer because there can be no sound in the vacuum of space, and the phenomenon was not so much an "explosion" as it was a dramatic unfolding of the dimensions of space and time . . . so the "Big Bang" was actually more of a "silent expansion."

host of explanations for other scientific observations (which is a hallmark of good science). The problem is that the equations don't work when the key variable—time—is set to zero. The closest we can get is 10^{-43} seconds after the Big Bang. So starting with an inconceivably dense, indescribably large and undefinably hot cosmic egg, we can say a lot about the universe. But we can't explain where that egg came from. At this point, we just have to believe that it existed. Somehow.

One recent explanation derives from quantum physics. As one compresses the universe into an ever smaller and smaller ball of energy—a singularity—we start to enter into an Alice-in-Wonderland world where the rules change dramatically. The certainty provided by Newtonian mechanics melts away and Heisenberg's uncertainty principle takes over. Before the beginning of time (if such an oxymoron can make sense), all possible outcomes existed together, and yet didn't actually exist, within that singularity. And then all of a sudden one of them became reality.

This is where the concept of fine-tuning comes in. Much has been made about how a long list of physical constants have taken on specific values, within a precision of many decimal places, which are required for life in our universe. If any of these were to have been even slightly different, life as we know it could not exist. These physical constants include the mass of the elementary particles (electrons, protons, neutrons), the gravitational constant, the speed of light, and a diverse array of other constants. Later in the evolution of the universe, we will add to that list many other factors such as the distance between the earth and the sun, and the size and placement of Jupiter.

That primal moment, the point at which all the physical constants took their precise values and the multitude of quantum possibilities crystalized into one reality, is a matter of faith. We can't prove it at this point of time. I'm not denying that it happened. Only pointing out that at this point in time we have to take it on faith.

Next, we come to another atheistic article of faith, the one pertaining to the origin of life. Despite all the bluster about how it was all inevitable, we have no concrete mechanism which explains how inanimate chemicals arranged themselves to produce the first living cell. Many hypotheses have been floated, but none have won the day. There have been numerous widely trumpeted announcements of a discovery which might explain one or another key step. I've avidly followed these out of keen interest, but have repeatedly been disappointed in two ways.

On the one hand, each of these major advances comprise baby steps in the entire process. It's like trying to convince someone that I can jump to the moon by casually demonstrating my ability to leap a few feet into the air. When the discrepancy in the distances involved is pointed out to me, I just repeat the effort again with every ounce of energy within me and thereby add a few more inches, and then make the point that with diligent training and practice I'll add yet a few more inches or even a foot. And after all that, I'll insist that because I've been able to demonstrate those three or four successive degrees of improvement, it is completely realistic to expect that eventually I'll be able to jump to the moon given enough time, training, and attempts.

Not so much.

One example (out of many) of this misplaced confidence pertains to nucleic acids forming spontaneously to give the first self-replicating system. Agreed, under certain carefully controlled conditions, nucleic acids can polymerize to form short-ish strands of DNA or RNA, and RNA has been shown to be capable of catalytic activity (albeit very weak and non-specific), but these demonstrations are only the equivalent of my best high-jump efforts, and a fully self-replicative system is the equivalent of the cow jumping over the moon in the night sky.

On the other hand, it isn't a simple matter of stacking up a long series of baby steps to arrive at the desired destination, in the manner that Lao-tzu (Chinese philosopher and founder of Taoism) intended when he wrote: "A journey of a thousand miles begins with a single step." This is because many of the steps are incompatible: the experimental conditions for one key step conflict completely with those of another step. For example, polymerization of many of the key molecules requires the removal of a water molecule, and yet other key steps can only occur in an abundance of water. In other words, in the proverbial small warm pond that Charles Darwin envisioned, or the deep sea hydrothermal vents which are now being considered, the chemical law of mass action dictates that the ubiquitous presence of water will make it far easier to add the water molecule (to break those polymers down) than to remove the water molecule. Certain key molecules can form spontaneously under acidic conditions, whereas other ones require basic conditions. Some require higher temperatures, while others denature or follow different reaction pathways at warmer temperatures.

So at this point, there is no unified party line on how life began on this planet. We just know that it did, and atheists believe it happened all by itself. I use the word "believe" simply because we have no proof that it can do so unassisted.

A very few atheists will concede to the concept of panspermia—the idea that life was seeded here by some alien intelligence—but will do so emphasizing heavily that this intelligence was something other than God. Their only reason for this proviso or exception is that they insist: "We cannot have God as an explanation; that kind of explanation is merely a God-of-the-gaps." And yet they'll acknowledge the possibility that an alien civilization could have been involved, one for which we have not a shred of evidence but which we might yet discover: an alien-of-the-gaps. This is another classic example of allowing one's worldview to influence one's interpretation of the data. Also, it simply moves back the goal-posts, since one then needs to account for how the aliens themselves came to be, so that too isn't an explanation.

If the challenge is raised "when/how did God come to be?" my response is that now we're talking about a category error.

Any naturalistic explanation must of necessity abide by natural laws. Agreed, we don't yet know all the laws of science, and we've frequently found ways to explain miraculous things by discovering new natural laws. For example, making ten tons of metal literally fly around the world by discovering the airfoil and the science of aerodynamics; or talking to people on the other side of the world by discovering electromagnetism and radio waves. Nonetheless, there will always be a requirement for naturalistic explanations to be grounded in natural laws: to ask how extraterrestrial aliens came to be is a perfectly legitimate question.

Supernatural explanations, on the other hand, are by definition exempt from this rule. They are super-natural: above the laws of nature. No doubt, many rationalists will insist that there is no such thing as the super-natural. But that too is a belief: it's simply a claim. We have no proof of it being impossible. The situation is much like the assumption that there is no such thing as the square root of a negative number. Until one brave mathematician postulated exactly that, and opened up a whole new field of mathematics which in turn was used to make entirely new discoveries and new technological applications.

My best explanation for how I, as a rational scientist, can accept the possibility of a Being having no beginning and doing things which seem impossible, involves an analogy I picked up from C. S. Lewis, who in turn derived it from a fictional story written by Edwin Abbott.[2] Before I proceed, let me inform or remind the reader that physicists today fully accept that there are more dimensions to our existence than simply the three spatial dimensions and time. I've searched for a definitive answer as to how many there are, and found a wide range of answers. But suffice to say that every physicist will agree that there are many more dimensions than simply forward-backward, left-right, up-down and time (which they will refer to as the x-, y-, z- and t-dimensions). In Abbott's fictional *Flatland*, there are only two spatial dimensions: x and y. The z-dimension has been squashed into imperceptibility (the same is said to have occurred for the many other dimensions of which we are not able to perceive, outside of mathematics and theoretical physics). In other words, the beings in Flatland live within a thin flat universe much like a page in a book, the edges of which go on in both directions into infinity. The only way to move around in Flatland is in the x- and y-directions; they have no practical understanding whatsoever of anything in the z-direction. Abbott then introduces a sphere—having three spatial dimensions, not just two—and imagines how that would interact with Flatland. But for theological reasons, I prefer to make my

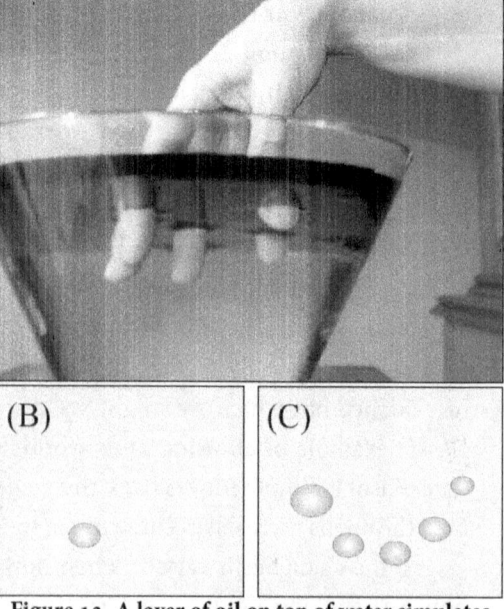

Figure 13. A layer of oil on top of water simulates a two-dimensional "Flatland". Inhabitants within that oil layer would only be aware of the parts of my 3-d hand that invade their 2-d universe (compare B and C).

2. Abbott, *Flatland*.

hand the three-dimensional object, rather than the sphere. You and I are fully able to operate in the z-dimension, although the Flatlanders can only perceive those parts of us which intersect with their x- and y-dimensions: the slice of our body that their x-/y- world cuts through.

Imagine I extend my finger down "out of the heaven" into Flatland: all of a sudden my finger appears in Flatland like a little circle that grows in circumference as I extend my finger deeper through the page. Seemingly out of thin air, I just suddenly appeared to them, albeit only in the form of an expanding circle. The other parts of me in the z-direction are invisible to the Flatlanders. And when I pull my finger back, I shrink and then suddenly disappear (figure 13B).

Next I reextend all four fingers and my thumb into Flatland, and from their perspective, I suddenly reappear, but now there are five of me that suddenly appear out of thin air in Flatland (Figure 13C)! And though they appear as five distinct and completely separate manifestations of me, they know somehow that all five are the same person. Five persons in one: a "quintity"? From my three spatial dimension experience, I can do things that Flatlanders can't comprehend. Think of the possibilities if there were a Being who could operate in all the dimensions of the universe, when physicists tell us there are at least ten dimensions!

Multiverse theory is another example of something I'll label as a belief that atheists choose to hold. Not that this theory needs to separate theists from atheists: there are also theists who choose to believe in this. But that's simply my point: we all, including atheists, can only choose to believe in this.

Multiverse theory postulates that our own universe is one of many different universes. In fact, there can be many very different kinds of universes, including ones with completely different physical constants. Many Hollywood movies play on the idea of other universes which are similar to our own but which have different timelines—other sequences of events—and the characters in their movies are trying to get back to the "right" timeline, or to restore the timeline back to what it "should be."

How many other universes could there be? One simply can't put a number on it: there is no way for us to ever find out. Anything that we can observe in one way or another is by definition within our universe. By extension, then, anything which is outside of our universe is by definition outside of the possibility of our observing it, let alone interacting with it.

It's all very heady stuff, and I'm aware of only two practical uses for this strange concept of multiverses outside of making Hollywood movies (in contrast to the immense usefulness of proposing the legitimacy of a square root of a negative number, for example).

First, it's a very useful naturalism-of-the-gaps. By that I mean that multiverse theory is very useful in "explaining" the problem of fine-tuning: if there is an infinite number of universes, each of which takes on a different set of physical constants, then one of them is bound to have the properties that our universe has. Problem solved!

Or the origin of life: if the chance of self-organization of inanimate chemicals into a living cell is so remote that fourteen billion years may not be enough to provide certainty that it will happen in one universe, then we can increase our chances by multiplying the number of universes, all of which are "trying" to form that life. Again: problem solved!

I for one don't see how explanations such as these are any better than saying, "God did it." They're just as pseudo-scientific—just as untestable and lacking explanatory power—than Intelligent Design. Anyone who embraces those concepts should stop railing against any God-of-the-gaps-like explanations. It's nothing more than materialism-of-the-gaps.

These beliefs are sometimes founded upon a presupposition. Yes, atheists too suffer from presuppositionism. That presupposition being the principle embodied within the metaphor of Occam's razor: that the simplest explanation, the one involving the fewest number of entities, must be the truth. Agreed, the probability is that this will more often than not be the truth. But this doesn't require it to be so. It could be possible for a more complicated mechanism to be the explanation for a certain observation or phenomenon. At the beginning of this book (p. 81), I gave an example of finding a tree in the middle of a forest, and agreed that the simplest explanation is that the tree was started by a seed dropped from one of the surrounding trees. But as we looked more closely at the tree, we realized that a much more complicated explanation was needed: that in fact the tree was started in another country and transported to this spot by grieving relatives.

In fact, from my point of view, the principle of Occam's Razor completely contradicts the atheistic explanations referred to above. How can an explanation which involves a nearly infinite number of universes, all trying out different values and parameters for a wide variety of physical constants and eventually producing sentient beings capable of exploring and understanding the universe, be compatible with Occam's Razor? Likewise for the origin of life and the succession of incredibly unlikely events required to produce the first functional living cell. As Keith Ward pointed out, isn't it more simple, more elegant, to propose a mechanism involving a single omnipotent Being motivated by love creating that universe? Admittedly, we have no evidence that God exists, but likewise we have no evidence of multiverses, of the spontaneous generation of substantially new genetic information, or of the self-generation of a living cell from abiotic raw materials.

How Can You Be So Sure?

A high school friend of mine sent me a link[3] to recordings from the Voyager space probe, and suggested I listen to them on a good sound system. Which I did. And was

---

3. Cole, "NASA's Voyager Probes."

blown away. Just listening to these "sounds" emanating from the cosmos unleashed a train of introspective thoughts. What the astrophysicists had done was record the electromagnetic radiation—vibrations in space—coming from the planets in our solar system by a satellite probe we'd lobbed out there thirty-five years ago, and then play those recordings back at a slower speed so we could hear them. "Sounds" recorded by NASA's Voyager probe as it zipped past Jupiter, Earth, Saturn, Uranus, and Neptune.

It's an amazing experience. Very much like the eerie background sounds in the movie *2001: A Space Odyssey*, or a bit like what you might hear in a Buddhist temple.

Listening to those recordings just brought home to me so strongly the point that there's so much going on around us about which we're unaware because we're so limited. Those planets have been "singing" away for millennia, and we've been completely deaf to it, and would continue to be so if some smart *Homo sapiens* hadn't developed a hearing aid for the rest of us.

Our ears hear only a fraction of the frequencies that make up sound. Apparently women can hear slightly higher frequencies than men, but dogs hear frequencies far beyond even them. In fact, ten times higher! And even a dog's ability is dwarfed by that of dolphins. In the other direction, elephants can hear frequencies far lower than we can, and do so through their feet! Peacocks make a sound with their tail feathers which is too low for us to hear, and they channel that sound like a spotlight by cupping their tail feathers into the shape of a satellite dish.

The same for our sense of vision. We see only a tiny fraction of the colors of light: the ones in a typical rainbow. Once again, animals can sense colors far above and below our range of perception. But even they can't sense the full breadth of the electromagnetic rainbow, which includes gamma rays and X-rays on the one end, and microwaves and radio waves on the other end.

The exact same things could be said about our sense of taste and smell (they're really the same thing), and I could use examples like blood hounds tracking a fugitive, drug-sniffer dogs, and a shark's ability to taste blood in water.

Sense of touch: ditto.

And it isn't just those classic five senses of ours which are so limited. Most of us can only conceive of four dimensions (three in space, one in time), and a minority of us (myself not included) might claim an understanding of certain other dimensions, but only in a mathematical sense. As I've said above, I can't find a definitive answer on how many dimensions there are, but the claims range up to eighteen!

And then there's our lifespan in the context of the great scheme of things. Despite our best efforts at keeping healthy, we last "just a blink of an eye."

We are just so puny and limited.

For this reason, I have to ask anyone who utterly discounts the existence of God: "How can you be so sure?" We are so limited. We perceive so little of what's going on all around us, and what has been going on for millennia. Our brains are incapable of containing a full understanding of the universe, and our feeble attempts at trying to do so are merely scratching the surface. Don't be so sure of what you can't know.

"I'm telling you, Larry, no Amoeba in the entire history of bacteriadom has ever found positive proof for humans. We've been to every corner of the pond. Put out our hairy feelers and chemical sensors, and not a single bug can ever point a pseudopod and say 'There! There they are!' Well, OK, every now and then some crackpot bug will claim to have encountered a human. But really, get a grip. No one really believes them. There's just no such thing as humans."

Seriously: how would a bacterium ever try to prove our existence, or claim to have a relationship with one of us? Even the brightest among them, using the most advanced technologies available to them, would be completely unable to reveal our existence, but would instead be totally at the mercy of us humans to reveal ourselves to them. What would the relationship between a four-dimensional being (*Homo sapiens*) and a ten-dimensional Being look like?

Agree to Disagree

The relationship between theists and atheists has too often been an unnecessarily antagonistic one, and sometimes an overly vicious one. True, we theists have sometimes brought that on ourselves by being insensitive and lacking tact when we share what we believe wholeheartedly to be true. But equally often, the combative response from atheists is a visceral one: it comes from deep inside. Perhaps both participants in this conflict are allowing that innate tribalism with which evolution has endowed us to control. Why not interrupt the either/or way of thinking and see it as a both/and situation?

Stephen Jay Gould, an extremely influential evolutionary biologist and popular writer believed that science and religion should be considered as two distinct domains of knowledge, referring to them as non-overlapping magisteria or NOMA. Separating science and faith may work much of the time, but sometimes we need both to get the complete picture. Atheists should consider the possibility that not all truth is found through reason and observation, just as much as theists should consider that not all truth is revealed in Scripture or tradition.

The astrophysicist Robert Jastrow recognized the complementarity of science and religion in our quest for truth. In his book *God and the Astronomers*,[4] he wrote: "At this moment it seems as though science will never be able to raise the curtain on the mystery of creation. For the scientist who has lived by his faith in the power of reason, the story ends like a bad dream. He has scaled the mountains of ignorance; he is about to conquer the highest peak; as he pulls himself over the final rock, he is greeted

---

4. Jastrow, *God and Astronomers*, 116.

by a band of theologians who have been sitting there for centuries." We're searching for the same thing—truth—and should ultimately arrive at the same answer.

Faith doesn't need to be a science-stopper; nor should science be a faith-stopper.

If it is possible to refer to the process of science and the body of knowledge it has produced as a quantifiable thing, then it is possible to say that the vast majority of both were produced by theists working in a theistic paradigm.

All of the world's oldest and most revered institutions of higher learning were founded by religiously motivated organizations: Yale, Oxford, Harvard, Princeton, Stanford, Columbia. The same can be said of many less recognized universities and academies.

Likewise, many of the most influential scientists and philosophers were theistic in their view: Plato, Aristotle, Socrates, Galileo Galilei, Johannes Kepler, Thomas Aquinas, René Descartes, Johannes Gutenberg, Christopher Columbus, Leonardo da Vinci, Nicholas Copernicus, Sir Francis Bacon, Blaise Pascal, Robert Boyle, Carolus Linnaeus, Gregor Mendel, Michael Faraday, Louise Pasteur, William Harvey, Max Planck, Werner Heisenberg, Albert Einstein, Alexander Graham Bell, John Polkinghorne (codiscoverer of the quark), Raymond Damadian (inventor of MRI), Francis Collins (former head of Human Genome Project), and hundreds of others. Apparently Sir Isaac Newton—the same one who gave us Newtonian mechanics, the wave theory of light, and calculus—wrote more about theological matters than scientific ones.

Conclusion

Atheists need to be careful about pointing fingers at the faith community. There is an expression that says when one points a finger at someone else, they're also pointing three fingers back at themselves. Atheists are too quick to dismiss people who have some form of faith, forgetting that they themselves often have a belief system of some sort (ideas which are not based on raw fact, but rather on hope, inference, expectation, and even peer pressure), and sometimes still try to hang on to some semblance of "spiritual" label. Many atheists are often guilty of a "naturalism-of-the-gaps" mentality, and have sometimes shown that they have vested interests in the outcome of a scientific question (the initial resistance to Big Bang cosmology, and the current resistance to ID theory, being two examples).

11

Standing on the Shoulders of Giants

WE *HOMO SAPIENS* HAVE continually built up ever greater bodies of knowledge and insight upon the work of our ancestors. Our vast amount of knowledge has been accreting like layers on the collective experience of billions of other people. Again, this is in part what sets us apart from the animals, and has helped us to dominate the planet (which was God's second command to us, after "be fruitful and multiply").[1] And now we're reaching out to the stars.

This building up of new ideas on top of old ones occurs at the biological level, too.

Evolution tells us that new genes arise from old ones: genes are duplicated and allowed to accumulate mutations until an entirely new and useful product arises. A simple change in one of the light-detecting genes leads to color vision, and a whole new world opens up to that animal.[2]

We also see this building up of new ideas in the form of anatomical structures. Again, in our eyes, our retina shows design modifications which compensate upon design flaws, greatly improving our vision in the process. There are many, many examples of genetic and anatomical breakthroughs which essentially amount to taking an already awesome idea and stepping it up a notch or pointing it in a new direction.

But it isn't only in scientific disciplines that we see this building up of new ideas on top of older ones. Another image or metaphor of ideas accumulating and building on one another is that of a diary or notebook. In the same way that an individual can keep these autographs as a storehouse of what they've learned—either in life, or in a particular academic subject—we can also do that as a species. As *Homo sapiens* evolved and expanded across the globe, they looked up and tried to understand the Great Being. Whole religions and theologies were born and died.

---

1. Gen 1:28.
2. Janssen, "Little Things in Life."

Some of them left behind only artefacts that now allow us to merely begin to understand their perspective on that compelling question: the Sumerians; those that built Stonehenge; the kingdom of Tuwana. In Asia, there is physical evidence for prehistoric religion going back to approximately five thousand years ago, during the late Neolithic era in the early Harappan period. In Turkey, amazing temple ruins have been found at Göbekli Tepe which date back to almost twelve thousand years ago.[3] In Siberia, researchers found frozen bodies and tools dating back to twenty-four thousand years ago, including "anthropomorphic Venus figurines, which are rare for Siberia but found at a number of Upper Palaeolithic sites across western Eurasia."[4]

Others left actual human representatives for us to dialogue with and record their stories: Zoroastrianism, Babylonians, Egyptians, Greeks, Romans, and others. In more recent history, we witnessed the appearance of new religions (from the mystic and polytheistic ones of the Far East such as Buddhism and Hinduism, to the Middle Eastern monotheistic Semitic religions of revelation like Judaism, Christianity, and Islam) and theologies (Catholicism, Protestantism, Sikhism, Mormonism). All of these and more trying to understand God, and devoting tremendous time, money, energy, and other resources toward that quest. I feel God has put that drive in us—"He has also set eternity in the human heart; yet no one can fathom what God has done from beginning to end"[5]—and each nugget of truth we've found has been inspired by his Spirit.

Among those gathering information about the divine were the ancient Jewish people, beginning with Abram and Sarai, who left the Babylonian world in their own pursuit of YWHW, the God of heaven. Their descendants gave us the Jewish religion, and its many writings. Those writings begin with a theology very different from the one Christians hold today. God is represented as often angry and destroying things. Vague references are made to a heaven, and none at all about a hell. Very few moral laws are given until we fast forward a couple thousand years to the story of Moses. And yet the biggest questions with which we still wrestle today are posed in those earliest of writings. In Genesis, the questions of how we came to be, and what is our purpose in life, are asked. Job, one of the oldest books of the Bible, if not the very oldest, addresses the question of why bad things happen to good people. As the centuries progress and the writings accumulate, we find more and more details of the Jewish understanding of God. His laws. His personal attributes, particularly his love and care. And this continues on to the representation given to us in Christ, who the Apostle Paul describes as the visible "image of the invisible God,"[6] and the writer of Hebrews as "the exact representation of his being."[7] Again, these are very different representa-

3. Mann, "Birth of Religion," 39.
4. Raghavan et al., "Upper Palaeolithic Siberian Genome," 87–91.
5. Eccl 3:11.
6. Col 1:15.
7. Heb 1:3.

tions from those given in the earliest Jewish writings. (Although looking back now, some will claim to be able to see glimpses and hints of Christ in the ancient writings, those glimpses and hints were subtle enough that they were completely missed by the authors and readers of that era.)

Many times I've alluded to these writings as a form of a diary of our species's life journey through this fundamental question: who is God? Each author of a book of Scripture, and each teacher which uses them, taking on at that moment in time the role of the author of *Homo sapiens*'s diary and adding another page. Or as a notebook which summarizes *Homo sapiens*'s various observations, glimpses, insights and hypotheses on the same question. We continue to build on all of those writings and reflections, even today.

But along the way, as we continue to learn more, we may need to adjust and readjust our thinking on a given matter as new insights emerge. Despite the tremendous power of the scientific revelations given to us by Sir Isaac Newton with respect to the laws of motion and the wave theory of light, both had to be markedly adjusted as we learned about general and special relativity, quantum theory and the particle theory of light. Likewise, the church had to adjust its view of ancient cosmology as the scientific evidence stacked up against the traditional understanding (ch. 2). Paul stood on Moses' shoulders and wrote about "the first Adam." But we can now stand on Paul's shoulders and, seeing further, be able to say, "Sorry, there's actually no first Adam." Doing so doesn't mean we claim any supremacy over Paul. As the sixteenth-century scholar Robert Burton put it, paraphrasing Bernard of Chartres, "A dwarf standing on the shoulders of a giant may see farther than a giant himself." Disagreeing with Paul also does not have to mean that we jettison our faith system either, although some adjustment may indeed be necessary (as was the case when the church finally accepted the scientific model of the cosmos; ch. 2).

So now we come to the present era, in which whole new bodies of scientific evidence are stacking up against the traditional understanding of our own human origins (chs. 5 and 6). Once again, this will require us to reconsider many theological points which we might have previously seen as immutable.

Where Are We At Right Now?

Science has moved so far beyond where it was in the days of Paul's writing about the consequences from the fall from grace of Adam and Eve in the garden of Eden, and far further beyond the time of the writers of Genesis which describe that specific event. In fact, we've moved so far beyond it that many believers find it hard to speak about theology and science in the same conversation. Trying to do so seems to inevitably divide those speaking. Scientific advances have also moved much of theology out of the realm of normal everyday experience for the average person.

With regards to the fundamental question "How did we get here?" there seems to be at least two diametrically opposing answers.

Looking at this question from one side of the table, with an open Bible as a reference and Christian tradition as our guide, we can point to several distinct touchpoints along the path describing the origin of humanity:

- we were created roughly six thousand years ago, in a Divine act completely separate from that which gave rise to the animals;
- there was one original couple, living in a garden somewhere in the vicinity of Mesopotamia;
- God took clay and created us in his image—we still don't understand exactly what the *imago Dei* is/was, but it might have been the spark which made us unique from the other animals—and breathed life into us;
- something happened in which there was a fundamental change in the human experience—a sharp break in the fellowship between God and man—a change which impacted agriculture, thorn-bearing plants, predation, and childbirth, and in the end we somehow died.

Rising up from that table, we walk around to the other side and take a seat beside several piles of evidence from various scientific disciplines, the most daunting of which comprise genetic and anthropological data. Looking at the very same question, but now from this new vantage point, we look down that path we had just moments ago drawn out on the table and find absolutely no confirmatory evidence for any of those touchpoints:

- the timeline is too short;
- the number of original humans is too few;
- the geographical location is incorrect;
- we aren't really all that different from the animals;
- death, thorn-bearing plants, predation, and hard childbirth have been with us for countless millions of years.

Instead, sitting on this second side of the table, one is compelled to draw out a completely different path, one at right angles to the first, to explain the origin of humans:

- we've been around for millions of years, having evolved in the genetic cradle of Africa from an ancestral type that we share in common with the other great apes;
- we were nearly wiped out as a species, but a few thousand of us hung on tenaciously and survived to dominate the planet;

- we have had many other genetic cousins and siblings who also came out of Africa and have since gone extinct.

These two different stories appear to be in complete opposition. Both can't be true.

Or can they?

At one point, scientists were viciously divided over how to explain the phenomenon of light. They too saw two diametrically opposed viewpoints, both supported by considerable bodies of experimental observations.

On the one hand, there was the particle theory of light. In the same way that matter is made up of tiny physical objects, so is light made up of something discrete and seemingly physical. Matter is composed of atoms, which in turn we learned were made up of yet smaller objects called electrons, neutrons and protons, but the physical entity which constituted light was unclear. Supporting this idea were observations made by one no less influential than Sir Isaac Newton, including the fact that light always travelled in straight lines when passed through prisms or bounced off of mirrors.

On the other hand, there was the wave theory of light, also backed by scientific observations, such as the negative interference obtained when light was sent through a diffraction grating (particles do not manifest this phenomenon).

Ultimately, physicists came to agree that neither theory was wrong: both were correct. Both were useful under different circumstances to explain different observations.

The same was true for mechanical theory. Newtonian physics, another contribution made by Sir Isaac, was hailed for centuries as a scientific breakthrough, and was used to explain the orbits of planets around the sun, the trajectories of cannonballs, and the flight of an arrow. More recently, it was used to send spacecraft to the moon. It was thought that there was no end to what Newtonian physics could do. Until scientists began to peer into the subatomic realm and saw that those physical laws no longer applied, and quantum physics took over. Position and trajectory no longer had any real meaning: it was possible to quantify one but not both. A given object could be everywhere and yet nowhere at the same time. Heisenberg's uncertainty principle took over.

Again, both theories and sets of laws are correct, but only under certain mutually exclusive conditions; and both account for different problems. We have the same situation today with relativity theory and quantum mechanics, which describe the universe at the unfathomably big scales and the invisibly small scales, respectively, and seem to be impossible to bring together into one grand unifying theory.

In a similar fashion, the scriptural description of the origin of mankind and the scientific explanation can both be correct, each being only useful to explain certain aspects of the story, but not sufficient to explain the whole story. One is not superior to the other. Each transcends the other. Each is essential to understand the question of our origin, albeit from different angles. One can explain *how* while the other can explain *why*.

The scriptural version was not intended to explain the physical details, the logistics, or the scientific mechanisms of how we humans came to be. Neither can the scientific version possibly resonate with our souls and give us insight into meaning ("Why am I here?"; "What is good?"). Both fail when pushed in those two distinct directions, just like the particle theory of light utterly fails to explain negative interference, and quantum mechanics cannot possibly be used to launch a Rosetta space probe to intercept a comet which would eventually come to brush past earth ten years later.

Looking Forward: Communication

One thing that is much needed is dialogue between the various theistic and scientific groups. It is true that there has already been much dialogue, but unfortunately too much of it has been antagonistic and confrontational in nature. We need to revisit many of our theological ideas without the preconception that science is "bent on destroying faith." Science and theology can be complementary: both merely seeking to discover truth. We've done it before, when we had to reconcile with the heliocentric theory.

We also need to fortify the bridges with the atheistic community, and stop viewing them as the enemy.

In this context, I applaud efforts such as a recent joint project between three major groups collectively representing millions of people:

- the National Association of Evangelicals (NAE)[8], which comprise more than forty-five thousand local churches from forty different denominations, and represents millions of Christians;
- the American Association for the Advancement of Science (AAAS),[9] whose Mission statement is "to advance science, engineering, and innovation throughout the world for the benefit of all people," and includes nearly two hundred and fifty affiliated societies and academies of science, serving ten million individuals world-wide;
- the American Scientific Affiliation (ASA)[10], who bring together thousands of Christian professionals engaged in the sciences.

---

8. National Association of Evangelicals, "About NAE."
9. American Association for the Advancement of Science, "About AAAS."
10. American Science Affiliation, "Who we are . . . What we do . . . And why."

The AAAS founded the Dialogue on Science, Ethics, and Religion (DoSER) Program in 1995.[11] Through this program, the AAAS, NAE and the ASA have collaborated on many projects and exchanges, including the publication of a series of essays written by leading evangelicals and scientists within a booklet entitled *When God and Science Meet: Surprising Discoveries of Agreement.*[12] This collaborative effort also received support from the John Templeton Foundation, whose Mission statement is to "serve as a philanthropic catalyst for discoveries relating to the Big Questions of human purpose and ultimate reality."[13]

The AAAS, through their DoSER program, have also partnered with the NAE, the John Templeton Foundation and with Rice University (Houston, Texas) to conduct a research endeavor entitled the Perceptions Project,[14] the purpose of which is to investigate the perceptions that members of their scientific and religious communities have about one another in order to "increase understanding and foster more productive dialogue about issues that are vital to the health of the nation."

The purpose of these massive collaborations is merely to build bridges and facilitate dialogue, and to seek truth.

We need more of this!

Looking Forward: Education

Readers should endeavor to become educated on subjects such as this in order to be relevant to mainstream society. Clearly, books are invaluable in this respect, but can become costly.

Perspectives on Science and Christian Faith[15] is a quarterly journal published by the ASA which features articles on all aspects of science and faith. These articles are written by scientists, philosophers and other academics, all of whom hold to orthodox Christian views.[16] More importantly, the articles are accessible without cost through the internet. There are many other such periodical publications.

The internet also provides an abundance of other resources in the form of articles, books, videos, blog posts, and other such websites. Sources for these include ICR, AiG, RTB, and the Biologos Foundation (see p. 11). However, one must bear in mind that these are administered and promoted by the members of the respective organizations, and are written by a relatively small number of individuals, some of whom may not have the appropriate credentials for the matter on which they write.

11. American Association for the Advancement of Science, "Dialogue on Science, Ethics, and Religion."
12. National Association of Evangelicals, "NAE Releases 'When God and Science Meet' Booklet."
13. John Templeton Foundation, "Mission."
14. American Association for the Advancement of Science, "Perceptions Project."
15. American Science Affiliation, "Perspectives on Science & Christian Faith."
16. American Science Affiliation, "About the ASA."

Moreover, the articles generally do not go through an arm's length peer-review. So although there is value in consulting these to get a sense of the range of thinking on a given matter, it is good to always keep in mind the authority and credentials of the one(s) providing the information.

We also need more communication and dialogue to occur at the local level, especially within churches. The purpose of this being to elevate the understanding of the layperson to the level appropriate to the twenty-first century. Too many of the average adherents have been educated on this subject from outdated resources. Too many have been home-schooled using those same outdated sources, and/or are home-schooling their own children in part to "protect" the latter from ideas that they would have otherwise received in the secular school system.

Too many young people have been raised in churches that prepared them poorly for entry in "the big bad world," teaching them bad science with the intent of defending the Bible. Those children then step onto high school, college and university campuses and are immediately overcome when confronted for the first time with modern science and world religions. The image in my mind is of the soldiers in the First World War emerging in waves from the trenches and immediately being struck down by machine gun fire.

One vigorous response against evolution theory from a high profile theologian went as follows: "That is the major weakness of the evolutionary creation viewpoint: it insists on a counterintuitive way of reading the biblical record so as to not read it for what it appears to actually say . . . the evolutionary creation viewpoint bases its conclusions on a full, unquestioning faith in secular evolutionary theory and on the concept of accommodation with regard to the written record in Scripture. Victims to this approach include the historicity of Genesis 1–11, instantaneous special creation as a divine act, divine revelation, and a consistent hermeneutic. Such casualties in the realm of biblical interpretation set the stage for even higher casualties among believers—especially among young people."[17] Curiously, I see the exact same criticism being applied in the opposite direction: the YEC perspective insists on a counterintuitive way of reading scientific data, bases its conclusions on a full, unquestioning faith in the biblical text, and sets the stage for intolerably high casualties among believers—especially among young people.

Many will resist vigorously the idea of educating our children on this subject, thinking that educating is the same as embracing. This doesn't need to be the case. One may still choose to believe differently, to hold on to a YEC viewpoint while considering the OEC or even atheistic arguments. Just like a medical doctor who knows all too well the association between cigarette smoking and lung cancer or pulmonary fibrosis, but chooses to smoke anyhow. But at least they do so having been properly informed on the matter. It is important to make our children aware of the issues and the consequences, and help them to develop their own reply to the challenge.

17. Barrick, *Four Views of Adam*, 84.

They're going to come across these ideas anyhow: we should help them prepare for the encounter.

Related to this subject of education is the controversial and divisive matter of trying to force the publicly funded school systems to teach Creationism (or to push Judeo-Christian laws into the legal system). To "give equal time for both sides." This incorrectly views there being only two sides on the matter. There are in fact many sides. Even within just the Christian community, there are numerous sides on the question of human origins: the YEC and OEC positions are only two extreme poles of a very broad spectrum, but this question actually has numerous poles going in several other dimensions. Then there are explanations endorsed by the Muslim community, which also comprises millions of individuals. And the Hindu version. The Aboriginal version. The Mayan and Ancient Egyptian versions. One can't possibly "give equal time for all sides." Instead, it comes down to a choice—a religious choice—as to which view to hold, and that kind of choice should occur in the home and the religious gathering place, not the publicly funded school system.

Besides, if one is going to require that we teach Creationism in biology classes because it's scriptural, wouldn't it equally be necessary to teach the kind of meteorology described in the book of Job to students in climatology classes?

Looking Forward: Unity

I distinctly remember an early phase in my life in which I was led to believe that other theists were eternally condemned because they didn't hold to the particular theology of the community of which I was part. Not only theists such as the Muslims, or Zoroastrians, but even other branches of Christianity such as the Roman Catholics; there was also a great cloud of uncertainty around other Protestants who were more liberal than we were. This was all based on certain interpretations of scriptural passages. I've long since left that way of thinking, but it persists in many other circles. For example, some will draw far-reaching conclusions based on whether or not one affirms penal substitutionary atonement, or believes in a conscious eternal suffering in hell. Those questions have almost become modern shibboleths. (In order to identify members of a competing tribe, certain ancient Israelites devised a plan based on a speech impediment endemic in that other tribe: "They said, 'All right, say 'Shibboleth.' If he said, 'Sibboleth,' because he could not pronounce the word correctly, they seized him and killed him at the fords of the Jordan. Forty-two thousand Ephraimites were killed at that time.")[18]

We absolutely need to move beyond this kind of tribalism (which, again, I see as an "appendix" left over from our evolutionary heritage). We can learn so much from each other, and can learn much about ourselves and our personal faith in dialoguing

---

18. Jdg 12:6.

with others of other faiths or even lack of faith. "As iron sharpens iron, so one person sharpens another."[19] Unity does not have to mean homogeneity.

For this reason, the reader should seek a better foundational understanding of other world religions. This should not be solely obtained third or fourth hand—for example, learning about Islam through a Christian teacher—as this will inevitably be distorted and lack personal insight. Instead, the reader should interact directly with those of other faiths, and ask them probing questions about their perspective. Not at all with the intent of debating which truth is more true. But simply to gain a new perspective on a vast multidimensional subject. Again, the only religion that Christ himself preached against was the hypocrisy of the Jewish leaders. He is never recorded as having said anything at all about the pantheon of gods which the Romans or Greeks of his time worshipped, nor about any religion practiced by any slaves of the occupying Romans (Africans, Celts, Gaul [France], Germanic tribes).

This does not have to be construed to be an endorsement of universalism.

Final Words

Only hours into our marriage and we were already fighting! And with knife in hand, no less! My beautiful wife and I, roughly thirty years ago. Standing in front of our wedding cake, posing for the classical picture while dozens of cameras were snapping pics.

As the knife touched the cake, I heard a distinct tapping sound. I leaned to my wife and said, "The cake is fake."

Her confident response: "No, my mom made it."

The scientist in me took a step back, and then I replied: "Listen," while I knocked on the top of the cake a few times, each time evoking a very distinct percussive reply.

Again: "No, my mom made it."

"But listen!"

"No, my mom made it."

"But *listen!*"

"No, my mom made it."

This went back and forth several times while the cameras continued to flash, and I got increasingly frustrated that she wasn't "listening to the data." We have a picture of the two of us examining the cake to make sense of what was going on.

In the end, we were both right. Her mom did bake the cake itself, but had a professional cake decorator put the icing on it, who in turn had put a plastic base on top of some of the layers to prevent the whole elaborate structure from collapsing in on itself.

Apparently the making was a slightly more complicated event than we had originally realized.

---

19. Prov 27:17.

Thirty years later, that kind of argument continues. Not between her and me, but between scientists and theologians, trying to make sense of how life came to be on earth. With the same sort of replies being exchanged: "Just look at the data" versus "No, God made it."

Many books have been written about the conflict between faith and science on the subject of origins, and specifically about the subject of human origins. I have referred to several of these throughout this book. In many cases, though, the authors begin with the Bible as the absolute authority on any subject (a laudable thing to do) and then work backwards to try to incorporate or accommodate the science and morality (which I find is too often an impossible thing to do). This forces them to certain conclusions about Adam and Eve, the origin of mankind, the fall and original sin, the flood, world religions, and so forth.

I now begin with the exceedingly well-established paradigm that humans have been around for hundreds of thousands of years and evolved from a common ancestor, and then move forward through time to see how we came to understand God. This approach leads me to very different conclusions about original sin, the atonement, authorship of Scripture, and world religions. One might expect that two different quests for ultimate truth, beginning from different positions and moving in different directions, should eventually arrive at the same destination. Perhaps we haven't waited long enough, and therefore haven't come to realize that the two seemingly opposite conclusions in fact point to the same thing (just like our current understanding of light being both a wave and a particle).

We have evolved. We are still evolving. Who knows what we will look like a hundred thousand years from now, if we're still around. That cautionary note comes not only from a religious upbringing which constantly reminded me of an imminent Armageddon, but also because of the constant warnings from doomsayers of how we're destroying the delicate balances of life on earth. A recent film documentary[20] on the life and scientific contributions of Dr. E. O. Wilson, a hugely influential sociobiologist and two-time Pulitzer Prize winning author, quoted him as saying: "I have learned that humanity's troubles are due substantially to the fact that we are a dysfunctional species. Dysfunctional! And why? Because we have Paleolithic emotions, we have medieval institutions, and on top of that we have developed God-like technology. And that is a dangerous mix."

As I promised to do at the beginning of this book, I have asked many questions, some of which may have offended certain readers. Please keep in mind that many of those were in fact posed as questions, rather than as statements. I may not necessarily hold the views that those questions allude to. But I am trying to stimulate thought and discussion. We have so much to learn, from each other and from the cosmos. I look forward keenly to new discoveries that tell us more about not just our origins, but also our destiny, and our relationship with the Divine.

20. Townsley and Schulze, "Ants and Men."

About the Author

DR. LUKE J. JANSSEN is a Professor in the Department of Medicine of a leading Canadian university, and holds a Bachelor's of Science degree in Biochemistry (1984), a Master's degree in Medical Science (1987) and a Doctorate of Philosophy in Medical Sciences (Physiology and Pharmacology) (1990). He has published 135 scientific research articles and book chapters, and has presented his research before many national and international scientific and pharmaceutical industrial audiences. He is currently carrying out medical research, in part using genetic techniques.

He has also served for over 30 years in various church ministries: these have included service on an elder's board, leadership of youth groups and young adult groups, teaching in Sunday school classes (children, youth, young adult and adult classes), and numerous ministries of service.

He is currently studying towards obtaining a graduate level theological degree.

He has also written a book detailing his own journey from a Fundamentalist background to one in which he has been able to dissolve its cognitive dissonance with modern scientific findings, questionable morality of certain Biblical stories, and world religions. Details of its contents can be found at his blog-site at https://lukejjanssen.wordpress.com/now-in-a-book/

Bibliography

Abbott, Edwin. *Flatland: A Romance of Many Dimensions*. London: Seely, 1884.

Alexander, Denis R. *Creation or Evolution: Do We Have to Choose?* Oxford: Monarch, 2008.

———. "How Does a BioLogos Model Need to Address the Theological Issues Associated with an Adam Who Was Not the Sole Genetic Progenitor of Humankind?" https://biologos.org/uploads/projects/alexander_white_paper.pdf.

American Association for the Advancement of Science. "About AAAS." http://www.aaas.org/about-aaas.

———. "Dialogue on Science, Ethics, and Religion." http://www.aaas.org/DoSER.

———. "Perceptions Project." http://perceptionsproject.org.

American Science Affiliation. "About the ASA." http://network.asa3.org/?page=ASAAbout#Who%20we%20are.

———. "Perspectives on Science & Christian Faith." http://network.asa3.org/?page=PSCF.

———. "Who We Are . . . What We Do . . . And Why." http://network.asa3.org/?page=ASAAbout.

Answers in Genesis. *Statement of Faith*. https://answersingenesis.org/about/faith.

Aquinas, Thomas. "First Part, Question 93." In *Summa Theologica of Saint Aquinas*, translated by Fathers of the English Dominican Province. 2nd and rev. ed. 1920. http://www.newadvent.org/summa/1093.htm.

Aristotle. *Politics*. I.i.9–12.

Arnason, U., et al. "Comparison between the Complete Mitochondrial DNA Sequences of *Homo* and the Common Chimpanzee Based on Nonchimeric Sequences." *Journal of Molecular Evolution* 42 (1996) 145–52.

Augustine of Hippo. "The Literal Meaning of Genesis." Translated by John Hammond Taylor. New York: Paulist, 1982. Passages retrieved from http://inters.org/augustine-interpretating-sacred-scripture.

Avarello, R., et al. "Evidence for an Ancestral Alphoid Domain on the Long Arm of Human Chromosome 2." *Human Genetics* 89 (1992) 247–49.

Barr, William R. "Life: Created in the Image of God." *Mid-Stream* 21 (1982) 473–84.

Barrett, Matthew, and Ardel B. Caneday, eds. "Adam, to Be or Not to Be." Introduction to *Four Views on the Historical Adam*. Grand Rapids: Zondervan, 2013.

Barrick, William D. "A Historical Adam: YEC View." In Barrett and Caneday, *Four Views on the Historical Adam*, 197–264.

Behe, Michael J. *Darwin's Black Box*. New York: Touchstone, 1996.

———. *The Edge of Evolution: The Search for the Limits of Darwinism*. New York: Free Press, 2007.

Bekoff, Marc. "Awareness: Animal Reflections." *Nature* 419 (2002) 255.
Belknap, Michael, and Tim Chaffney. "Reimagining Ark Animals." *Answers in Genesis*, December 13, 2015. https://answersingenesis.org/noahs-ark/reimagining-ark-animals.
Bellah, Robert N. *Religion in Human Evolution: From the Paleolithic to the Axial Age*. Cambridge: Belknap of Harvard University Press, 2011.
Berger, L. R., et al. "*Homo naledi*, a New Species of the Genus *Homo* from the Dinaledi Chamber, South Africa." *eLife* 4 (2015) e09560.
Bersaglieri, Todd, et al. "Genetic Signatures of Strong Recent Positive Selection at the Lactase Gene." *American Journal of Human Genetics* 74 (2004) 1111–20.
Bertazzo, S., et al. "Fibres and Cellular Structures Preserved in 75-Million-Year-Old Dinosaur Specimens." *Nature Communications* 6 (2015) 7352.
BioLogos. "What We Believe." https://biologos.org/about-us.
Brauch, Manfred T. *Abusing Scripture: The Consequences of Misreading the Bible*. Downers Grove: InterVarsity, 2009.
Brooke, John Hedley, and Geoffrey Cantor. "The Contemporary Relevance of the Galileo Affair." In *Reconstructing Nature: the Engagement of Science and Religion*, 106–40. New York: Oxford, 1998.
Brown, P., et al. "A New Small-Bodied Hominin from the Late Pleistocene of Flores, Indonesia." *Nature* 431 (2004) 1055–61.
Browne, Janet. "Noah's Flood, the Ark, and the Shaping of Early Modern Natural History." In *When Science and Christianity Meet*, edited by David C. Lindberg and Ronald L. Numbers, 111–38. Chicago: University of Chicago Press, 2003.
Brueggemann, Walter. *Genesis*. Interpretation. Atlanta: John Knox, 1982.
Cadena, E. A., and J. F. Parham. "Oldest Known Marine Turtle? A New Protostegid from the Lower Cretaceous of Colombia." *Paleobios* 32 (2015) 1–42.
Cahill, Thomas. *The Gifts of the Jews: How a Tribe of Desert Nomads Changed the Way Everyone Thinks and Feels*. New York: Doubleday, 1998.
Cairns, David. *The Image of God in Man*. London: SCM, 1953.
Cajal, Ramón y. *Recuerdos de mi vida: Historia de mi labor scientífica*. Vol. 2, p. 76, Madrid: Moya, 1917.
Cann, R. L., et al. "Mitochondrial DNA and Human Evolution." *Nature* 325 (1987) 31–36.
Chimpanzee Sequencing and Analysis Consortium. "Initial Sequence of the Chimpanzee Genome and Comparison With the Human Genome." *Nature* 437 (2005) 69–87.
Clarke, Errnest G. *The Wisdom of Solomon: A Commentary*. Cambridge: Cambridge University Press, 1973.
Cole, Tom. "NASA's Voyager Probes at 35: Listen to the Music of the Planets." http://www.radiotimes.com/news/2012-10-24/nasas-voyager-probes-at-35-listen-to-the-music-of-the-planets.
Collins, C. John. "Adam and Eve as Historical People, and Why It Matters." *Perspectives on Science and Christian Faith* 62 (2010) 147–65.
———. *Did Adam and Eve Really Exist? Who They Were and Why You Should Care*. Wheaton, IL: Crossway, 2011.
———. "A Historical Adam: Old Earth Creation View." In Barrett and Caneday, *Four Views on the Historical Adam*, 143–96.
———. *Science and Faith: Friends or Foes?* Wheaton, IL: Crossway, 2003.
Collins, Francis. *The Language of God: A Scientist Presents Evidence for Belief*. New York: Free Press, 2007.

Crew, Bec. "Catch the Wave: Decoding the Prairie Dog's Contagious Jump-Yips." *Scientific American*, January 7, 2014. http://blogs.scientificamerican.com/running-ponies/catch-the-wave-decoding-the-prairie-doge28099s-contagious-jump-yips.

Damrosch, David. *The Buried Book: The Loss and Rediscovery of the Great Epic of Gilgamesh*. New York: Holt, 2007.

Dangel, A. W., et al. "Complement Component C4 Gene Intron 9 Has a Phylogenetic Marker for Primates: Long Terminal Repeats of the Endogenous Retrovirus ERV-K(C4) Are a Molecular Clock of Evolution." *Immunogenetics* 42 (1995) 41–52.

Danielson, Dennis R. "The Great Copernican Cliché." *American Journal of Physics* 69 (2001) 1029–35.

Dao, Anh H. "Human Tails and Pseudotails." *Human Pathology* 15 (1984) 449–53.

Darwin, Charles. *On the Origin of Species by Means of Natural Selection, or the Preservation of Favoured Races in the Struggle for Life*. London: Murray, 1859.

———. *The Descent of Man and Selection in Relation to Sex*. Akron, OH: Werner, 1875.

Davies, Robertson. *Tempest-Tost*. Toronto: Irwin, 1951.

Dawkins, Richard. *The Blind Watchmaker*, London: Penguin, 1988.

———. *The God Delusion*. New York: Bantam, 2006.

———. *The Greatest Show on Earth: The Evidence for Evolution*. New York: Free Press, 2009.

Dembski, William A. *Intelligent Design: The Bridge between Science and Theology*. Downers Grove: InterVarsity, 1999.

DeWitt, David. "It's an Ape . . . It's a Human . . . It's . . . It's . . . a Missing Link! Detailed Analysis of *Australopithecus sediba* Presents Problems for Evolution." *Answers-in-Genesis*, September 13, 2011. https://answersingenesis.org/human-evolution/australopithecus-sediba.

Diamond, Jared. *The World until Yesterday: What Can We Learn from Traditional Societies?* New York: Penguin, 2012.

Dirks, Paul H. G. M., et al. "Geological and Taphonomic Context for the New Hominin Species *Homo naledi* from the Dinaledi Chamber, South Africa." *eLife* 4 (2015) e09561.

DoHarris L., et al. "Molecular Motors: How to Make Models That Can Be Used to Convey the Concept of Molecular Ratchets and Thermal Capture." *Advances in Physiological Education* 35 (2011) 213–18.

Dubrow, Terry J. "Detailing the Human Tail." *Annals of Plastic Surgery* 20 (1988) 340–44.

ENCODE Project Consortium. "An Integrated Encyclopedia of DNA Elements in the Human Genome." *Nature* 489 (2012) 57–74.

Eiberg, Hans, et al. "Blue Eye Color in Humans May Be Caused by a Perfectly Associated Founder Mutation in a Regulatory Element Located within the HERC2 Gene Inhibiting OCA2 Expression." *Human Genetics* 123 (2008) 177–87.

Enns, Peter. *The Evolution of Adam: What the Bible Does and Doesn't Say about Human Origins*. Grand Rapids: Brazos, 2012.

———. *Inspiration and Incarnation: Evangelicals and the Problem of the Old Testament*. Grand Rapids: Baker Academic, 2005.

———. "7 Problems with a Recent Evangelical Defense of the Historicity of Genesis 1–11." *Patheos*, May 26, 2015. http://www.patheos.com/blogs/peterenns/2015/05/7-problems-with-a-recent-evangelical-defense-of-the-historicity-of-genesis-1-11.

Falk, D., et al. "Brain Shape in Human Microcephalics and *Homo floresiensis*." *Proceedings of the National Academy of Science USA* 104 (2007) 2513–18.

Fisher, S. E., and C. Scharff. "FOXP2 as a Molecular Window into Speech and Language." *Trends in Genetics* 25 (2009) 166–77.

Franze, Kristian, et al. "Müller Cells Are Living Optical Fibers in the Vertebrate Retina." *Proceedings of the National Academy of Sciences* 104 (2007) 8287–92.

Fu, Q., et al. "An Early Modern Human from Romania with a Recent Neanderthal Ancestor." *Nature* 524 (2015) 216–19.

George, Andrew. *The Epic of Gilgamesh: The Babylonian Epic Poem and Other Texts in Akkadian and Sumerian*. Translated and with introduction by Andrew George. New York: Penguin, 2000.

Geyer, John. *The Wisdom of Solomon: Introduction and Commentary*. London: SCM, 1963.

Gibbons, Ann. "A New Kind of Ancestor: *Ardipithecus* Unveiled." *Science* 326 (2009) 36–40.

Gishlick, Alan. "Baraminology." *Reports of the National Center for Science Education* 26 (2006) 17–21.

Gosse, Philip. *Omphalos: An Attempt to Untie the Geological Knot*. London: Van Voorst, 1857.

Greene, Mott T. "Genesis and Geology Revisited: The Order of Nature and the Nature of Order in Nineteenth-Century Britain." In *When Science and Christianity Meet*, edited by David C. Lindberg and Ronald L. Numbers, 139–60. Chicago: University of Chicago Press, 2003.

Gregory, W. K. "*Hesperopithecus* Apparently Not an Ape Nor a Man." *Science* 66 (1927) 579–81.

Haarsma, Deborah, and Loren Haarsma. *Origins: A Reformed Look at Creation, Design and Evolution*. Grand Rapids: Faith Alive, 2007.

Hagelberg, E., et al. "DNA from Ancient Mammoth Bones." *Nature* 370 (1994) 333–34.

Haile-Selassie, Yohannes, et al. "Late Miocene Teeth from Middle Awash, Ethiopia, and Early Hominid Dental Evolution." *Science* 303 (2004) 1503–1505.

Ham, Ken. "We'll Find a New Earth within 20 Years." *Answers in Genesis*, July 20, 2014. https://answersingenesis.org/blogs/ken-ham/2014/07/20/well-find-a-new-earth-within-20-years.

———. "Were You There? Evidence for Creation." http://www.icr.org/article/670/88.

———. "Were You There? How Can Anyone Know What Happened in the Beginning if None of Us Were There? Ken Ham, AiG–U.S., Explains." *Answers in Genesis*, July 15, 2011. https://answersingenesis.org/the-word-of-god/were-you-there.

Handt, O., et al. "Molecular Genetic Analyses of the Tyrolean Ice Man." *Science* 264 (1994) 1775–78.

Harlow, Daniel C. "Creation according to Genesis: Literary Genre, Cultural Context, Theological Truth." *Christian Scholar's Review* 37 (2008) 163–98.

Heidel, Alexander. *The Gilgamesh Epic and Old Testament Parallels*. Chicago: University of Chicago Press, 1946.

Higham, T., et al. "Testing Models for the Beginnings of the Aurignacian and the Advent of Figurative Art and Music: The Radiocarbon Chronology of Geißenklösterle." *Journal of Human Evolution* 62 (2012) 664–76.

Holsinger-Friesen, Thomas. *Irenaeus and Genesis: A Study of Competition in Early Christian Hermeneutics*. Journal of Theological Interpretation Supplement 1. Winona Lake, IN: Eisenbrauns, 2009.

Hooke, S. H. *In the Beginning*. Clarendon Bible Old Testament 6. Oxford: Clarendon, 1947.

Huff, C. D., et al. "Mobile Elements Reveal Small Population Size in the Ancient Ancestors of *Homo sapiens*." *Proceedings of the National Academy of Science* 107 (2010) 2147–52.

Hughes, J. F., and J. M Coffin. "Evidence for Genomic Rearrangements Mediated by Human Endogenous Retroviruses during Primate Evolution." *Nature Genetics* 29 (2001) 487–89.

Institute for Creation Research. "Principles of Scientific Creationism." http://www.icr.org/tenets.

———. "Who We Are." http://www.icr.org/who-we-are.

Ijdo, J. W., et al. "Origin of Human Chromosome 2: An Ancestral Telomere-Telomere Fusion." *Proceedings of the National Academy of Sciences* 88 (1991) 9051–55.

Janssen, Luke J. "Little Things in Life." https://lukejjanssen.wordpress.com/2016/04/02/the-little-things-in-life.

———. *Reaching into Plato's Cave: Bringing the Bible into the 21st Century*. Charleston, SC: Kindle, 2014.

Jastrow, Robert. *God and the Astronomers*. Toronto: McLeod, 1992.

Jern, P., et al. "Divergent Patterns of Recent Retroviral Integrations in the Human and Chimpanzee Genomes: Probable Transmissions between Other Primates and Chimpanzees." *Journal of Virology* 80 (2006) 1367–75.

Kappelman, John, and John G. Fleagle. "Age of Early Hominids." *Nature* 376 (1995) 558–559.

Keith, M. C., and G. M. Anderson. "Radiocarbon Dating: Fictitious Results with Mollusk Shells." *Science* 141 (1963) 634–37.

Krause, J., et al. "The Complete Mitochondrial DNA Genome of an Unknown Hominin from Southern Siberia." *Nature* 464 (2010) 894–97.

Labin, A. M., and E. N. Ribak. "Retinal Glial Cells Enhance Human Vision Acuity." *Physical Review Letters* 104 (2010) 158102.

Lamoureux, Denis. "Ancient Astronomy." Episode 3 of *The Bible & Ancient Science*. http://www.ualberta.ca/~dlamoure/wlas3/index.html.

———. "Ancient Biology." Episode 4 of *The Bible & Ancient Science*. http://www.ualberta.ca/~dlamoure/wlas4/index.html.

———. "Beyond the Cosmic Fall and Natural Evil." *Perspectives on Science and Christian Faith* 68 (2016) 44–59.

———. *Evolutionary Creation: A Christian Approach to Evolution*. Eugene, OR: Wipf & Stock, 2008.

———. "No Historical Adam: Evolutionary Creation View." In Barrett and Caneday, *Four Views on the Historical Adam*, 37–88.

Lander E. S., et al. "Initial Sequencing and Analysis of the Human Genome." *Nature* 409 (2001) 860–921.

Lennox, John C. *Seven Days That Divide the World: The Beginning according to Genesis and Science*. Grand Rapids: Zondervan, 2011.

Leroi-Gourhan, A. "Shanidar et Ses Fleurs." *Paléorient* 24 (1998) 79–88.

Lewis, C. S. *The Problem of Pain*. London: Bles, 1943.

Li, H., and R. Durbin. "Inference of Human Population History from Individual Whole-Genome Sequences." *Nature* 475 (2011) 493–97.

Lindberg, David C. "Galileo, the Church, and the Cosmos." In *When Science and Christianity Meet*, edited by David C. Lindberg and Ronald L. Numbers, 33–60. Chicago: University of Chicago Press, 2003.

———. "The Medieval Church Encounters the Classical Tradition: Saint Augustine, Roger Bacon, and the Handmaiden Metaphor." In *When Science and Christianity Meet*, edited by David C. Lindberg and Ronald L. Numbers, 33–60. Chicago: University of Chicago Press, 2003.

Livingstone, David R. *Adam's Ancestors: Race, Religion, and the Politics of Human Origins.* Baltimore: Johns Hopkins University Press, 2008.

Lubenow, Marvin. "The Neanderthals: Our Worthy Ancestors." *Answers-in-Genesis*, November 29, 2006. https://answersingenesis.org/human-evolution/neanderthal/the-neandertals-our-worthy-ancestors.

MacArthur, John. *The Battle for the Beginning: Creation, Evolution, and the Bible.* Nashville: Nelson, 2001.

Mack, S. J., et al. "Common and Well-Documented HLA Alleles: 2012 Update to the CWD Catalogue." *Tissue Antigens* 81 (2013) 194–203.

Madueme, Hans, and Michael Reeves, eds. *Adam, the Fall and Original Sin: Theological, Biblical, and Scientific Perspectives.* Grand Rapids: Baker Academic, 2014.

Mann, Charles C. "The Birth of Religion." *National Geographic*, June 2011, 39–59.

Marchi, E., et al. "Neanderthal and Denisovan Retroviruses in Modern Humans." *Current Biology* 23 (2013) R994–95.

Markey, Sean. "Monkeys Show Sense of Fairness, Study Says." *National Geographic News*, September 17, 2003. http://news.nationalgeographic.com/news/2003/09/0917_030917_monkeyfairness.html.

Mayer, J., et al. "Human Endogenous Retrovirus K Homologous Sequences and Their Coding Capacity in Old World Primates." *Journal of Virology* 72 (1998) 1870–75.

McDonald, Bob. "Dinosaur Fossils Preserve Blood Cells." http://www.cbc.ca/radio/quirks/quirks-quarks-for-june-13-2015-1.3111320/dinosaur-fossils-preserve-blood-cells-1.3111394.

McGrath, Gavin Basil. "Soteriology: Adam and the Fall." *Perspectives on Science and Christian Faith* 49 (1997) 252–63.

McGuckin, John Anthony. *The Westminster Handbook to Origen.* Louisville: Westminster John Knox, 2004.

McPherron, Shannon P., et al. "Evidence for Stone-Tool-Assisted Consumption of Animal Tissues Before 3.39 Million Years Ago at Dikika, Ethiopia." *Nature* 466 (2010) 857–60.

Mendez, F. L., et al. "An African American Paternal Lineage Adds an Extremely Ancient Root to the Human Y Chromosome Phylogenetic Tree." *American Journal of Human Genetics* 92 (2013) 454–59.

Menton, David, and John UpChurch. "Who Were Cavemen? Finding a Home for Cavemen." *Answers-in-Genesis*, April 1, 2012. https://answersingenesis.org/human-evolution/cavemen/who-were-cavemen.

Meyer, M., et al. "A High-Coverage Genome Sequence from an Archaic Denisovan Individual." *Science* 338 (2012) 222–26.

Meyer, Matthias, et al. "A Mitochondrial Genome Sequence of a Hominin from Sima de lo Huesos." *Nature* 505 (2014) 403–6.

Meyer, Stephen C. *Signature in the Cell: DNA and the Evidence for Intelligent Design.* New York: HarperOne, 2009.

Middleton, J. Richard. *The Liberating Image: the* Imago Dei *in Genesis 1.* Grand Rapids: Baker/Brazos, 2005.

Miller, Webb, et al. "The Mitochondrial Genome Sequence of the Tasmanian Tiger (*Thylacinus cynocephalus*)." *Genome Research* (2009) 213–20.

Minns, Denis. *Irenaeus.* Washington, DC: Georgetown University Press, 1994.

Mitchell, Elizabeth. "Denisovan Gene Gave Tibetans Their High-Altitude Tolerance." *Answers-in-Genesis*, July 19, 2006. https://answersingenesis.org/genetics/denisovan-gene-gave-tibetans-their-high-altitude-tolerance.

———. "Does Hugh Ross Believe in Soulless Ancient Humans?" *Answers-in-Genesis*, April 20, 2015. https://answersingenesis.org/human-evolution/origins/does-hugh-ross-believe-in-soulless-ancient-humans.

———. "*Homo erectus* Grew Teeth Like Modern Humans, Not Like Chimps." *Answers-in-Genesis*, February 2, 2013. https://answersingenesis.org/human-evolution/homo-erectus-grew-teeth-like-modern-humans-not-chimps.

———. "Is *Homo floresiensis* a Legitimate Human 'Hobbit' Species?" *Answers-in-Genesis*, August 19, 2014. https://answersingenesis.org/human-evolution/homo-floresiensis/homo-floresiensis-legitimate-human-hobbit-species.

———. "Is *Homo naledi* a New Species of Human Ancestor? News to Know." *Answers-in-Genesis*, September 12, 2015. https://answersingenesis.org/human-evolution/homo-naledi-new-species-human-ancestor.

Montgomery, David. *The Rocks Don't Lie: A Geologist Investigates Noah's Flood*. New York: Norton, 2012.

Morris, Henry M. *The Genesis Record: A Scientific and Devotional Commentary on the Book of Beginnings*. Grand Rapids: Baker, 1976.

Morwood, M. J., and W. L. Jungers. "Conclusions: Implications of the Liang Bua Excavations for Hominin Evolution and Biogeography." *Journal of Human Evolution* 57 (2009) 640–48.

National Association of Evangelicals. "About NAE." http://nae.net/about-nae.

———. "NAE Releases 'When God and Science Meet' Booklet." http://nae.net/nae-releases-when-god-and-science-meet-booklet.

National Geographic News. "Grizzly-Polar Bear Hybrid Found—But What Does It Mean?" http://news.nationalgeographic.com/news/2006/05/polar-bears_2.html.

National Park Service. "Wolf Restoration Continued." http://www.nps.gov/yell/learn/nature/wolfrestorationinfo.

Niimura, Y., and M. Nei. "Evolution of Olfactory Receptor Genes in the Human Genome." *Proceedings of the New York Academy of Sciences* 100 (2003) 12235–40.

Orlando, L., et al. "Recalibrating *Equus* Evolution Using the Genome Sequence of an Early Middle Pleistocene Horse." *Nature* 499 (2013) 74–78.

Osborn, Eric. *Irenaeus of Lyons*. Cambridge: Cambridge University Press, 2001.

Osborn, Henry Fairfield. "*Hesperopithecus*, the First Anthropoid Primate Found in North America." *Science* 55 (1920) 463–65.

Owen, James. "5 Surprising Facts about Otzi the Iceman." *National Geographic*, September 10, 2015. http://news.nationalgeographic.com/news/2013/10/131016-otzi-ice-man-mummy-five-facts.

Pääbo, Svante. "Molecular Cloning of Ancient Egyptian Mummy DNA." *Nature* 314 (1985) 644–45.

———. *Neanderthal Man: In Search of Lost Genomes*. New York: Basic (Perseus), 2014.

Pääbo, Svante, and Allan C. Wilson. "Polymerase Chain Reaction Reveals Cloning Artefacts." *Nature* 334 (1988) 387–88.

Palagi, Elisabetta, et al. "Rapid Mimicry and Emotional Contagion in Domestic Dogs." *Royal Society Open Science* 2 (2015) 150505.

Palagi, Elisabetta, et al. "Yawn Contagion in Humans and Bonobos: Emotional Affinity Matters More Than Species." *Peer Journal* 2 (2014) e519.

Pew Research Center. "America's Changing Religious Landscape." http://www.pewforum.org/2015/05/12/americas-changing-religious-landscape.

———. "Scientists and Belief." http://www.pewforum.org/2009/11/05/scientists-and-belief.

Pinkston, William S. *Biology for Christian Schools*. Greensville, SC: Bob Jones University Press, 1991.

Pinnock, Clark H. "Climbing Out of a Swamp: The Evangelical Struggle to Understand the Creation Texts." *Interpretation* 43 (1989) 143–55.

Plantinga, Alvin. *Where the Conflict Really Lies: Science, Religion, and Naturalism*. New York: Oxford University Press, 2011.

Polkinghorne, John. *Quarks, Chaos & Christianity*. London: SPCK, 2005.

Population Reference Bureau. *How Many People Have Ever Lived on Earth?* http://www.prb.org/Articles/2002/HowManyPeopleHaveEverLivedonEarth.aspx.

Prüfer, K., et al. "The Complete Genome Sequence of a Neanderthal from the Altai Mountains." *Nature* 505 (2014) 43–49.

Raghavan, M., et al. "Upper Palaeolithic Siberian Genome Reveals Dual Ancestry of Native Americans." *Nature* 505 (2014) 87–91.

Raichlen, David A., et al. "Laetoli Footprints Preserve Earliest Direct Evidence of Human-Like Bipedal Biomechanics." *Public Library of Science One* 5 (2010) e9769.

Ramm, Bernard. *Special Revelation and the Word of God*. Grand Rapids: Eerdmans, 1961.

Ramsey, C. B., et al. "A Complete Terrestrial Radiocarbon Record for 11.2 to 52.8 kyr BP." *Science* 338 (2012) 370–74.

Rana, Fazale. "Chromosome 2: The Best Evidence for Evolution?" *Reasons-to-Believe*, June 1, 2010. http://www.reasons.org/articles/chromosome-2-the-best-evidence-for-evolution.

———. "Neanderthal Brains Make Them Unlikely Social Networkers." *Reasons-to-Believe* April 1, 2013. http://www.reasons.org/articles/neanderthal-brains-make-them-unlikely-social-networkers.

Rana, Fazale, and Hugh Ross. *Who Was Adam? A Creation Model Approach to the Origin of Man*. Colorado Springs: NavPress, 2005.

Range F., et al. "The Absence of Reward Induces Inequity Aversion in Dogs." *Proceedings of the National Academy of Science* 106 (2008) 340–45.

Reader, John. *Missing Links: The Hunt for Earliest Man*. Boston: Little, Brown, 1981.

———. *Missing Links: In Search of Human Origins*. New York: Oxford, 2011.

Reasons to Believe. "Hominids." http://www.reasons.org/rtb-101/hominids.

———. "Our Mission: Engage & Equip." http://www.reasons.org/about/our-mission.

Reich, David, et al. "Genetic History of an Archaic Hominin Group from Denisova Cave in Siberia." *Nature* 468 (2010) 1053–60.

Reimer, P. J. "Atmospheric Science: Refining the Radiocarbon Time Scale." *Science* 338 (2012) 337–38.

Reus, K., et al. "HERV-K(OLD): Ancestor Sequences of the Human Endogenous Retrovirus Family HERV-K(HML-2)." *Journal of Virology* 75 (2001) 8917–26.

Riddle, Mike. "Chapter 7: Doesn't Carbon-14 Dating Disprove the Bible?" https://answersingenesis.org/geology/carbon-14/doesnt-carbon-14-dating-disprove-the-bible.

Roffman, Itai, et al. "Preparation and Use of Varied Natural Tools for Extractive Foraging by Bonobos (*Pan Paniscus*)." *American Journal of Physical Anthropology* 158 (2015) 78–91.

Romano, C. M., et al. "Demographic Histories of ERV-K in Humans, Chimpanzees and Rhesus Monkeys." *Public Library of Science* 2 (2007) e1026.

Ross, Hugh. *More than a Theory: Revealing a Testable Model for Creation*. Grand Rapids: Baker, 2012.

Rowlands, Mark. *Can Animals Be Moral?* New York: Oxford, 2012.

Rubin, M., and Taylor, D. W. "Radiocarbon Activity of Shells from Living Clams and Snails." *Science* 141 (1963) 637.

Sala, Nohemi, et al. "Lethal Interpersonal Violence in the Middle Pleistocene." *Public Library of Science One* 10 (2015) e0126589.

Sanders, A. R., et al. "Genome-Wide Scan Demonstrates Significant Linkage for Male Sexual Orientation." *Psychological Medicine* 45 (2015) 1379–88.

Sanders, Paul. "Missing Link in Hebrew Bible Formation." *Biblical Archaeology Review* 41 (2015) 46–52.

Scally, A., et al. "Insights into Hominid Evolution from the Gorilla Genome Sequence." *Nature* 483 (2012) 169–75.

Seibert, Eric A. *Disturbing Divine Behavior: Troubling Old Testament Images of God*. Minneapolis: Augsburg Fortress, 2009.

Shipman, Pat. *The Invaders: How Humans and Their Dogs Drove Neanderthals to Extinction*. Cambridge: Belknap of Harvard University Press, 2015.

Shreeve, Jamie. "This Face Changes the Human Story. But How?" *National Geographic*, September 10, 2015. http://news.nationalgeographic.com/2015/09/150910-human-evolution-change.

Smith, John Clark. *The Ancient Wisdom of Origen*. Cranbury, NJ: Associated University Presses, 1992.

Steinhuber, S., et al. "Distribution of Human Endogenous Retrovirus HERV-K Genomes in Humans and Different Primates." *Human Genetics* 96 (1995) 188–92.

Stolzenburg, Will, writer. *Lords of Nature: Life in a Land of Great Predators*. Film, 57 mins. Directed by Karen Anspacher-Meyer. 2009.

Stringer, Chris. "Human Evolution: Small Remains Still Pose Big Problems." *Nature* 514 (2014) 427–29.

Stringer, C. B., and Andrews P. "Genetic and Fossil Evidence for the Origin of Modern Humans." *Science* 239 (1988) 1263–68.

Sullivan, John Edward. *The Image of God: The Doctrine of St. Augustine and Its Influence*. Dubuque, IA: Priory, 1963.

Surin, Kenneth. "Atonement and Christology." *Neue Zeitschrift für Systematische Theologie und Religionsphilosophie* 24 (1982) 131–49.

Templeton Foundation. "Mission." https://www.templeton.org/who-we-are/about-the-foundation/mission.

Tenneson, Michael, et al. "A New Survey Instrument and Its Findings for Relating Science and Theology." *Perspectives on Science and Christian Faith* 67 (2015) 200–202.

Torrents, D., et al. "A Genome-wide Survey of Human Pseudogenes." *Genome Research* 13 (2003) 2559–67.

Townsley, Graham, and Shelley Schulze. *Of Ants and Men*. PBS International. 2015. http://pbsinternational.org/programs/e-o-wilson-of-ants-and-men.

Tripolitis, Antonia. *Origen: A Critical Reading*. New York: Lang, 1985.

Vannucci, R. C., et al. "Craniometric Ratios of Microcephaly and LB1, *Homo floresiensis*, using MRI and Endocasts." *Proceedings of the National Academy of Science USA* 108 (2001) 14043–48.

Vesalius, Andreas. *De Humani Corporis Fabrica*. Book 1. Padua School of Medicine, 1543.

Walton, John H. "A Historical Adam: Archetypal Creation View." In Barrett and Caneday, *Four Views on the Historical Adam*, 89–142.

———. *The Lost World of Adam and Eve: Genesis 2–3 and the Human Origins Debate*. Downers Grove: IVP Academic, 2015.

———. *The Lost World of Genesis One*. Downers Grove: InterVarsity, 2009.

———. "Response from the Archetypal View." In Barrett and Caneday, *Four Views on the Historical Adam*, 68–69.

Whitcomb, John C., and Henry M. Morris. *The Genesis Flood: The Biblical Record and Its Scientific Implications*. Philadelphia: Presbyterian & Reformed, 1961.

White, T. D., et al. "*Australopithecus ramidus*, a New Species of Early Hominid from Aramis, Ethiopia." *Nature* 371 (1994) 306–12.

Wilcox, David L. "A Proposed Model for the Evolutionary Creation of Human Beings: From the Image of God to the Origin of Sin." *Perspectives on Science and Christian Faith* 68 (2016) 22–43.

Yang, H., et al. "Phylogenetic Resolution within the Elephantidae Using Fossil DNA Sequence from the American Mastodon (*Mammut americanum*) as an Outgroup." *Proceedings of the National Academy of Sciences* 93 (1996) 1190–94.

Young, Davis A. "The Antiquity and the Unity of the Human Race Revisited." *Christian Scholar's Review* 24 (1995) 380–96.

Yunis, Jorge J., et al. "The Striking Resemblance of High-Resolution G-Banded Chromosomes of Man and Chimpanzee." *Science* 208 (1980) 1145–48.

Yunis, Jorge J., and Om Prakash. "The Origin of Man: A Chromosomal Pictorial Legacy." *Science* 215 (1982) 1525–30.

Zevit, Ziony. "Was Eve Made from Adam's Rib—or His Baculum?" *Biblical Archaeology Review* 41 (2015) 33–35.

———. *What Really Happened in the Garden of Eden?* London: Yale University Press, 2013.

Subject Index

Abbott, Edwin, 278
Abel, 37, 157, 234, 252
Abram / Abraham
   leaving Ur, 21, 25, 62, 199, 285
   carrying clay tablets, 199
   renaming of, 21, 25, 199
   forefather of Jesus, 197, 201
   descendants, 26, 43, 62, 249
   father of Judaism, Christianity and Islam, 199, 202
   favored by God, 247
   challenged God, 271
accommodation, 8, 171–173, 251–252, 266–267
Adam
   made from dirt, 44, 45, 174, 229
   Adam's rib used to create Eve, 41, 192, 229
   naming animals and people, 79, 174
   being named by God, 174–175
   father of Semitic races, 202–204, 225, 249
   representative of mankind, 13, 173–174, 191–192, 203, 225, 252, 267, 268, 274
   origin of sin and death in mankind, 156, 157, 168, 171, 182, 189, 215, 226, 234, 271
   first Adam … last Adam, 15, 37, 152, 173, 252, 286
   refurbished hominid, 192, 267–268
   Y-chromosomal Adam, 127–130
   Caucasian, 65–67, 213
   Neolithic farmer, 192, 285
agency detection, 222, 232, 235
Akkadian civilization, 20, 43, 61, 72, 154
   also see Ancient Near East, literature
Alexander, Denis, 17, 192
alleles, 133, 141–142, 180
American Association for the Advancement of Science (AAAS), 289
American Scientific Affiliation, 289
Ancient Near East(ern), 13, 20, 21, 25, 26, 27, 61, 72, 154, 189, 199, 206, 223

Göbekli Tepe, 285
   literature, 43, 61, 190
   also see Akkadian civilization, Babylonian civilization, Egyptian civilization, Enuma Elish, Epic of Gilgamesh, Greek civilization, Sumerian civilization, Mesopotamia, Zoroastrianism
Answers-in-Genesis, 8, 10, 181, 215, 258, 262–264, 290
   AiG's Creation Museum, 10, 181, 262
   also see Ken Ham
anthropology, 6, 70, 83–85, 103, 110–112, 191, 203
Anthropopithecus, 108–109
apologetics, 8, 9, 17, 272–273
Apparent Age (Theory of), 56, 168, 259–260
Aquinas, Thomas, 283
   on the *imago Dei*, 207, 208, 210
archaeology, 61, 69, 72, 114, 115, 203
Ardipithecus, 113, 124, 148
Aristarchus of Samos, 30
Aristotle, 30, 283
   on cosmos, 30
   on difference between animals and humans, 218
atonement, 186, 231, 255, 271, 274
Augustine of Hippo,
   on days of creation, 155, 250, 259
   on *imago Dei*, 208
   on original sin, 225
   on theology versus science, 38, 272–273
   on the trinity, 209
   all rational creatures are of Adam, 213, 235
   on Hebrew as first language, 67
Australopithecus, 109–113, 124, 132, 148
   afarensis, 112
   africanus, 109, 113, 124
   robustus, 113
   boisei, 111–113
   anamensis, 113
   bahrelghazali, 113

SUBJECT INDEX

Australopithecus (cont.)
  deyiremeda, 113
  garhi, 113
  sediba, 113
  footprints, 120–121
  also see Lucy
Babylonian
  civilization, 21, 24, 43, 61, 62, 72, 285
  King's List, 21
  also see Ancient Near Eastern literature
Bacon, Sir Francis, 127, 244
baramin, 7, 181
Barrick, William D., 17, 194
Behe, Michael, 17
Bellah, Robert, 235
Bellarmine, Cardinal Robert, 34–35, 102, 190
Berekiah, 252
Berger, Lee, 117
Bernard of Chartres, 39, 218, 286
Bible, *see Scripture; Council of Nicea*
Big Bang, 13, 68, 72, 176, 179, 214, 251, 258, 266–267, 275, 276, 283
Biologos, 11, 290, *also see Collins, Francis*
Brahe, Tycho, 34
broken image-bearers, broken world, 176–177
Buckland's Dilemma, 58
Buckland, Reverend William, 58
Busk, George, 106
Cahill, Thomas, 25
Cain, 66, 157, 189, 200, 234, 238, 247, 252, 266,
Cajal, Ramony, 89
Calvin, John, 34
Campanella, Tommaso, 215, 235
cancer, 100, 102, 152, 159, 169, 170, 187, 271, 274, 291
Catholic Church, 155, 285
  on cosmology, 34–36
  on original sin, 226, 228
  on Vesalius, and number of ribs in men, 41
  on trial of Galileo, 34–36
  on North American aboriginals, 65–66, 213
  on *imago Dei*, 65, 207–209, 213
centromere, 100, 133, 137
  also see chromosome
Christian denominations
  Anglican, 58, 274
  Assemblies of God, 9
  Calvinist, 34, 66, 186, 231
  Coptic, 231
  Fundamentalism, 4, 5, 15, 67, 72, 155, 243, 257, 258, 269
  Greek Orthodox, 10, 226, 228, 246

Prosperity Message, 238, 239
  Lutheran, 34, 155
  Mormonism, 13, 198, 285
  Protestant, 9, 34, 36, 155, 226, 228, 259, 285
  snake handlers, 12
  Westboro Baptist Church, 12
  also see Catholic Church
chromosome, 87, 89–90, 133
  inversion, 132, 136, 137–139
  rearrangement, 136
  staining, 89–90, 134
  also see centromere, telomere, karyotype
clay
  tablets, 61–62, 127, 199, 227, 236
  humans, 43, 44, 104, 190, 208, 228
cognitive dissonance, 2, 4, 6, 13, 40, 265
Collins, C. John, 17, 191, 194, 268
Collins, Francis, 8, 11, 17, 283
  also see BioLogos Foundation
common design, 131, 144, 162, 163
common descent, 11, 72, 131, 138, 143, 144, 149, 162, 163, 167, 168, 220
Compartmentalism, 8, 9, 282
Complementarism, 8, 9
Concordism, 8, 9
Copernicus, 31, 33, 34, 283
Council of Nicea, 240, 246
Crick, Francis, 100
Cro-Magnon, 148
Cuozzo, Jack, 261
Cuvier, Georges, 60
D'Armate, Salviano, 89
Dart, Raymond, 109–110, 124
Darwin, Charles, 45, 277
  Darwinism, Darwinian model, 10, 11, 72, 76, 163, 263, 264
  Origin of Species, 45, 104, 105, 123
  small, warm pond, 277
David, 10, 247
  author of Psalms, 27, 42
  numerology of name, 201
  ruler of Israel, 43, 62
Davies, Robertson, 4, 5
Davis, Peter, 111
Dawkins, Richard, 8, 163, 229
  on his definition of biology, 69, 262
Dawson, Charles, 121
Day, Michael, 111
death, 151, 156–159, 170, 227, 254
Dembski, Michael, 18
Denisovans, 115–116, 124, 126, 132, 148
Dennett, Daniel, 8
Diamond, Jared, 210

SUBJECT INDEX

diary,
- as analogy for DNA, 91
- as analogy for scripture, *see Scripture, diary*

dinosaurs, 4, 10, 103, 159, 254, 262, 271

DNA, 87–88
- transcription, 88
- translation, 87
- mutations, 97, 131, 151, 156–157
- *also see chromosome, ENCODE Project, Human Genome Project, junk DNA*

dome, *see firmament*

Dubois, Eugene, 108–109, 124

Duncan, Isabelle, 266

Earth
- four corners, 32
- three-layered, 20, 26–28, 249
- unmoveable, 27, 28, 34
- *also see geocentrism, heliocentrism, Flat Earth*

eclipse, 23

Egyptian civilization, 20, 21, 24, 25, 43, 61, 62, 72, 199, 285, 292
- *also see Ancient Near Eastern literature*

Einstein, Albert, 283

Elijah, 232, 247

elohim, 21, 154, 195, 247

ENCODE Project, 100

Enns, Peter, 17, 153, 274

Enoch, 203, 232, 247, 246

Enuma Elish, 43, 190, 228

enzyme
- digest milk, 178
- gulono-γ-lactone oxidase, 144
- metabolic, 96
- mutations in, 97
- olfactory receptor, 144–145, 164
- reverse transcriptase, 98, 100
- vitamin C, 144, 164, 220
- *see vision, rhodopsin*

Epic of Gilgamesh, 21, 24, 60–63, 72, 190, 202, 227, 246
- *also see Ancient Near Eastern literature*

epicycles, 30

erector pili, 132, 138, 163, 220, 270

ex nihilo, 228

fallen creatures, fallen world, 15, 176–177

firmament, 26, 27, 36, 182, 190, 242, 249, 261

flat earth, 16, 26, 27, 31, 32, 71, 80, 185, 190, 242, 251, 273

Flatland, 278

Flood
- Babylonian account, 21, 61–62, 127
- account in Genesis, 4, 57–60, 62
- Jesus referring to, 37, 171, 173, 252
- Peter referring to, 37, 171, 251, 252
- Noah's Flood, 27, 37, 57–60, 171, 247
- caused genetic changes, 182, 254, 261
- geological deformation, 40
- layering of bodies, 59, 72
- *also see representative pairs,*

fossilization, 77, 84, 124–125, 185

Franklin, Rosalind, 100

Galen, 41

Galileo, 31, 34–36, 283

Gap Theory, 66, 266

Garden of Eden
- fall in the, 10, 11, 67, 230, 232, 271, 273
- location, 21, 196, 202, 287

gene therapy, 101, 102, 189, 223, 234

genetic bottleneck, 128–130, 193, 268

genome:
- human, 100, 131, 133, 139
- Neanderthal, 101, 131, 139, 146
- Denisovan, 101, 131, 139, 146
- chimpanzee, 131, 134, 139, 146
- gorilla, 131, 134, 139
- quagga, 101
- mammoth, 101
- *also see Human Genome Project*

genotype, 7, 180

geocentrism, geocentric theory, 16, 30–31, 33–35, 161, 190, 240

God-of-the-gaps, 70, 277, 280

goose-bumps, *see erector pili*

Gosse, Henry Philip, Theory of Apparent Age, 57

Gould, Stephen J., 8, 183, 282
- punctuated equilibrium model, 183
- *also see NOMA*

Greek civilization, 29–30, 33, 36, 156, 236, 285

Haarsma, Deborah and Loren, 17

Hades, *see Hell*

Ham, Ken, 8, 10, 56, 181, 235
- *also see Answers-in-Genesis*

hairs on skin, *see erector pili*

hamartia (sin), 186, 230

Harris, Sam, 8

Hartle, James, 68

Hartsoeker, Nicolas, 44

Hawking, Stephen, 68, 275

Heisenberg Uncertainty Principle, 276, 283, 288

heliocentrism / heliocentric theory, 30, 33–37, 92, 151, 154, 251, 265, 289

hell, 27, 38, 251, 285

Hesperopithecus haroldcookii, 123

Hitchcock, Edward, 76

Hitchens, Christopher, 8

309

hoax
    Piltdown Man, 121–123
    Nebraska Man, 123
    Apparent Age, 56–57, 168, 259
    Creationists and 20,000 year fossil, 53–54
    living mollusks, 52–53
hobbit, *see Homo floriensis*
Homo
    antecessor, 119, 147
    diluvii testis, 60
    erectus, 108–109, 112, 114–115, 121, 124, 126, 132, 147, 213, 263
    ergaster, 109, 119
    floresiensis, 114, 124, 126, 147
    habilis, 112, 126, 132, 147
    heidelbergensis, 120, 139, 147
    naledi, 67, 116–119, 124–126, 147, 184, 195, 263
    rhodesiensis, 120
homosexuality, 12, 175, 223, 249
Hovind, Kent, 56
Hoyle, Fred, 68
Human Genome Project, 11, 283
human leukocyte antigen complex, 141
Huxley, T. H., 104, 130
image of God; *imago Dei*, 204–215, 274
    definition, 15, 168, 204–205
    ancient Hebrew view, 37, 204–207
    in Apocryphal period, 207
    in Apostolic period, 207
    in Patristic period, 65, 208–209
    in modern era, 209–211, 214, 287
    Christ as ultimate figure, 204, 285
    implications, 211–215
    in aboriginals, 65
    in non-human species, 66, 196, 204, 212, 214, 263, 268
    and evolution, 168–169
infertility, 43, 138
Institute for Creation Research, 9, 261, 290
Intelligent Design, 7, 68, 132, 138, 163–168, 269–271
Irenaeus, 208
Jastrow, Robert, 282
Java Man, 108–109, 114, 124,
Jefferson, Thomas, 5
Jehoida, 172, 252
Job
    on weather, 63, 292
    on human origins, 43–45
    on adversity/testing, 243, 247
    on justice, 271, 285
    on creation, 27, 28, 251
    on God, 168, 205
    on sin, 226

John Templeton Foundation, 290
Josephus, 197
junk DNA, 99–100
Kanzi, 217, 218
karyotype, 90, 134–136, 137, 139
Kepler, Johannes, 34, 283
keystone species, 159, 160
Keith, Arthur, 109, 122
Kingdom of Tuwana, 285
Kitzmiller versus Dover Trial, 175
Koko, 217, 218
La Peyrère, Isaac, 66, 203
    on pre-Adamites, 66–67
Laetoli footprints, 120, 262
Lamarck, Jean-Baptiste, 76
Lamoureux, Denis, 8, 18, 43, 171, 193, 194
Lao-tzu, 277
Leakey, Louis and Mary, 110–113, 120, 124, 217
Le Gros Clark, 122
Leibniz, Gottfried Wilhelm, 38
Lennox, John C., 17
Leviathan, 173, 251
Lewis, C. S.
    on Adam, 192
    Flatland, 278
    on redemption, 238
Lincoln, Abraham, 232
Linnaeus, Carl, 79, 180, 215, 283
Livingstone, David, 64, 212
Lucy, 110, 112, 114, 124, 126
Luther, Martin
    on the geocentric theory, 34
    on the Purchase of Indulgences, 155
MacArthur, John, 17
Madhava of Sangamagrama, 39
Magellan, Ferdinand, 32
Magi, 236
McGrath, Gavin Basil, 192
Membracidae 255
Mesopotamia, 16, 20–21, 43, 64, 188, 198, 205, 287
    also see Ancient Near East
Meyer, Stephen C., 18
microevolution / macroevolution, 7, 153, 185–186, 264
Middleton, J. Richard, 17, 205–209, 212
missing link
    existence/absence of, 45, 74–76, 108, 121, 124–125, 152, 184, 261, 263
    branching nature of evolution, 73–76, 125
    transitional forms, 76–78, 184
    also see fossilization
mitochondria
    origins of, 91–92

parental inheritance of, 93-94
energy generation in cell, 92
mitochondrial DNA, 93-94, 101, 127-130, 146
Montgomery, David, 17, 57-58
Morris, Herbert William, 266
Morris, Henry,
founder, Institute for Creation Research, 9
on the *imago Dei*, 209
Morwood, Mike, 114-115
Moses, 20, 195, 196, 199, 233, 285
educated in Egypt, 25, 62
challenging God, 271
stone tablets, 199
establishing temple practices, 233-234
also see Scripture, authorship, Moses; also see Pentateuch
multiverse theory, 279
mustard seed, 171, 242
myth, 7, 154, 176, 189, 227
National Association of Evangelicals, 289
Neanderthal, 15, 105-108, 116, 126, 132, 148
discovery, 72, 105
description, 105, 106-107
tool-making, 108, 188
burial practice, 108, 119, 188-189, 210, 213, 236
imago Dei, 66-67, 196
leaving Africa, 130, 146
as Nephilim, 195
see genome, Neanderthal
Nebraska Man, 123
Nephilim, 4, 60, 195, 227
Newton, Sir Isaac
religious writings 269, 283
scientific discoveries, 38-39, 269, 286, 288
standing on shoulders of giants, 38, 218
Noah, 25, 232, 233, 247
as author of Scripture, 25
Noah's Ark, 10, 57-60, 179, 181, 264
days of, 173, 195, 251
righteous man, 233, 247
offering, 233-234
father of mankind, 263
noble elements (Greek), 33, 37
NOMA / non-overlapping magisteria, 8, 282
Occam's Razor, 280
OEC, 4, 9, 11, 13, 20, 190, 192, 196, 257, 291
oral record; oral transmission, 198-201
Origen, 155, 208, 250, 259
original sin, 15, 65, 225-231
os baculum, 41-42
out-of-Africa hypothesis, 146, 148, 287
Pääbo, Svante, 145
paleontology, 6, 57, 73, 86, 120, 189, 243, 262

panspermia, 69, 269, 277
Pascal, Blaise, 236, 283
Paul (Apostle),
encourages study, 14
third heaven, 26, 172, 173, 251
on gender equality, 249, 253
on hell, 28
on death and redemption, 156, 170, 193, 271
on imago Dei, 207, 285
on original sin, 225
in Athens, 236, 238, 252
on inspiration of Scripture, 240
Pentateuch, 8, 20, 21, 25, 62, 72, 234, 240, 247, 248
also see Scripture, authorship
Peter
referring to flood, 37, 171, 251, 252
vision, 238
Apocalypse of, 246
Gospel of, 246
Pew Forum on Religion and Public Life, 13, 236
phenotype, 7, 178, 180
Phasmatodea, 254
Phycodurus eques, 254
Phyllium celebicum, 254
physical constants, 68, 259, 260, 276, 278-280
Piltdown Man, 121-123
Pit of bones, 116, 119
Pithecanthropus, 109, 124
"a plain reading of Scripture," 4, 18, 26, 29, 33, 38, 40, 45, 46, 83, 150, 242, 243, 250, 258
plagiarism, 1-2, 72, 131, 149
Plantinga, Alvin, 17
Plinians, 64, 103, 212, 215, 235
also see pre-Adamites
Pliny the Elder, 64, 212-213
Polkinghorne, John, 18, 283
Population Reference Bureau, 233
pre-Adamites, pre-humans, 15, 64-68, 266, 267
Isabelle Duncan, 266
Isaac La Peyrère, 66-67, 203
Pliny the Elder, 64, 103, 211-214, 235
also see Plinians
predation, 159-160, 169, 222, 254, 255, 271
presuppositionalism, 18, 45, 112, 156, 191-192, 194, 268, 274, 280
provirus / proviral sequences, 98-99, 131, 142-144, 162
pseudogene, 95-96, 131, 144-145, 162, 164
Pythagorus, 29

quantum mechanics, 288, 289
radioactive decay, 46–47
radiometric dating, 45–57, 85, 114, 261
    carbon-14, generation, 48, 56
    errors, outliers, 50–51
    errors, underestimation, 53–55
    errors, overestimation, 52–53
    errors, contamination, 51–52, 53–54, 55–56
    errors, marine reservoir effect, 53
    corroborated by other techniques, 49–50
Rana, Fazale (Fuz), 17, 102, 192, 195
RATE group, 54–56
Reasons-to-Believe, 8, 10, 192, 263, 266, 290
recurrent laryngeal nerve, 163, 270
redemption, 15, 65
representative pairs, 58, 181–182, 264
retina, *see vision, retina*
retrograde motion, 30
retrovirus, 98–99, 143
rhodopsin, *see vision, rhodopsin*
Ross, Hugh, 8, 10, 17, 195
    also see Reasons-to-Believe,
Sarah / Sarai, 21, 25, 26, 43, 62, 285
    also see Abraham/Abram
Scheuchzer, Johann, 59
Scopes Monkey Trial, 175
Scripture
    authority, 10, 11, 12, 36, 240–243, 259
    authorship, God, 10, 12, 198, 241, 244
    authorship, Moses, 21, 25, 62, 191, 198, 244, 248
    authorship, scribes, 20, 21, 25, 62, 191
    authorship, patriarchs, 25, 200
    censoring, 5, 13
    dietary laws, 223, 247, 248, 249, 251
    divine inspiration, 5, 10, 11, 12, 16, 26, 240–243
    human inspiration, 13, 16, 25, 26, 42, 62, 154, 196, 241–242, 251, 257
    inerrancy, infallibility, 7, 10, 12, 15, 16, 259
    interpretation, 5, 15, 18, 40, 81, 185, 242, 243, 244, 250, 273
    is/contains God's words, 16, 243
    non-Biblical Christian texts, 246
    different number of books of Bible, 10, 246
    inconsistencies and contradictions, 11, 198, 241
    contradicted by science, 36, 40–68, 151
    as the human diary/textbook, 244–247, 257, 265, 284, 286
    also see oral transmission, Pentateuch; Council of Nicea
Scientific Method, 50, 80–83
    also see Occam's Razor

selfishness, 216, 221, 223, 224, 229, 231, 239
SETI Project, 69
Shaaffhousen, Hermann, 105–106, 124
Sheol, *see* Hell
Silver, Daniel Jeremy, 244
Sima de los Huesos, *see Pit of Bones*
single nucleotide polymorphism, 95
slavery, 65–66, 67, 172, 223, 242, 248, 253
slippery slope, 196, 257
Smith, George, 72
Smith, Grafton Elliott, 106
Sodom and Gomorrah, 247
Soejono, Raden, 114
special creation, 11, 137, 138, 151, 152, 161–168, 188, 195, 219–220
species
    definition, 179–180
    classification, *see* Linnaeus, Carl
stars, 23, 24, 26, 30, 34, 236
stellar parallax, 34
Stillingfleet, Edward, 274
Stonehenge, *see world religions*
Stott, John, 192
subjugation of women, 172, 198, 238, 242, 247, 249, 253
Sullivan, Louis H., 83
Sumerian civilization, 20, 24, 30, 43, 61, 72, 190, 211, 227, 285
sun
    rising/setting, 28, 30, 37–38
    returns through underground tunnel, 23, 24
    going backwards, 28
    drives weather, 62
    also see heliocentrism, geocentrism,
Tacitus, 197
Taung child, 109
taxonomy, *see Carl Linnaeus*
telomere, 100, 133, 137
    also see chromosome
Theophilus of Antioch, 212
theory, 7, 175–176
third heaven, 26, 172, 173, 251
Thomas (disciple), 246, 272
thorns, 169, 182, 255, 287
Thylacine wolf, 80, 86, 146, 170
Tobias, Philip, 112
tools,
    animal use, 217–218
    Neanderthal use, 108, 119, 188–189, 210, 213, 236
Tower of Babel, 4, 21, 227, 247
transitional form, *see missing link*
Tree of Knowledge, 156, 159, 182, 210, 216, 225, 230

SUBJECT INDEX

Tree of life
   spiritual/metaphorical, 76, 157, 158, 227
   biological, 4, 45, 76, 85, 86, 102, 126, 149, 160, 229, 269
   paleontological, 76
   its fruit, 157, 158, 192, 225, 226, 230
tribalism, 220–221, 242, 232, 265, 292
Twain, Mark, 3
Ut-napishtim, 61, 62, 227
Valentinians, 207
Van Leeuwenhoek, Anton, 89
Venus
   planet, 35
   statues, 119, 285
Vesalius, Andreas, 41
Viracocha, *see world religions*
vision, 281, 284
   retina, 165, 270
   rhodopsin, 97, 281
vitellin, 97, 144, 164
Walton, John, 17, 172, 193, 206, 209
Ward, Keith, 280
Washoe, 217, 218
waters above/below, 26, 27, 28, 182, 267
Watson, James, 100
Weiner, Joseph, 122
Wellhausen, Julius, 72
White, Timothy D., 113
Wilberforce, Samuel, 104, 130
Wilson, E. O., 294
Wise, Kurt, 8

world religions
   human search for Great Being, 15, 210, 222, 235, 244, 246, 284
   Babylonian, 24, 61, 227
   Buddhism, 3, 239, 240, 285
   Confucianism, 240, 244
   Greek, 33, 36, 156, 161, 173, 236, 237, 285, 293
   Hinduism, 240, 244, 285, 292
   Islam, 12, 198, 202, 237, 239, 240, 244, 259, 285, 292
   Jainism, 244
   Judaism, 199, 202, 226, 233–234, 244, 285
   Kingdom of Tuwana, 285
   Mayan, 24, 61, 292
   polytheism/monotheism, 25
   Roman, 236, 237, 285, 293
   Sikhism, 239, 285
   Stonehenge, 24, 285
   Taoism, 244, 277
   Viracocha, 237
   Zoroastrianism 3, 24, 236, 244, 285, 292
   also see Christian denominations
Wright, Frank Lloyd, 83
Y-chromosome, 94, 127–130
YEC, 4, 9, 10, 12, 54, 67, 72
Young, Davis A., 195, 267
Zechariah, 172, 252
Zevit, Ziony, 41
Zinjanthropus boisei, 111

Scripture Index

Genesis

1:1	266
1:2–2:3	266
1:16	154
1:24	204
1:26	154, 204
1:26–7	204
1:27	205
1:28	203, 211, 284
1:29–30	254
1:30	220
2:7	219
2:9	157, 158, 216
2:10–14	202
2:15	211
2:16	157, 175
2:16–17	158
2:17	156, 157, 216, 225
2:19	175
2:19–20	79
2:20	175
3:10	157
3:17	175
3:20	175
3:21	175
3:22	158
4:3–4	234
4:5	157
4:8	157, 252
4:13–14	157
4:21	189
5	202, 203
5:1	204, 228
5:3	157, 204
5:5	157
5:24	232
6:2	203
6:4	60
6:8–9	232
6:17	220
7:11	27
7:15	220
7:21	220
7:22	220
8:20	234
9:6	204, 212
10	202
11	202, 203
11:5	247
11:30	43
15:7	21, 25, 199
17	62
18:20–21	247
18:23–33	271
20:17–8	43
25:21	43
29:31	43
30:1–3, 22	43

Exodus

15:3	205
21:20–21	253
32:11–13	271
33:21–23	168, 205

Leviticus

11:10–12	248, 249
15:19–22	248
19:27	248
21:18	248
21:20	248
25:44–46	253
27:3–7	253

Numbers

21:14	246

Deuteronomy

10:14	26
15:12–18	253
21:10–14	198, 248
21:18–21	248
32:22	28

Joshua

3:16	226
10:13	28, 246
24:2	21, 25

Judges

12:6	292
13:2	43

1 Samuel

1:2 and 6	43
2:8	27

2 Samuel

6:23	43
22:9	168, 205

1 Kings

8:27	26
11:41	246
14:29	246

2 Kings

2:11	232
2:23–5	198

1 Chronicles

1:1	225
16:29–30	27, 28, 34
29:29	246

2 Chronicles

2:6	26
6:18	26
9:29	246
12:15	246
21:4–26:23	201
24:20	172, 252

Nehemiah

9:6	26

Job

9:6	27
10:8–9	43, 44
11:8	28
31:33	226
31:35	271
37:3	63
4–5	63
10	63
11	63
15–16	63
37:18	27
38:6	63
22	63
24	63
25	63
38:22–25	153
38:29	63, 206
41:1	251
41:18–21	251
41:20	168, 205

Psalms

8:4–5	178
8:6–8	211
18:8	168, 205
19:1	210
19:5–6	28
30:3	27
40:8	42
74:13–14	251
75:3	27
90:10	261
91:4	206
93:1	27, 28, 34
95:3	205
96:10	27, 28, 34
104:5	27, 28, 34
104:26	251
119:90	27, 28, 34
137:8–9	241, 251
139:13	44
139:14	270
139:15	43

Proverbs

27:17	293
31:6–8	243

Ecclesiastes

1:5	28
3:9–10	215
3:11	216, 285
3:18–19	216
8:15	243
10:19	243

Isaiah

11:6–7	255
11:12	32
27:1	251
29:16	43
35:9	255
38:8	28
40:18	204
40:22	32
54:1	44
64:6	231
64:8	43
65:25	255
66:13	206

Jeremiah

1:5	44
17:9	231
18:1–6	43

Ezekiel

1:5	204
1:10	204
1:13	204
1:16	204
1:22	204
1:26	204
1:28	204
7:2	32
8:2	204
10:1	204
10:10	204
10:21	204
10:22	204

Hosea

6:7	226

Matthew

1:8	201
1:17	201
4:2	241
4:18	241
5:45	38
5:27–28	248
5:31–32	248
5:33–34	248
5:38–39	248
5:43–44	248
5:45	38
5:48	177
7:12	229
7:27	15
8:4	249
8:24	241
9:16–17	155
10:6	173
11:23	28
13:1–9	256
15:11	248
15:17–20	248
15:24	173, 249
19:4–6	37, 152, 171, 174
19:8–9	248
22:20	204
23:25–26	225
23:35	37, 172, 252
23:37	206
24:27–29	172
24:36	172, 241
24:37–38	37, 171
24:38–39	252

Mark

1:32	38
1:44	249
4:6	38
4:18	241
4:31	171, 242
4:38	241
5:9	172, 241
6:38	172, 241
7:20–23	224
9:21	241
10:6	252
10:6–8	37
10:6–9	152, 174
12:16	204
14:36	272
16:2	38

Luke

1:7 and 36	44
4:40	38
5:14	249
6:27–31	248
9:51–56	238
10:15	28

317

Luke (cont.)

11:51	37
14:25–33	14
17:26–27	37, 171
20:24	204
24:45–47	246

John

1:1–18	241
2:4	249
3:1–21	157
4	238
7:6	249
7:10	249
10:16	238
12:24	172, 242
14:6	238, 239
20:25	272

Acts

7:22	62
10	238
15:5	173
15:20	154, 173
17:16–34	152, 236, 238
17:26	37, 171, 174, 252
19:35	204
23:6	252

Romans

5:12–19	37, 152, 156, 157, 171, 225, 251, 271
6:1–2	177
6:6	225, 230
7:18–25	177
8:13	177
8:19–22	177, 230
8:29	204, 207
11:4	204
12:2	177
13:11	153

1 Corinthians

5:7	225
8:4	154
8:9	16
10:23–33	154
11:7	204, 207
11:13–16	172, 249
15:21–22	37, 152, 171, 173, 225, 251, 271
15:36–37	242
15:45	15
15:49	204, 207

2 Corinthians

2:15	153
3:18	204, 207
5:17	225
12:2	26, 38, 172, 251

Galatians

3:28	249
4:27	44
5:12	243
5:19–21	239

Ephesians

3:6	249
4:22	230
4:24	207
4:26	38
6:5	172

Philippians

1:8	42
2:10	28, 38

Colossians

1:15	204, 285
3:5	177
3:9	225, 230
3:10	204

1 Timothy

2:13–15	37, 152, 171, 173

2 Timothy

2:15	14
3:16	240
4:18	153

Titus

2:2	230

Hebrews

1:3	207, 285
5:8–9	241
11:7	171
11:11	43

James

1:4	177
1:11	38
3:9	212

1 Peter

1:9	153
3:15	258
3:20	37, 171

2 Peter

2:5	37, 171, 251, 252

Revelation

7:1	32
12:14	204
12:15	204
14:9	204
14:11	204
15:2	204
16:2	204
19:20	204
20:4	204
20:8	32
22:2	158

Additional writings

Sirach 17:3	207
Wisdom 2:23	207
2 Esdras 8:44	207